Comparative aspects of neuropeptide function

Studies in neuroscience *No. 11*

Series editor Dr William Winlow, *Dept. of Physiology, University of Leeds, LS2 9NQ, UK*

Neuroscience is one of the major growth areas in the biological sciences and draws both techniques and ideas from many other scientific disciplines. *Studies in neuroscience* presents both monographs and multi-author volumes drawn from the whole range of the subject and brings together the subdisciplines that have arisen from the recent explosive development of the neurosciences.

Studies in neuroscience includes contributions from molecular and cellular neurobiology, developmental neuroscience (including tissue culture), neural networks and systems research (both clinical and basic) and behavioural neuroscience (including ethology). The series is designed to appeal to research workers in clinical and basic neuroscience, their graduate students and advanced undergraduates with an interest in the subject.

1. **The neurobiology of pain**
   ed. A. V. Holden *and* W. Winlow
2. **The neurobiology of dopamine systems**
   ed. W. Winlow *and* R. Markstein
3. **Working methods in neuropsychopharmacology**
   ed. M. H. Joseph *and* J. L. Waddington
4. **Growth and plasticity of neural connections**
   ed. W. Winlow *and* C. R. McCrohan
5. **Aims and methods in neuroethology**
   ed. D. M. Guthrie
6. **The neurobiology of the cardiorespiratory system**
   ed. E. W. Taylor
7. **The cellular basis of neuronal plasticity**
   ed. A. Bulloch
8. **Higher order sensory processing**
   ed. D. M. Guthrie
9. **Function and dysfunction in the basal ganglia**
   ed. A. J. Franks, J. W. Ironside, R. H. S. Mindham, R. J. Smith, E. G. S. Spokes *and* W. Winlow
10. **Neurobiology of *Mytilus edulis***
    ed. George B. Stefano
11. **Comparative aspects of neuropeptide function**
    ed. Ernst Florey *and* George B. Stefano
12. **Neuromuscular transmission: basic and applied aspects**
    ed. Angela Vincent *and* Dennis W.-Wray
13. **Simpler nervous systems**
    ed. D. A. Sakharov *and* W. Winlow
14. **Nociceptive afferent neurones**
    David C. M. Taylor *and* Friedrich-Karl Pierau
15. **Signal molecules and behaviour**
    ed. W. Winlow, D. V. Vinogradova *and* D. L. Sakharov

Comparative aspects of neuropeptide function

edited by Ernst Florey

Faculty of Biology, University of Konstanz

and George B. Stefano

Multidisciplinary Center for the Study of Aging, State University of New York

Manchester University Press

Manchester and New York

Distributed exclusively in the USA and Canada by St. Martin's Press

Published by Manchester University Press
Oxford Road, Manchester M13 9PL, UK
and Room 400, 175 Fifth Avenue,
New York, NY 10010, USA

Distributed exclusively in the USA and Canada
by St. Martin's Press, Inc.,
175 Fifth Avenue, New York, NY 10010, USA

British Library cataloguing in publication data
Comparative aspects of neuropeptide function.
1. Animals. Nervous system. Neuroactive peptides
I. Florey, Ernst II. Stefano, George B. III. Series
591.188

Library of Congress cataloging in publication data
Comparative aspects of neuropeptide function/edited by Ernst Florey
and George B. Stefano.
p. cm.—(Studies in neuroscience: no. 11)
Based on a meeting held in Konstanz, Germany in 1989 and sponsored
by the Deutsche Forschungsgemeinschaft and others.
ISBN 0-7190-3298-9 (hardback)
1. Neuropeptides—Congresses. 2. Comparative neurobiology—
Congresses. I. Florey, Ernst. II. Stefano, George B., 1945–
III. Deutsche Forschungsgemeinschaft. IV. Series.
[DNLM: 1. Arthropods—physiology—congresses. 2. Helminths—
physiology—congresses. 3. Hormones—physiology—congresses.
4. Mollusca—physiology—congresses. 5. Neuroimmunology—
physiology—congresses. W1 ST927K no. 11/WL 104 C7367 1989)
QP552.N39C65 1990
591.1'88—dc20
DNLM/DLC
for Library of Congress 90-6326

ISBN 0 7190 3298 9 *hardback*

Printed in Great Britain by
Biddles Ltd., Guildford and King's Lynn

Contents

This book is dedicated to

BERTA SCHARRER

in recognition of her inspiring work, both as a scientist and as a teacher. Her pioneering research, carried out in conjunction with that of her eminent husband, Ernst Scharrer, opened the road that finally led to the discovery that peptides made and secreted by nerve cells are the major agents which mediate and modulate the interactions between nerve cells, and between nerve cells and other kinds of cells, tissues and organs. For more than six decades Professor Scharrer has been at the forefront of research. Today she is again leading on to new perspectives in a field which commands increasing attention: the chemical interaction between neurons and cells of the immune system by means of neuropeptides.

Foreword

The discovery of endogenous opioid peptides in the mammalian brain has led to an almost explosive development of research into the structure, synthesis and functional role of neuropeptides. The role of peptide hormones had been recognized long ago. In fact, the first hormones to be discovered were peptides. It was also known that the neurohormones produced by so-called neurosecretory cells are, by and large, polypeptides. The discovery of opioid peptides (first named enkephalins and endorphins), however, focussed attention on the fact that numerous types of nerve cells contain, and evidently release peptides as chemical messengers, and that these neuropeptides qualify as transmitter and modulator substances. Even those classical hormones, known as secretin (first described in 1902), gastrin (first reported in 1833) and insulin (discovered in 1927) turned out to be genuine peptides which are mainly (secretin, gastrin) or also (insulin) produced by neurons.

As a consequence of the new findings, the field of endocrinology has lost much of its exclusiveness: even the concept of neurosecretion has changed its meaning since it was found that the production and secretion of neuropeptides is a general feature of most, if not all, neurons. On the other hand, the concepts of neuronal function have undergone a major transformation as it is now recognized that neurons, by secreting certain neuropeptides, control the differentiation and growth of many kinds of cells, tissues and organs, and that neurohormones can initiate different kinds of behaviour.

The discovery of peptide receptors in blood cells (immunocytes, haemocytes), opens the area of neuroimmunology, and the discovery that haemocytes manufacture neuropeptides now adds a new dimension to our ideas about the control of brain function by the immune system.

Advances of our knowledge of the mechanisms of gene expression now permit new insights into the molecular control of peptide synthesis.

In view of the magnitude of current research in neuropeptides, it may seem presumptuous to attempt a synthesis by bringing together a relatively small group of researchers from different specialities of neuropeptide research in an effort to assess current concepts, and to create new perspectives. Indeed, there have been numerous reviews and monographs covering various aspects of neuropeptide structure and function, and there have been several successful symposia – and symposium publications – addressed to the same topic. We felt, however, that there was a need to counteract increasing specialization by providing a forum for an exchange of knowledge and of ideas between representatives of the diverse disciplines concerned with neuropeptides. The fields represented range from psychiatry to zoology, from research in development to investigations of ageing, from molecular genetics to behaviour, and from structural chemistry to immunology. Discussion was intensive and productive. In no small way it influenced the preparation of the manuscripts to be incorporated in this book.

We are grateful to all participants for their contributions in the form of formal lectures, of lively discussions, and of the chapters they prepared for this book. We thank the sponsors for their encouragement and financial support: the Deutsche Forschungsgemeinschaft, the National Science Foundation, the National Institutes of Health, the Paul Martini Stiftung, the Freunde und Förderer der Universität Konstanz, Boehringer–Ingelheim International, Bogdan Enterprises, Bristol–Myers, Byk Gulden-Lomberg, Glaxo, Lion Informational Systems, Morrell Instrument Company, Searle, Upjohn Company, Wellcome GmbH, Wyeth–Ayerst, and Carl Zeiss, Inc.

The meeting was held in the Sporthotel Höri on the shores of Lake Constance. We are in the dept of Ms. Grace Y. Chapman for her untiring efforts on our behalf. We thank the staff for friendly hospitality and every kind of co-operation. Manchester University Press deserve credit for their efficient handling of the editing of this book.

Konstanz, New York, August 1990

ERNST FLOREY
GEORGE B. STEFANO

Contributors

Robin A. Barraco, Department of Physiology, Wayne State University, School of Medicine, 540 E. Canfield, Detroit, MI 48201, USA.

Eckehard Baumann, Department of Animal Physiology, Friedrich-Schiller-Universität, Erbertstrasse 1, O-6900 Jena, FRG.

Harry H. Boer, Faculty of Biology, Free University, P.O. Box 7161, 1007 MC Amsterdam, The Netherlands.

Patrick Cadet, Multidisciplinary Center for the Study of Aging, State University of New York/College at Old Westbury, Old Westbury, NY 11568, USA.

David O. Carpenter, Wadsworth Center for Laboratories and Research, New York State Department of Health and School of Public Health, Albany, NY 12237, USA.

Joseph F. Cubells, Department of Neuroscience, Albert Einstein College of Medicine, Bronx, NY 10461, USA.

Silvia De Biasi, Dipartimento di Fisiologia e Biochimica Generali, Sezione di Istologia e Anatomia Umana, Università degli Studi di Milano, via Celoria 26, 20133 Milano, Italy.

Heinrich Dircksen, Institut für Zoophysiologie der Universität Bonn, Endenicher Allee 11–13, W-5300 Bonn 1, FRG.

Ernst Florey, Biology Faculty, University of Konstanz, Postfach 5560, W-7750 Konstanz, FRG.

Bernard Fournier, Laboratoire de Neuroendocrinologie, URA CNRS 1138, Université de Bordeaux I, 33405 Talence Cédex, France.

Yuko Fujisawa, Faculty of Integrated Arts and Sciences, Hiroshima University, Hiroshima 730, Japan.

Josiane Girardie, Laboratoire de Neuroendrocrinologie, URA CNRS 1138, Université de Bordeaux I, 33405 Talence Cédex, France.

Tetsuya Ikeda, Faculty of Integrated Arts and Sciences, Hiroshima University, Hiroshima 730, Japan.

Cornelis Janse, Faculty of Biology, Vrije Universiteit, P.O. Box 7161, 1007 MC Amsterdam, The Netherlands.

Eve W. Johnson, Department of Microbiology, University of Texas Medical Branch, Galveston, TX 77550, USA.

Janine L. Kallen, Zoological Laboratory, Faculty of Sciences, Catholic

University of Nijmegen, Toernooiveld 6525 ED, Nijmegen, The Netherlands.

Yoshimi Kamatani, Suntory Institute for Bioorganic Research, Osaka 618, Japan.

Tomoko Kanda, Faculty of Integrated Arts and Sciences, Hiroshima University, Hiroshima 730, Japan.

Martin Kavaliers, Division of Oral Biology, Faculty of Dentistry, University of Western Ontario, London, Ontario, Canada N6A 5B7.

Ron M. Kerkhoven, Faculty of Biology, Free University, P.O. Box 7161, 1007 MC Amsterdam, The Netherlands.

Kah Hwi Kim, Department of Physiology, Gifu University School of Medicine, Tsukasa-machi 40, Gifu 500, Japan.

Tibor Kiss, Balaton Limnological Research Institute, P.O. Box 35, H-8237 Tihany, Hungary.

Yasuo Kitajima, Faculty of Integrated Arts and Sciences, Hiroshima University, Hiroshima 730, Japan.

Gábor L. Kovàcs, Central Laboratory, Markusovsky Teaching Hospital, H-9701 Szombathely, Hámán Kató u.28, Hungary.

Ichiro Kubota, Suntory Bio Pharma Tech Center, Gunma 370-05, Japan.

Yoshihiro Kuroki, Faculty of Integrated Arts and Sciences, Hiroshima University, Hiroshima 730, Japan.

Michael K. Leung, Department of Chemistry, State University of New York/College at Old Westbury, Old Westbury, NY 11568–0210, USA.

Jonathan Lundy, Department of Chemistry, State University of New York/College at Old Westbury, Old Westbury, NY 11568–0210, USA.

Maynard H. Makman, Departments of Biochemistry and Molecular Pharmacology, Albert Einstein College of Medicine, Bronx, NY 10461, USA.

Margaret M. Mason, Department of Pathology, University of New Mexico, Albuquerque, NM 87131, USA.

Hiroyuki Minakata, Suntory Institute for Bioorganic Research, Osaka 618, Japan.

Stacia B. Moffett, Department of Zoology, Washington State University, Pullman, WA 99164, USA.

David B. Morton, Department of Zoology, NJ-15, University of Washington, Seattle, WA 98195, USA.

Yojiro Muneoka, Faculty of Integrated Arts and Sciences, Hiroshima University, Hiroshima 730, Japan.

Harald Murck, Physiologisches Institut, Universität Marburg, W-3550 Marburg, FRG.

Ronald J. Nachman, Western Regional Research Center, Agricultural Research Service, US Department of Agriculture, Berkeley CA 94720, USA.

Kyosuke Nomoto, Suntory Institute for Bioorganic Research, Osaka 618, Japan.

Heinz Penzlin, Department of Animal Physiology, Friedrich-Schiller-Universität, Erbertstrasse 1, O-6900 Jena, FRG.

Anthony M. Perks, Department of Zoology, University of British Columbia, Vancouver, BC, Canada V6T 2A9.

Martine Picquot, Laboratoire de Neuroendocrinologie, URA CNRS 1138, Université de Bordeaux I, 33405 Talence Cédex, France.

David J. Prior, Department of Biological Sciences, Northern Arizona University, Flagstaff, AZ 86011–5640, USA.

Jacques Proux, Laboratoire de Neuroendocrinologie, URA CNRS 1138, Université de Bordeaux I, 33405 Talence Cédex, France.

Katalin S.-Rózsa, Balaton Limnological Research Institute, P.O. Box 35, H-8237 Tihany, Hungary.

János Salánki, Balaton Limnological Research Institute, P.O. Box 35, H-8237 Tihany, Hungary.

Berta Scharrer, Multidisciplinary Center for the Study of Aging, State University of New York/College at Old Westbury, Old Westbury, NY 11568, USA.

Juan Sinisterra, Multidisciplinary Center for the Study of Aging, State University of New York/College at Old Westbury, Old Westbury, NY 11568, USA.

Eric M. Smith, Department of Psychiatry, University of Texas Medical Branch, Galveston, TX 77550, USA.

Joachim Stangier, Institut fur Zoophysiologie der Universität Bonn, Endenicher Allee 11–13, W-5300 Bonn, FRG.

George B. Stefano, Multidisciplinary Center for the Study of Aging, State University of New York/College at Old Westbury, Old Westbury, NY 11568, USA.

Hiroshi Takeuchi, Department of Physiology, Gifu University School of Medicine, Tsukasa-machi 40, Gifu 500, Japan.

James W. Truman, Department of Zoology, NJ-15, University of Washington, Seattle, WA 98195, USA.

Marcel van der Roest, Faculty of Biology, Vrije Universiteit, P.O. Box 7161, 1007 MC Amsterdam, The Netherlands.

Dennis E. Van Epps, Applied Sciences, Baxter Healthcare Corporation, Round Lake, IL 60073, USA.

François Van Herp, Zoological Laboratory, Faculty of Sciences, Catholic University of Nijmegen, Toernooiveld 6525 ED, Nijmegen, The Netherlands.

Jan van Minnen, Faculty of Biology, Free University, P.O. Box 7161, 1007 MC Amsterdam, The Netherlands.

Laura Vitellaro–Zuccarello, Dipartimento di Fisiologia e Biochimica

Generali, Sezione di Istologia e Anatomia Umana, Universita degli Studi di Milano, via Celoria 26, 20133 Milano, Italy.

Christian Walther, Physiologisches Institut, Universität Marburg, W-3550 Marburg, FRG.

Ian G. Welsford, Department of Biological Sciences, Northern Arizona University, Flagstaff, AZ 86011–5640, USA.

W. C. Wildering, Faculty of Biology, Vrije Universiteit, P.O. Box 7161, 1007 MC Amsterdam, The Netherlands.

Anchalee Yongsiri, Department of Physiology, Gifu University School of Medicine, Tsukasa-machi 40, Gifu 500, Japan.

Klaus E. Zittlau, Physiologisches Institut, Universität Marburg, W-3550 Marburg, FRG.

Ernst Florey

Introduction

The challenge of neuropeptides

Since the pioneering work of Ernst and Berta Scharrer on the phenomenon and functional significance of neurosecretion, it emerged that most of the secretory products are peptides and that these neuropeptides are signal molecules produced not only by neurons which show the characteristic symptoms of true secretory cells but by most, if not all, neurons. Indeed, they are even produced by non-neuronal cells, especially by haemocytes. They serve in signal transmission and thus in the control of cellular behaviour. The investigation of these events on the molecular scale has become a major challenge of pharmacology.

Pharmacology has been defined as a science concerned with the effects of chemical agents on living systems. Pharmacology can also be regarded as an extension of physiology in as much as pharmacological research is nothing but physiological research carried out with the aid of molecular tools. Pharmacology can also be regarded as molecular physiology.

Biology has entered the molecular and even atomic dimension. Pharmacology, as a branch of biology, is thus moving within the wide range of molecular biology, and its methodology encompasses all the new concepts of biochemistry, biophysics and molecular genetics. This is especially true for the field of neuropharmacology which has become a major aspect of neurobiology, the most active area of biology today.

Like physiology, the science of pharmacology is distinguished among the other experimental, biological sciences in that it retains a strong organismic outlook. It makes use of the extensive knowledge of molecular biology to solve the mysteries of the functioning of entire organisms. The genetic aspects of the synthesis of neuropeptides within nerve cells make sense only when we consider the role these neuropeptides play within the context of specific operations and behaviours of the whole organism. The analysis of the effects of neuropeptides on receptor molecules or on single ionic channels is worthwhile because it explains how neurons accomplish their sophisticated control of the cells they either innervate directly or affect by humoral mechanisms.

Neuropeptides, with their nearly limitless number of combinations of amino acid sequences, offer an almost unimaginable range of specific interactions with special receptor molecules, carriers, channels and enzymes. The recognition that nerve cells communicate with other cells (including other neurons) by means of the controlled release of such peptides implies that, in addition to the intercellular communication by the ubiquitous nerve impulses with their stereotypic sameness, the nervous system is capable of a nearly infinite variety of specific interactions. In the hand of the neuropharmacologist, specific peptides become powerful tools for the manipulation and study of cells, tissues and organisms.

The knowledge gained through neuropharmacological research of peptide effects finds immediate application in clinical medicine, where it generates new perspectives for the explanation and cure of specific diseases. Today, medicine is reaping the benefits of knowledge gained from research into the mode of action of small molecule transmitters such as acetylcholine, nor-epinephrine, dopamine, 5-hydroxytryptamine, gamma-amino butyric acid, glycine and L-glutamate. Pharmacological interventions with various forms of mental disease, with Parkinsonism, muscular dystrophy, epilepsy, heart disease, or even with such simple entities as motion sickness have thus found a rational basis. The application of knowledge gained in the field of neuropeptide pharmacology promises an even greater range of medical competence.

Biology, of course, is more than medicine. Its object is the understanding of life on our planet earth. Biology is systematic knowledge of living systems. Over the centuries this science has passed through the stage of mechanism, and gone through a phase in which life was understood as a strictly physicochemical process. During recent decades, biologists have discovered the dimension of cybernetics. We now realize that life is a hierarchy of living systems, each with organismic properties, each with its own level of complex interactions. Each level of organization interacts with any other level of organized life.

The concept of organism

We recognize that all organisms that populate our planet are genetically linked, as are all cells of a given organism. In a sense each eukaryont cell is a composite of symbionts that compete for survival. Mitochondria with their own DNA are symbionts of the nucleus, they may compete for the same amino acids when it comes to synthesizing proteins, and they are subject to control and even to destruction by other organelles (e.g. lysosomes) that inhabit the same cell. The cells of the organism interact in many ways: they cooperate, they compete, and they interact like predators and prey. Within a given ecosystem the organisms that constitute the system interact again as symbionts and parasites, as predators and prey. Cells, organisms, and ecosystems, when looked at over appropriate spans of time appear to be in a steady-state condition. Indeed, it is only because of this apparent steady state that we can recognize them as cells, as organisms and as ecosystems. Viewed over more extended periods of time, the cells change and the organisms change in form and function – just as entire ecosystems become transformed with time. When we consider cells we recognize growth, differentiation, and death. When we consider organisms we see differentiation and development, we see ageing and death.

Steady-state conditions exist only over limited time spans and they can be recognized only when low spatial resolution is employed. When we talk of 'the heart' or 'the kidney' we have in mind such a steady-state system. When we talk of a rat or a pond snail we refer to an organism in such a steady-state condition. But this condition, as we all know, is an illusion. Looked at over a briefer span of time, and with higher spatial resolution, a steady state is no longer steady. There are innumerable processes, fluctuations of synthesis and degradation, of forward and backward movement, of expansion and contraction, of growth and decay – the apparent steady state exists only because all these events are linked by complex feedback and feedforward causation. The sum, or better, the integral, of these interactions is the cell, the organism, the ecosystem.

Unless we look at individual events under the aspect of this integral, these events are meaningless. The so-called molecular approach in biology derives its meaning from the perspective of the system under consideration. The ionic currents moving through membrane channels are meaningless unless they are viewed within the context of specific cell function; and cell functions are meaningless unless they are viewed within the context of the organism. No cellular system within an organism can be understood by itself – if by 'understanding' we mean recognition of significance. Its meaning becomes evident only when seen in the context of the next higher level of organization in the hierarchy.

Neuropeptides as chemical signal molecules are significant when their functional role is considered. Their genesis, action and degradation become meaningful when viewed in the context of the life of the entire organism, and in many cases only when viewed in the context of the entire ecosystem.

Neuropeptide research ranges from molecular genetics and the cellular level up through that of the interactions of entire organisms. The organismic approach is essential for the understanding of the role of neuropeptides in the interactions between the nervous system and 'immunocytes'.

New perspectives and questions

Many new problems can now be recognized and challenge future research: (1) 'Classic' (small molecule) transmitters, such as acetylcholine or biogenic amines, are synthesized mainly in the nerve terminal. This synthesis is governed by the arriving nerve impulses. The amount of transmitter within a given terminal represents a steady state between synthesis and removal (through release and/or metabolism). The amount of releasable transmitter is thus determined by these two processes which are located in the nerve terminals. Neuropeptides, on the other hand, are synthesized in the cell soma. Axoplasmic transport carries them to the release sites of the nerve terminals. Does release also occur from the soma? What mechanisms are involved in the control of synthesis? Do nerve impulses effect increased synthesis? Do presynaptic receptors for transmitter and modulator substances mediate effects on the rate of peptide synthesis? What mechanisms control the rate of neuropeptide transport from soma to terminals? Is there feedback from the terminal, telling the soma when to step up synthesis, or to stop sending more peptide to an already loaded terminal?

(2) Some – perhaps all – neurons have multiple release sites. Is the release at these sites synchronized or independent? If independent, what mechanism initiates release at a given site?

(3) The release of 'classic transmitters' is either spontaneous (molecular release, quantal release), or impulse-coupled (quantal release). Both forms of release are chemically controlled. Release has several components that are subject to regulation, among them synthesis, translocation (mobilization), packaging, and fractional release from a 'releasable store'. What agents control these various processes in the case of neuropeptide release?

(4) Neuropeptides are the result of gene activation. What is the cause of gene expression? The situation is likely to be complicated: expression of a neuropeptide may involve gene expression of one or more enzymes which split the pre-(pro-)peptide, and these processes may involve one or more activators. Peptide production thus involves multiple control points. What are the controlling agents; how do they interact?

(5) Neuropeptides interact with more or less specific receptors. Such receptors are found not only on neuronal membranes but, among others, on immunocytes (haemocytes). What factors control the synthesis (expression) of these receptors, their incorporation into the cell membrane and their internalization?

(6) Haemocyte behaviour is, most likely, controlled by other humoral factors as well. If 'stress' causes a large increase in the number of circulating haemocytes, most of the haemocytes obtained by puncture after 'handling stress' may be in a condition that differs from that of 'unstressed' haemocytes. To what extent and in what role are neuropeptides involved?

(7) Haemocytes not only have receptors for neuropeptides, they also manufacture and, presumably, release neuropeptides. Haemocytes can attach to endothelium. Attached haemocytes may release neuropeptides

which pass through the endothelium into the extracellular tissue fluid. How is this release controlled? What is its role? The cerebral cortex of mammals is highly vascularized. The pattern of neuronal columns matches that of capillaries; thus each column may be provided with its own capillary. Peptide-releasing, attached haemocytes may be in a unique position to achieve modulation of neuronal activity. We need experiments to explore this.
(8) Haemocytes can be regarded as floating synaptosomes. They may be part of a reflex arc that involves both neural and haemal cells. In addition, haemocytes have been shown to invade nerve tissue and, in some cases, to become transformed into microglia. What is the role of these cells? Do they respond to, or manufacture neuropeptides? The possibility of haemocytes and/or microglia modulating brain function is a tantalizing prospect.

The chapters included in this volume attend to some of these questions. But they also raise others and offer a challenge for future, imaginative research. The chapters are grouped in sections for greater ease of orientation. In part, the reason is one of taxonomy: because of their natural, evolutionary relationship, molluscs, or arthropods, or mammals, have special features which make comparisons within the taxonomic group easier than comparisons between groups. Nonetheless it is obvious that cross-comparisons are possible. Indeed, the parallels of neuropeptide functions in molluscs and mammals, or mammals and arthropods are striking. Most astounding is the correspondence between the immune systems of mammals and molluscs – as illustrated in the chapters of the last section of the book.

It is obvious that this book does not cover the entire field of neuropeptide research. Its strength lies in the interdisciplinary approach and the vista it generates – and this, we feel, justifies its publication.

Part I

General aspects

Echiuroid studies

1 *Anthony M. Perks*

Neurosecretion: the beginnings

For most of us the concept of neurosecretion started with the classical studies of the Scharrers, around the 1930s (Scharrer, 1928, 1930, 1932; Scharrer and Scharrer, 1940, 1954). However, there were shadowy yet colourful figures in the past who, in their own way, heralded these important concepts.

Our understanding probably started with Galen (*c.* AD 180), who seems to have recognized the pituitary gland, perhaps in the course of his duties as physician to the least successful gladiators of Pergamum (Clendening, 1942; Greene, 1970). However, it was Andreas Vesalius who left us the first known diagram, in his *De Humani Corporis Fabrica, libri septum*, published in 1555 (Fig. 1.1, upper diagram). He shows a gland, apparently supplied by fluids from the infundibular funnel of the brain, and equipped with four ducts for carrying secretions outwards; their final destination was not clear, but there was a general belief, stemming from Galen's time, that the fluids passed on into the nose and throat (Zuckerman, 1954; Hunter and Macalpine, 1963; Dewhurst, 1980). This idea had never received experimental support, and the first test had to await Thomas Willis, in 1664.

In recent years the events which led up to, and surrounded, the work of Thomas Willis have been brought together by Dewhurst (1980). During the English Civil War, Willis left his farm near Bagley Wood, and went down into Oxford to fight for the King. It is probable that he saw action. However, when the war was lost he turned to medicine, and seems to have obtained his medical degree with great rapidity (six months), mainly due to the influence of another Royalist, Thomas Clayton, the Regius Professor of Medicine. His success as a physician was considerable. It may well have been helped by his marriage to a lady whose father, Samuel Fell, was the Vice Chancellor, and whose brother, John, became Bishop of Oxford. It was undoubtedly helped by his enviable reputation for raising the dead (one Ann Green). He is still remembered for the Circle of Willis, for his pioneer neurology, *Cerebri anatome*, of 1664 (illustrated by Sir Christopher Wren, who later slipped into designing cathedrals; see Feindel, 1965), and for his remarkable comment that, in the malady known as the Pissing Evil (diabetes), the urine was of a sweet taste (Willis, 1674, 1675). However, he must also be remembered for his

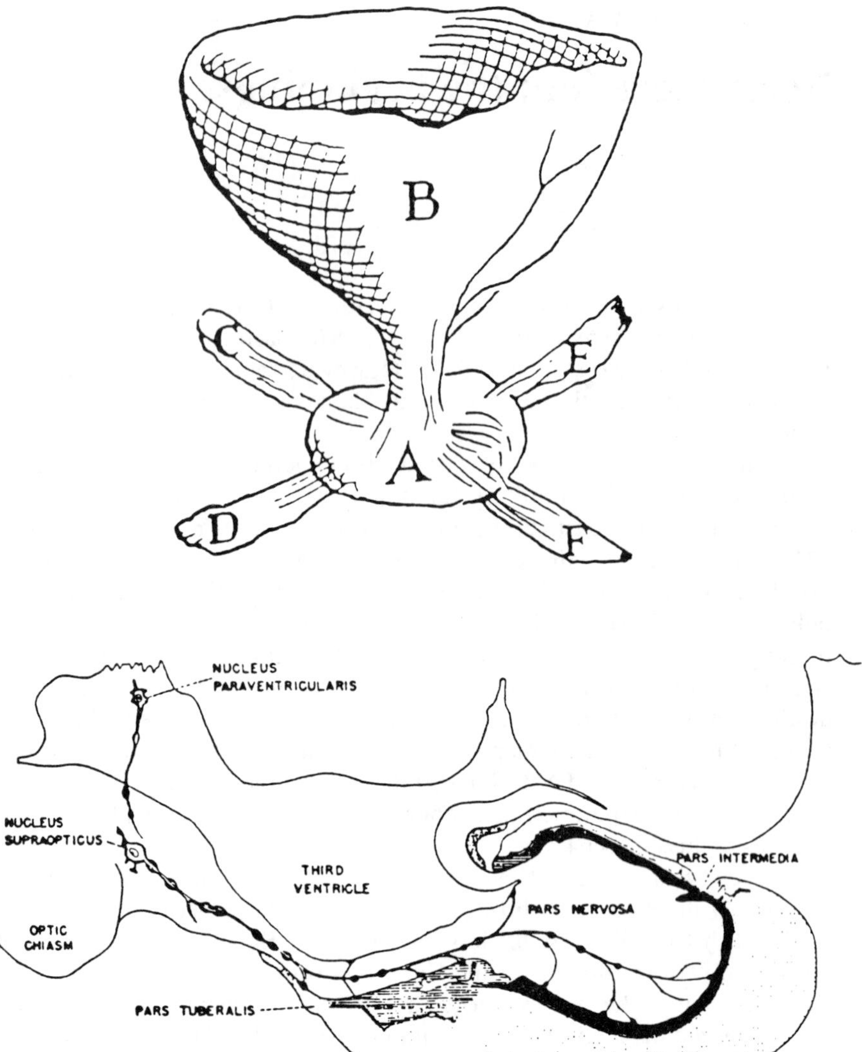

Fig. 1.1. Neurosecretory systems, ancient and modern. *Upper diagram*: the pituitary and infundibular funnel, according to Vesalius, 1555: A, the pituitary gland; B, the infundibulum; C, D, E and F, ducts leading from the pituitary. *Lower diagram*: the hypothalamic–hypophysial system, according to Bargmann and Scharrer, 1951.

experiments on the pituitary gland: in 1663 he syringed milk, then ink, into the pituitary fossa of a calf, and found that the fluids emerged in the jugular veins, and not in the nose and throat (Lower's description to Boyle, 4 June 1663; Dewhurst, 1980). However, Willis's conclusions were not clear. Although he appeared to accept the results in his *Cerebri anatome* (Pordage's translation, pp. 104–5; Feindel, 1965), his lectures, delivered about the same time, cling to the old ideas, mainly on clinical grounds (Dewhurst, 1980, pp. 140–1). Dewhurst has concluded that Willis had difficulty in ridding himself of the old theories. By this narrow margin, Thomas Willis missed pre-eminence as the Father of Neuroendocrinology. However, his assistant, Sir Richard Lower, was convinced by the results.

Sir Richard Lower had worked for Willis as a student. However, in his *Cerebri anatome* of 1664, Willis describes Sir Richard as a 'most learned Physician and highly skillful Anatomist ... the edge of whose Knife and Wit I willingly acknowledge ... as also his indefatigable Industry, and unwearied Labour' (quoted from Hunter and Macalpine, 1963, and Dewhurst, 1980). Richard Lower became a successful physician; he attended King Charles II during his last illness, and he ministered to his successor, James II, although the king remarked that he 'did him more mischief than a troop of horse' (Cunningham, 1892; Hunter and Macalpine, 1963). He also attended the famous orange-seller and mistress of the king, Nell Gwynne, with whom he had a splendid relationship, and from whom he picked out all the gossip of the court (Cunningham, 1892; Bevan, 1969). On the academic side he made one of the most important observations in the history of physiology when he showed that dark venous blood became bright red after passing through the lungs (Lower, 1669; see Franklin, 1932). Soon afterwards he turned his attention to neurosecretion. In 1669, Lower published *De Catarrhis*, and there he described the earlier experiment with Thomas Willis. He writes: 'Thus, if milk or some black substance is squirted into these openings with a syringe, it penetrates at once on each side into the jugular veins. But none of this tincture will appear around the palate, nostrils, mouth, throat or larynx. In this way, in the case of a calf, all fluid coming from the cerebrum returns wholly into the veins'. He adds: 'I have recently found this also confirmed by experiment on the human cranium' (quoted in translation from Hunter and Macalpine, 1963). Clearly, Sir Richard Lower accepted the concept that the secretions of the brain found their way into the pituitary, and then on into the blood stream. Clearly, with his experimental approach, he foreshadowed our understanding of neurosecretion.

Later, there was some general support for Sir Richard from Swedenborg (see Zuckerman, 1954; Hanṣtröm, 1957), but after this almost regal start there followed a relatively 'dark age'. A number of investigators regarded the posterior pituitary as little more than a connective tissue appendage of the brain (Müller, 1871; Toldt; 1888; Prenant, 1896). It was not until the beginning of this century that Ramon y Cajal (1911) finally established the essentially neural nature of this structure. It was then left to the Scharrers, together with Bargmann, to take the vital step which showed the neurosecretory nature of the hypothalamic–hypophysial system (Fig. 1.1, lower diagram), and extended the basic concepts into both vertebrate and invertebrate physiology (Scharrer, 1928, 1930, 1932; Bargmann and Scharrer, 1951; Scharrer and Scharrer, 1954). In so doing, they changed the vague ideas of Thomas Willis and Sir Richard Lower into the clearest of modern terms.

Acknowledgements

We are glad to thank Dr Berta Scharrer and the *American Scientist* for permission to use the lower diagram of Fig. 1, and Christine Ruddock for help concerning the locality of Thomas Willis.

References

Bargmann, W. and Scharrer, E. (1951). The site of origin of the hormones of the posterior pituitary. *Am. Sci.*, **39**, 255–9.

Bevan, B. (1969). *Nell Gwyn*. Robert Hale, London, p. 190.

Clendening, L. (1942). *Source Book of Medical History*. Dover Publications, New York, pp. 41–51.

Cunningham, P. (1892). *The Story of Nell Gwyn and the sayings of Charles II*. W. W. Gibbings, London, p. 224.

Dewhurst, K. (1980). *Thomas Willis's Oxford Lectures*, Sanford Publications, Oxford, p. 182.

Feindel, W. (1965). *Thomas Willis; The anatomy of the brain and nerves*. McGill University Press, Montreal, p. 104.

Franklin, K. J. (1932). De corde by Richard Lower, with introduction and translation. *Early Science in Oxford*, edited by R. T. Gunther, Vol. 9. Oxford.

Greene, R. (1970). *Human Hormones*. World University Library, McGraw-Hill, New York and Toronto, p. 256.

Hanström, B. (1957). Reflections on the secretory role of the brain. *Nova Acta Soc. Sci. Upps*. [4], **17**, 1–12.

Hunter, R. and Macalpine, I. (1963). *Richard Lower, "De Catarrhis, 1672"*. Dawsons of Pall Mall, London, p. 29.

Lower, R. (1669). Tractatus de corde. London; cited from Franklin, K. J., 1932, De corde by Richard Lower, with introduction and translation. *Early Science in Oxford*, edited by R. T. Gunther, Vol. 9. Oxford.

Müller, W. (1871). Über Entwickelung und Bau des Hypophysis und des Processus infundibuli cerebri. *Jena Z. Med. Naturw.*, **6**(3), 201–11.

Prenant, A. (1896). Elements d'embryologie d l'homme et des vertèbres, Vol. 2. Cited by Gentes, L., in *Trav. Lab. Soc. sci. Arcachon*, 1907, 2nd année, pp. 129–275.

Ramon y Cajal, J. (1911). *Histologie du Système nerveux*, Vol. 2. Malione, Paris.

Scharrer, E. (1928). Die Lichtempfindlichkeit blinder Elritzen (Untersuchungen über das Zwischenhirn der Fische. I). *Z. vergl. Physiol.*, **7**, 1–38.

Scharrer, E. (1930). Über sekretorisch tätige Zellen im Thalamus von *Fundulus heteroclitus* L. (Untersuchungen über das Zwischenhirn der Fische. II). *Z. vergl. Physiol.*, **11**, 767–73.

Scharrer, E. (1932). Die Sekretproduktion im Zwischenhirn einiger Fische (Untersuchungen über das Zwischenhirn der Fische. III). *Z. vergl. Physiol.*, **17**, 491–509.

Scharrer, E. and Scharrer, B. (1940). Secretory cells within the hypothalamus. *Res. Publs. Ass. Res. nerv. ment. Dis.*, **20**, 170–94.

Scharrer, E. and Scharrer, B. (1954). Hormones produced by neurosecretory cells. *Recent Prog. Horm. Res.*, **10**, 193–240.

Toldt, C. (1888). Lehrbuch der Gewebelehre 3. Aufl. Cited by Gentes, L., in *Trav. Lab. Soc. sci. Arcachon*, 1907, 2nd année, pp. 129–75.

Vesalius, A. (1555). De humani corporis fabrica, libri septum. Cited from Zuckerman, S. (1954), The secretions of the brain. Relation of hypothalamus to pituitary gland. *Lancet*, **266**, 739–43.

Willis, T. (1664). *Cerebri Anatome*. Martyn and Allestry, London. Translated by S. Pordage, 'The anatomy of the brain', in *The Remaining Works of Dr. Thomas Willis*, 1681; see Feindel, W. 1965 (above).

Willis, T. (1674, 1675). *Pharmaceutice rationalis sive diatriba de medicamentorum operationibus in humano corpore*, 2 Vols. Oxford and London.

Zuckerman, S. (1954). The secretions of the brain. Relation of hypothalamus to pituitary gland. *Lancet*, **266**, 739–43.

2 *George B. Stefano*

Stereospecificity as a determining force stabilizing families of signal molecules within the context of evolution

2.1. On the evolution of chemical signal systems

It has become evident that intercellular communication is mediated primarily by chemical signal molecules. During the course of evolution, organisms in which this form of communication developed appear to have increased their chances of survival and thus passed this trait on to their descendants. A plausible explanation for the emergence/dominance of this mechanism of communication can be based on its inherent level of sophistication, as noted not only by synaptic molecules that can enter into intercellular communication, but hormonal ones as well. A further advantage of this method is that it is not limited by spatial requirements. Mechanisms which employ direct contact, by their nature, require a great deal of contact space, whereas the only requirement of chemical communication is scaled-down space for receptors. This also allows for a greater diversity of the signal molecules and their corresponding receptors, and the same chemical communication mechanism would permit synaptic growth and plasticity. The end-result of such chemical communication mechanisms would be a higher degree of sophistication and detailed information transfer, which allows for a greater number of behavioural characteristics (including afferent hormonal influences) to enhance an organism's chance for survival in a changing environment. If indeed this system was favoured, it can be predicted that the organisms accumulating the greatest diversity of cellular communication would eventually begin to control their environment. Thus, these organisms would expand our concept of natural selection. Also, it could be predicted that the anatomy of the chemical communication mechanism during evolution would be diversified. That is, the distance between the origin of the signal molecule and its receptor does not have to be fixed. In other words, the closer the origin of a signal molecule is to its target receptor, the quicker the response. Therefore, at the extremes it is possible to have ongoing long-term communication (hormonal) which is not essential for immediate action and immediate short-term communication (synaptic).

In order to understand the significance and mechanisms of conserving information in DNA we should consider the following: molecular evolution, in part, concerns itself with determining the genetic basis of natural selection. Changes in the genetic code, if favourable, will make an organism better-prepared to cope with its environment. One of the most dominant changes in DNA that may occur is the point mutation (base addition, removal or change). In recent years other mechanisms of DNA alteration have been found to exist (DNA transposition, unequal cross-over and reverse transcriptase). Many exogenous as well as endogenous factors exist that can induce these errors.

Many of the protein molecules active in this communication appear to be derived from larger polypeptide gene products. This is certainly true of the various opioid peptides (for a recent review on the subject see Leung and Stefano, 1987). Briefly, the endogenous opioids with the lowest molecular weights are the pentapeptides, Met-enkephalin and Leu-enkephalin. The prefixes indicate the difference between these two enkephalins on the last residue. The first four residues, Tyr–Gly–Gly–Phe, are identical for both compounds. The higher molecular weight bioactive opioids in general contain one of the two enkephalin sequences at their N-terminals. In the case of peptide E and F of proenkephalin, they each contain an additional copy of the enkephalin sequence.

All known opioid peptides may be classified into three families: the pro-opiomelanocortin (POMC), the proenkephalin and the prodynorphin families. This classification is based upon the macromolecular precursors from which these neuropeptides are derived. These precursors all contain at least one copy of an enkephalin sequence. Interestingly, nearly all the opioid peptides in the three precursors are bracketed by paired basic amino acids on the N- and C-terminals. Recently, with the use of cDNA techniques, the nucleotide sequences of the genes as well as the amino acid sequences for all three precursors have been determined. Although there is no direct evidence, the tremendous sequence homology among the precursors does strongly suggest a common origin. In addition, there is considerable similarity among the genes of the precursors in the arrangements of the coding.

The peptides of the POMC family are derived from a single macromolecular precursor. POMC is an interesting molecule in that its processed products include hormones such as ACTH, LPH, a-MSH, and B-MSH as well as the opioid peptide, β-endorphin. B-endorphin possesses a very strong analgesic effect and contains the Met-enkephalin sequence at its N-terminal. The evolutionary and physiological significance of the structural association between β-endorphin and the other opioid peptides remain unclear at this point. However as POMC appears to be processed to different final products in the anterior and neurointermediate lobes of the pituitary gland, it is possible that POMC may serve different functions in different tissues. In POMC, β-endorphin represents the last 31 amino acid sequence of the precursor. The two amino acids at its N-terminal are Lys and Arg. Thus, β-endorphin is released by neuropeptide processing enzyme with trypsin-like activity. The absence of a basic paired amino acid after the Met-enkephalin in β-endorphin makes it an unlikely precursor for Met-enkephalin. In addition, immunocytochemistry studies have shown that β-endorphin and enkephalin do not coexist in the same regions of the brain.

The cDNA analyses of a number of intermediate-size peptides in bovine adrenal chromaffin cells led to the complete identification of the macromolecular precursor proenkephalin. These studies raised the question whether the enkephalins are the only 'intended' final products of the precursor. This is especially true since some of the 'intermediates' do exist at a relatively high level in the tissues and have very strong opioid activities. Nevertheless, the primary structure of proenkephalin showed that of the six copies of Met-enkephalin sequence, and one copy of Leu-enkephalin sequence, only two Met-enkephalin sequences are not bracketed at the N- and C-terminal by paired basic amino acids. They are Met-enkephalin-Arg^6-Phe^7 and Met enkephalin-Arg^6-Gly^7-Leu^8. Thus, enkephalins can clearly be generated from proenkephalin neuropeptide processing enzymes. In addition, time course studies with bovine chromaffin granules showed that with time the proenkephalin intermediates in the granules are processed to Met- and Leu-enkephalin as well as the heptapeptide and octapeptide, and their relative proportions are also found to remain constant.

The ratio of Met-enkephalin sequence to Leu-enkephalin sequence in proenkephalin is 6:1. On the other hand, prodynorphin contains three copies of Leu-enkephalin sequence, but no Met-enkephalin sequence. Interestingly, the Leu-enkephalin sequences of prodynorphin are also bracketed by paired amino acid sequences. Similar to proenkephalin, prodynorphin in the tissues appears to be processed first to various dynorphins which have potent opioid activities and contain the Leu-enkephalin sequence. Here again, Leu-enkephalin may only be one of the 'intended' products of prodynorphin.

Thus, these large gene products can be specifically cleaved into smaller active or non-active components. For example, 240 amino acids are incorporated into POMC, which may then be cleaved into smaller biologically active peptides whose structure can vary depending on the tissue and species. Apparently, this processing also occurs in protozoa (Le Roith *et al.*, 1982) and molluscs (Leung and Stefano, 1987). In addition to the coding regions of DNA (exons) these larger polypeptide-producing genes contain non-coding regions (introns). It has been hypothesized that introns (found only in eukaryotic genes) may function as 'spacers', contain codes for enzymatic cleavage, allow for proper folding of the smaller peptide and/or allow for different combinations of the smaller peptides to possibly make a slightly larger one. An additional function for introns can be proposed: introns may be viewed as structures which have the ability to increase an organism's life span because they can act as 'damage buffers'. Eukaryotic cells contain 10 times the amount of genes as prokaryotic cells, as well as having 100 times the amount of DNA. If genetic mutations occur they will most likely affect intron regions. Organisms that have these regions would therefore stand a better chance of survival. In short, introns may be viewed as a naturally occurring anti-ageing mechanism.

Eukaryotic genes tend to have numerous copies of certain exons, as is the case of methionine enkephalin in proenkephalin A. These repetitions of important exons also may be regarded as a mechanism to ensure signal fidelity in older organisms. Again, if an important signal molecule is going to be used during an organism's life span precautions have to be incorporated into stabilizing the molecule. By repeating the sequence of a biologically important molecule over and over the chances of retaining a good copy are enhanced. If this molecule is hit by misfortune its replicates will take its place. Thus, molecular redundancy of key sequences may also be regarded as another anti-ageing mechanism.

Other phenomena which are equally important are the mechanism of simultaneous expression of the signal molecule and the receptor in different cell types. It is known that the same signal system can be used in different or the same ways in different phyla. What causes the precise expression of both complementary systems? This dual expression certainly suggests the existence of a functional interaction of these two 'separate' aspects of the same signal system. It would be interesting to speculate that since both 'up' and 'down' regulation of a receptor population can be regulated by the concentration of the signal molecule, the signal molecule itself can induce the presence of its receptor in a distant cell (non-specific pinocytosis coupled to DNA disinhibition or initiation, etc., activating a dormant recognition system).

2.2. **Stabilization of a signal system within evolution**

The presence of biologically active neuropeptides in invertebrates, which are comparable to those of vertebrates, has been known for a considerable period of time (Frontali and Gainer, 1977; Haynes, 1980; Scharrer, 1967, 1978). The recent upsurge of interest in the diverse roles and modes of operation of these molecules has sparked a search for their evolutionary history. While several reports on the occurrence of endogenous opioids in submammalian vertebrates have become available (Audigier *et al.*, 1980), comparable data in invertebrates are emerging. They consist of the demonstration in certain invertebrate ganglia of either opioid peptides or their specific receptor sites (review – Leung and Stefano, 1987).

Immunocytochemical methods have revealed enkephalin-like materials in the nervous systems of a cephalopod (*Octopus vulgaris*, Martin *et al.*, 1979), a gastropod (*Achatina fulica*, Van Noorden *et al.*, 1980), a pelecypod (*Mytilus edulis*, Stefano and Martin, 1983) and a leech (*Haemopis marmorata*, Zipser, 1980). Enkephalin and β-endorphin-like products have been demonstrated in the cerebral ganglion of the earthworm, *Lumbricus terrestris* (Alumets *et al.*, 1979) of *Mytilus edulis*, and an α-endorphin-like peptide in the suboesophageal ganglion of another lumbricid, *Dendrobaena subrubicunda* (Remy and Dubois, 1979). The latter compound seems to occur also in the suboesophageal ganglion of a lepidopteran larva, *Thaumetopoea pitocampa* (Remy *et al.*, 1978). In addition, ACTH and β-endorphin-like amino acid sequences were detected immunologically in protozoa as part of a high-molecular-weight macromolecule (LeRoith *et al.*, 1982).

The presence of an opiate receptor mechanism in the central nervous system of the marine mollusc, *Mytilus edulis*, was suggested by a rise in dopamine level following intracardiac administration of exogenous Met- and Leu-enkephalin, an effect reversible by naloxone (Stefano, 1982). Similar results were obtained in the freshwater bivalve *Anodonta cygnea* and the land snail *Helix pomatia* (Osborne and Neuhoff, 1979). The first actual demonstration of high-affinity opiate-binding sites in an invertebrate ganglion was accomplished by Stefano *et al.* (1980) in *Mytilus edulis*. The biochemical characteristics of this system, analysed in detail by Kream *et al.* (1980), have been found to parallel those of mammalian systems. With respect to specific binding sites in insects, early indications are the studies by Pert and Taylor (1980), and Edley *et al.* (1982), who showed that suspensions prepared from *Drosophila* heads avidly bind [^{3}H]Leu-enkephalin and the opioid ligand [^{3}H]diprenorphine. Recently, specific high-affinity binding sites for a synthetic enkephalin analog, D-Ala^2-Met^5-enkephalinamide (DAMA), were demonstrated in the cerebral ganglia and midgut of the insect *Leucophaea* (Stefano and Scharrer, 1981). The results strongly suggest the presence, in this invertebrate, of opiate receptors that are confined to certain areas of the nervous tissue. Again these opioid receptors were found to resemble those described in mammalian systems.

Enkephalin immunoreactivity has been demonstrated in the pedal, cerebral and visceral ganglia and various connectives of *Mytilus edulis* (Stefano and Martin, 1983) in which previous studies have established the presence of opiate-binding sites which appear to interact with dopaminergic neural elements (Stefano *et al.*, 1982). Recently, Met- and Leu-enkephalin and Met-enkephalin-Arg^6-Phe^7 were isolated and identified by HPLC analysis of acid extracts of *M. edulis* pedal ganglia (Stefano and Leung, 1986). Taken together, these results demonstrate that the structure, receptor characteristics, and presynaptic modulatory activities of these neuropeptides have been conserved throughout evolution. This in turn leads one to suggest that mechanisms exist to maintain certain intercellular signal molecules during evolution.

Furthermore, we now know that opioid peptides are involved with immuno-autoregulatory mechanisms involving specific immunocytes in *Mytilus edulis* (Stefano *et al.*, 1989a,b). DAMA can initiate both cell adhesion as well as migration in invertebrate immunocytes and human granulocytes. These activities are quite similar to that which has been reported to occur in mammals (Stefano, 1989). Thus, further strengthening the argument that information has been retained during evolution concerning not only a signal family but its function as well.

Why should these signalling molecules and their apparent systems/mechanisms be retained relatively intact during the course of evolution? In order to answer this question we must briefly review some basic principles of intercellular signalling. It should also be noted at this time that the same may apply to intracellular signal systems. Major requirements of a compound to be established as a signalling molecule, be it as a neurotransmitter or a hormone, are (1) presence of the molecule in a particular cell, (2) its release from that cell upon appropriate stimulation, (3) high-affinity, stereospecific binding to a receptor on the target cell, (4) a specific, physiological effect of the molecule on the effector cell, and (5) a specific inactivation mechanism.

In peptidergic signal systems, these characteristics are directly gene-determined. The enzymes that synthesize and process such signal molecules must be present in their cells of origin and the information to produce these enzymes resides, obviously, in the DNA of the cell. This is also true of the stereoselective receptor molecules found on the target cell as well as all stereoselective components of a given intercellular communication system. The entire sequence, from synthesis of the signal molecule to its inactivation, is based on sequential stereospecific events, including, in another cell, receptor recognition. Therefore, the components of the system had to evolve simultaneously in order for the system to be operational. Compatible structural conformations had to be found in the synthesizing enzymes, the signal molecule, the receptor molecule, and the inactivation enzymes. Such conformational 'matching' of molecules within each signal system is difficult and time-consuming to achieve. In addition, to be operational within an organism, all the components had to be expressed simultaneously. Thus evolutionary changes had to occur on the corresponding genes of the components within a particular signal system if large-scale changes were to take place. The conformational complexity and rigidity of the 'match' among the sequential components for a given signal system would thus seem to exert a determining influence during evolution to maintain the conformational integrity of the signal system given the degree of difficulty in obtaining it originally

(Stefano, 1984, 1986). Thus, 'ancient' communication systems, e.g. opioids, would tend to remain relatively intact in increasingly complex animal phyla, especially the structure or conformation of the bioactive portions of the molecules themselves.

This principle of conservation does not preclude events that may lead to an old signal system being used in a new functional capacity. In summary, the determining force during evolution which appears to maintain signal systems may well be the number of highly precise stereospecific conformational matching events associated with intercellular communication mechanisms.

This same consideration may also be applied to intracellular communication systems being maintained throughout evolution. Certainly cAMP appears to occur in all organisms. In addition, the primary structure of calmodulin, a multifunctional intracellular messenger, is believed to be retained since mammalian antibodies appear to react with extracts from coelenterates (Cheung, 1980).

The list of 'mammalian type' signalling molecules in 'simpler' organisms is steadily growing. In prokaryotes, chorionic gonadotropin-like material has been detected (Acevedo *et al.*, 1978; Stefano, 1988). In protozoans not only have similar mammalian-type neuropeptides been detected, but opioids have been shown to alter feeding behaviour, an effect inhibited by naloxone, the opiate receptor blocker (Josefsson and Johansson, 1979). Taken together, the evidence indicates that signal effector and receptor communication mechanisms may be present in unicellular organisms. This in turn suggests that the origin of signal systems may have occurred during prokaryotic development. Indeed, many of these systems may have started out as intracellular communication mechanisms. Thus, the term 'neuropeptides' may be totally erroneous, even in vertebrates, since many serve and are found in locations other than neural tissues.

Aside from those in unicellular organisms, the list of mammalian-type peptides found in various invertebrates is quite impressive (Stefano, 1988). Both molluscs and insects have been shown to possess high-affinity opioid receptors (Stefano *et al.*, 1980; Stefano and Scharrer, 1981). Met-enkephalin, Leu-enkephalin, and Met-enkephalin-Arg^6-Phe^7 have now been demonstrated in invertebrates (Leung and Stefano, 1987). The endogenous opioid system appears to be strongly involved in regulating dopamine metabolism in molluscs (Stefano, 1982). In *M. edulis* this system appears to be part of the regulatory system associated with central regulation of cilioinhibition in the organism's visceral ganglia (Aiello *et al.*, 1986) and in modulating anterior byssus retractor activity (Bianchi and Wang, 1986). Thus in molluscs opioid mechanisms have been firmly established.

If one examines all the information available to date regarding opioids in invertebrates, a complex picture emerges. There is evidence for variations in post-translational processing in different molluscs (Martin *et al.*, 1986). There is also preliminary evidence for the existence of the three mammalian-type gene product precursors (or one or two large polypeptides) based on the various smaller opioid substances reported to exist. Opiomelanocortin products also have been identified in molluscs (Martin *et al.*, 1986). β-endorphin is quite potent in displacing [^{3}H]DAMA binding in molluscs (Kream *et al.*, 1980). Proenkephalin products have been reported in leech extracts (Flanagan and Zipser, 1986), and dynorphin is a potent binding agent in molluscan neural tissue extracts (Leung and Stefano, 1987). Proenkephalin products have been detected in molluscs, e.g. Met-enkephalin-Arg^6-Phe^7, and it seems present in an amidated form (Voigt and Martin, 1986). As a result quite a complex opioid picture emerges in invertebrates.

The accumulating evidence also strongly indicates that this is not a 'static' system in invertebrates, but is one that is highly variable in regard to quantitative changes, even within the same species and in the same tissue. Ageing variations have been shown to occur in regard to opioid levels and high-affinity opioid binding site densities (Leung and Stefano, 1987) as also noted in mammals (Codd and Byrne, 1980). Seasonal variation in opioid binding densities occurs in *M. edulis*, (Stefano and Leung, 1986). In addition, seasonal changes apply to the endogenous opioid levels in the pedal ganglia of *M. edulis* (Stefano and Leung, unpublished). These variations manifest themselves in changing ratios of the three detected opioids, including periods of time when they are below our level of detection. Seasonal opioid variations also occur in leeches (Flanagan and Zipser, 1986) and in the anterior byssus retractor muscle response to morphine (Bianchi and Wang, 1986). Clearly, the opioid systems in invertebrates are complex and show striking similarities with those of mammals, thus indicating a common evolutionary trend.

2.3. **How can change or diversification be introduced into this conservative concept of signal system evolvement?**

It would seem as though change has occurred in various signal system 'families'. Even though substance P can exert physiological effects in invertebrates, attempts to isolate it biochemically have proven unsuccessful (Kream *et al.*, 1986). However, closely related molecules, the tachykinins, have been found in lower life forms (physaelamine in amphibians and eledoisin in Octopus). Given the rather conservative nature of signal system evolvement noted in this reveiw, how can one account for the diversified signal molecules found in higher organisms? It is likely that the answer lies in the concept of conformational matching. The most significant aspect of conformational recognition within the stereospecific components of a signal system is by definition its functional shape. Substitution of amino acids would or could be tolerated only if the characteristic conformation of the signal system component (enzymes, receptor, messenger molecule, etc.) retains its functional shape, within very narrow limits, at each recognition event. Thus, an amino acid (nucleotide level) may be substituted and the resulting point mutation be retained and incorporated into the signal system as long as the molecule maintains its proper conformation. I would speculate that the pliable nature of a signal molecule, especially peptides, can be modified at a recognition step by the normal microenvironmental factors (pH, ionic composition, etc.). This therefore implies that there exists a degree of conformational 'give'. It would be possible to initiate diversification of particular signal systems within this narrow 'give' range. Given the small amount of 'play' allowed, change would still occur at a very slow pace. The immunoreactive histochemical studies have certainly demonstrated that strong structural similarities exist with regard to peptides of vertebrate origin (in which they were first discovered) being found in invertebrates. Thus, signal system 'families' are centralized around an essential conformation.

The conservative characteristic of signal system evolution can also be noted in the basic structure of neurons and of nervous systems in general. All neurons can generate action potentials and then propagate them to a point where they are coupled to a secretion event. Nervous systems in total are built around three aspects of function: sensory, integrative and motor. Endocrine systems also have similar characteristics. If anything, there is a basic theme in all systems that is modified to be distinctive. As a result of the basic theme an organism requires less DNA to be different, only messages that can differ slightly to put a variation on the central theme. For example, secretion tends to involve certain events that are constant regardless of the tissue involved (Ca^{2+} dependency, etc.).

In summary, the data on invertebrate opioids indicate that post-translational variation exists in addition to post-translational modification processes. There is diversity in the number of opioid products presumably originating from one or more common precursors (Stefano and Leung, 1986). The opioid system is dynamic and varying with age and seasons. Both ageing (Stefano and Leung, 1986) and seasonal variations have been shown to exist in mammals, although the literature in this regard is quite sparse. Variations in the processing of opiomelanocortin are known to occur in different higher vertebrate neural tissues (Smyth, 1986). This demonstrates flexibility and tissue specificity in processing mechanisms, as also noted in invertebrates (Martin *et al.*, 1986).

Since many opioid characteristics are common to both invertebrates and vertebrates, their 'ancient' status appears to be established. Again, we must ask why this signal system would remain intact, especially the shorter molecules. I believe the answer in part is in the need for conformational matching of the stereospecific nature of these systems, as mentioned earlier. Each signal molecule contains a critical area responsible for the initiation of biological activity, and the conformation of this critical area is the most important aspect of the molecule. For example, only the first 24 of 39 amino acids of ACTH are required for its full activity and this important region is stable among various mammals. However, at the C-terminal and (25–39) variation of amino acid composition occurs and seems tolerable. Another interesting pattern emerges when we examine ACTH neurochemistry and induced cellular and organismal behaviour of this peptide. Firstly, POMC can be processed differently depending on its anatomical location. ACTH and beta-endorphin are the predominant end-products in the anterior pituitary of the rat. In the intermediate lobe of the pituitary and brain POMC is processed to CLIP, alpha-MSH and β-endorphin or the N-terminal fragments alpha- and gamma-endorphin (review see Civelli *et al.*, 1987). Interestingly, each of these smaller peptides has its own distinct biological activity. Furthermore, the number of biologically active POMC-derived peptides that can be generated by additional processing (phosphorylation, acetylation, glycosylation and sulphation) have the ability to alter biological activity in a equally specific manner (Civelli *et al.*, 1987). Secondly, POMC-derived peptides appear to be involved in numerous behaviour-modulating functions. These peptides can influence aggression, grooming, nerve cell regeneration, sexual behaviour, pain, immobility, sedation, learning, memory, mood, social behaviour, motivation, attention, maternal behaviour, addiction, temperature regulation, food intake, immune and neuroimmune activities (Clement-Jones and Besser, 1987; Stefano, 1989). Thus, these related peptides are quite large in their sphere of influence as well as the potential number of active signal molecules and their corresponding receptors. Given this great wealth of communication material how can we account for all the information being maintained during evolution and its apparent yet subtle diversity?

In examining ACTH we may find the clues and answers to these questions. The ACTH tetrapeptide (4–7) is the functional core of the ACTH molecule since it is the smallest unit to have ACTH behavioural activity (Grevin and De Wied, 1973). The biologically active part of the molecule for inducing grooming is also 4–7 (Gispen and Isaacson, 1986) whereas nerve regeneration is 6–10 of ACTH (Bijlsma *et al.*, 1983). Opiate antagonistic effects can be manifested in the derivative ACTH-7–10 (Plomp and Van Ree, 1978). Additionally, animal tissues appear to be able to process these ACTH peptides to biologically smaller peptides (Civelli *et al.*, 1987). In one study in particular, depending on the pH, different *in vitro* products emerged, demonstrating the potential of the physiological environment to modify immediate end-product processing (Wang *et al.*, 1983). Additionally, β-endorphin (1–31) displays opioid activity, whereas 1–27 and 1–26 are competitive antagonists that do not have this activity (Akil *et al.*, 1981). Clearly, the stereospecific (conformational) structure of the resulting peptide is critical for its agonist as well as antagonist properties.

Thus, this limited review of the data suggests that large opioid precursors (maybe all neuropeptide precursors) carry not only potentially active ligands but endogenous antagonists as well. This may not be as surprising as it may sound, since an agonist must not only bind but be able to initiate activity at a stereospecific receptor. Thus, the potential to generate an antagonist is present in the agonist since both bind. Cleaving the portion of the molecule that initiates activity would surely create an antagonist. Again, vital information is multiplied and conserved in DNA, since it has such great potential for diversity of expression. This phenomenon may also be critical in stabilizing signal system families (agonists and antagonists) during the course of evolution.

If this serves as an example, we should expect to find more short chain sequences of biologically important molecules or parts of these molecules to be identical in most phyla, whereas variation may be found in the larger signalling molecules, especially in regions distant from the active 'centre'. Somatostatin also appears to be highly conserved in that 35 of the 42 nucleotides coding for it are conserved in rat and angler fish (Goodman *et al.*, 1982). These organisms are believed to have diverged in evolution about 400 million years ago. The same appears to be true for gastrin/CCK, as noted in a recent report (Hansen *et al.*, 1987), and insulin (LeRoith *et al.*, 1980). Indeed, these modifications may be considered to be the evolutionary changes or 'advances'. In brief, over long periods of time, additions to essential 'key' conformations may occur, increasing the efficiency of a particular signal system. Thus, as in the case of the opioid substances, it may be more proper to speak of a 'family' of compounds. The establishment of such a family lies in the rigidity of the conformational components and events occurring simultaneously in the involvement of vital biochemical reactions. Therefore, conformational matching becomes a determining force and not a selective force in establishing a basic signal system that can be enriched as time goes on.

Acknowledgments

The author was in part supported by Grants NIH-MBRS 08180 and ADAMHA-MARC 17138, the Long Island Community Foundation, Inc., Hoffmann LaRoche, the Upjohn Company and the State University of New York Research Foundation. The major portion of this report comes from *Comp. Biochem. Physiol.*, **90C**, 287–294. The author wishes to acknowledge thoughtful comments from B. Scharrer.

References

Acevedo, H. F., Slifkin, M., Pouchet, G. R. and Pardo, M., (1978). Immunochistochemical localization of a choriogonadotropin-like protein in bacteria isolated from cancer patients. *Cancer,* **41**, 1217–29.

Aiello, E., Hager, E., Akiwumi, C. and Stefano, G. B. (1986). An opioid mechanism modulates central and not peripheral dopaminergic control of ciliary activity in the marine mussel *Mytilus edulis. Cell. Molec. Neurobiol.*, **6**, 17–30.

Akil, H., Young, F., Watson, S. J. and Coy, D. H. (1981). Opiate binding properties of naturally occurring N- and C-terminus modified beta-endorphins, *Peptides*, **2**, 289–92.

Alumets, J., Hakanson, R., Sundler, F. and Thorell, J. (1979). Neuronal localization of immunoreactive enkephalin and B-endorphin in the earthworm. *Nature (Lond.)*, **279**, 805–7.

Audigier, Y., Deprat A. M. and Cros, J. (1980). Comparative study of opiate and enkephalin receptors on lower vertebrates. *Comp. Biochem. Physiol.*, **67**, 191–4.

Bianchi, C. P., and Wang, Z. (1986) Morphine enhancement of the cholinergic response of anterior byssus retractor muscle of *Mytilus edulis*. In: Stefano, G. B. (ed.): *Handbook of Comparative Opioid and Related Neuropeptide Mechanisms*, vol. 2. CRC Press, Boca Raton, FL, pp. 59–64.

Bijlsma, W. A., Schotman, P., Jennekens, F. G. I., Gispen, W. H. and De Wied, D. (1983). The enhanced recovery of sensorimotor function in rats is related to the melanotropic moiety of ACTH/MSH neuropeptides. *Eur. J. Pharmacol.*, **92**, 231–6.

Cheung, W. Y. (1980). Calmodulin plays a pivotal role in cellular regulation. *Science*, **207**, 19–27.

Civelli, O., Douglass, J. and Herbert, E. (1987). Proopiomelanocortin: a polyprotein at the interface of endocrine and nervous systems. In: Udenfriend, S. and Meishofer, J. (eds), *Opioid Peptides: Biology, Chemistry and Genetics*. Academic Press, New York, pp. 70–95.

Clement-Jones, V. and Besser, G. M. (1987). Opioid peptides in humans and their clinical significance. In: Udenfriend, S. and Meishofer, J. (eds), *Opioid Peptides: Biology, Chemistry and Genetics*. Academic Press, New York, pp. 324–90.

Codd, E. E. and Byrne, W. L. (1980). Seasonal variation in the apparent number of H-naloxone binding sites. In: Way, E. L. (ed.), *Endogenous and Exogenous Opiate Agonists and Antagonists*. Pergamon Press, New York, pp. 67–71.

Edley, S. M., Hall, L., Herkenham, M. and Pert, C. B. (1982). Evolution of striated opiate receptors. *Brain Res.*, **249**, 184–38.

Flanagan, T. and Zipser, B. (1986). Opioid-peptide and substance P immunoreactivity in cytological preparations and tissue homogenates of the leech. In: Stefano, G. B. (ed.): *Handbook of Comparative Opioid and Related Neuropeptide Mechanisms*, vol. 1. CRC Press, Boca Raton, FL, pp. 165–80.

Frontali, N. and Gainer, H. (1977). Peptides in invertebrate nervous system. In: Gainer, H. (ed.): *Peptides in Neurobiology*. Plenum Press, New York, pp. 259–71.

Gispen, W. H., and Isaacson, R. L. (1986). Excessive grooming in response to ACTH, In: de Wied, D., Gispen, W. H. and van Wimersma Greidanus, Tj. B. (eds), *Neuropeptides and Behavior*, vol. 1. Pergamon Press, Oxford, pp. 273–312.

Goodman, R. H., Jacobs, J. W., Dee, P. C. and Habener, J. R. (1982). Somatostatin-28 encoded in a cloned cDNA obtained from a rat medulary thyroid carcinoma. *J. Biol. Chem.*, **257**, 1156–9.

Grevin, H. M. and De Wied, D. (1973). The influence of peptides derived from corticotropin (ACTH) on performance. Structure–activity studies, *Prog. Brain Res.*, **39**, 429–42.

Hansen, G. N., Hansen, B. L. and Scharrer, B. (1987). Gastrin/CCK-like immunoreactivity in the corpus cardiacum-corpus allatum complex of the cockroach *Leucophaea maderae. Cell. Tiss. Res.*, **248**, 595–8.

Haynes, L. W. (1980). Peptide neuroregulation in invertebrate. *Prog. Neurobiol.*, **15**, 205–23.

Josefsson, J. O. and Johansson, P. (1979). Naloxone reversible effect of opioid on pinocytosis in *Amoeba proteus. Nature (Lond)*, **78**, 283–92.

Kream, R. M., Zukin, R. S. and Stefano, G. B. (1980). Demonstration of two classes of opiate binding sites in the nervous tissue of the marine mollusc *Mytilus edulis*. Positive homotropic cooperativity of lower affinity binding sites. *J. Biol. Chem.*, **255**, 9218–24.

Kream, R. M., Leung, M. K. and Stefano, G. B. (1986). Is there authentic substance P in invertebrates? In: Stefano, G. B. (ed.). *Handbook of Comparative Opioid and Related Neuropeptide Mechanisms*, vol. 1. CRC Press, Boca Raton, FL., pp. 65–72.

LeRoith, D., Shiloach, J., Roth, J. and Lesniak, M. A. (1980). Evolutionary origins of vertebrate hormones: substances similar to mammalian insulins are native to unicellular eukaryotes. *Proc. Natl. Acad. Sci., USA.*, **77**, 6184–8.

LeRoith, D., Liotta, A. S., Roth, J., Shiloach, J., Lewis, M. E., Pert, C. B. and Krieger, D. T. (1982). ACTH and β-endorphin-like materials are native to unicellular organisms. *Proc. Natl. Acad. Sci., USA*, **79**, 2086–90.

Leung, M. K. and Stefano, G. B. (1987). Comparative neurobiology of opioids in invertebrates with special attention to senescent alterations. *Prog. Neurobiol.*, **28**, 131–59.

Martin, R., Frosch, D., Weber, E. and Voigt, K. H. (1979). Met-enkephalin-like immunoreactivity in a cephalopod neurochemical organ. *Neurosci. Lett.*, **15**, 253–7.

Martin, R., Haas, C. and Voigt, K.-H. (1986) Opioid and related neuropeptides in molluscan neurons. In: Stefano, G. B. (ed.), *Handbook of Comparative Opioid and Related Nueropeptide Mechanisms*, vol. 1. CRC Press, Boca Raton, FL., pp. 49–64.

Osborne, N. N., Cuello, A. C. and Dockray, G. J. (1982). Substance P and cholecystokinin-like peptides in *Helix* neurons and cholecystokinin and serotonin in a giant neuron. *Science*, **216**, 409–11.

Osborne, N. N. and Neuhoff, V. (1979). Are there opiate receptors in the invertebrates? *J. Pharm. Pharmacol.*, **31**, 481.

Pert, C. B. and Taylor, D. (1980). Type 1 and type 2 opiate receptors: a subclassification scheme based upon GTP's differential effects on binding. In: Way, E. L. (ed.), *Endogenous and Exogenous Opiate Agonists and Antagonists*. Pergamon Press, New York, pp. 87–94.

Plomp, G. J. J. and Van Ree, J. M. (1978). Adrenocorticotrophic hormone fragments mimic the effect of morphine *in vitro. Br. J. Pharmacol.*, **64**, 223–7.

Remy, C. and Dubois, M. P. (1979). Localization par immunofluorescence de peptides analogues a l'endorphine dans les ganglions infraoesophagiens du lombricide Dendrobaena subrubicunda Eisen. *Experientia*, **35**, 137–8.

Remy, C., Girardie, J. and Dubois, M. P. (1978). Présence dans le ganglion sous-oesophogien de la chenille processionnaire du pin de cellules révélées en immunofluorescence par un anticorps anti-β-endorphine. *C.R. Acad. Sci. (Paris)*, Ser. D., **286**, 651–3.

Scharrer, B. (1967). The neurosecretory neuron in neuroendocrine regulatory mechanisms. *Am. Zool.*, **7**, 161–8.

Scharrer, B. (1978). Peptidergic neurons: facts and trends. *Gen. Comp. Endocrinol.*, **34**, 50–62.

Smyth, D. G. (1986). Flexibility in the processing of beta-endorphin. In: Stefano, G. B. (ed.), *Handbook of Comparative Opioid and Related Neuropeptide Mechanisms*, vol. 1. CRC Press, Boca Raton, FL, pp. 37–40.

Stefano, G. B. (1982). Comparative aspects of opioid-dopamine interaction. *Cell. Mol. Neurobiol.*, **2**, 167–78.

Stefano, G. B. (1984). Evolutionary considerations regarding signal systems. Symposium SUNY/College at Old Westbury, Old Westbury, New York, Abs. G.1.

Stefano, G. B. (1986). Conformational matching: a possible evolutionary force in the evolvement of signal systems. In: Stefano, G. B. (ed.), *Handbook of Comparative Opioid and Related Neuropeptide Mechanisms*, vol. 2. CRC Press, Boca Raton, FL, pp. 271–7.

Stefano, G. B. (1988). The evolvement of signal systems: conformational matching a determining force stabilizing families of signal molecules. *Comp. Biochem. Physiol.*, **90C**, 287–94.

Stefano, G. B. (1989). Role of opioid neuropeptides in immunoregulation. *Prog. Neurobiol*, **33**, 149–59.

Stefano, G. B. and Leung, M. K. (1986). Opioid aging and seasonal variations in invertebrate ganglia: evidence for an opioid compensatory mechanism. In: Stefano, G. B. (ed.), *Handbook of Comparative Opioid and Related Neuropeptide* Mechanisms, vol. 2. CRC Press, Boca Raton, FL., pp. 199–209.

Stefano, G. B. and Martin, R. (1983). Enkephalin-like immunoreativity in the pedal ganglion of *Mytilus edulis* (Bivalvia) and its proximity to dopamine-containing structures. *Cell. Tiss. Res.*, **230**, 147–54.

Stefano, G. B. and Scharrer, B. (1981). High affinity binding of an enkephalin analog in the cerebral ganglion of the insect *Leucophaea maderae* (Blattaria). *Brain Res.*, **225**, 107–14.

Stefano, G. B., Kream, R. M. and Zukin, R. S. (1980). Demonstration of stereospecific opiate binding in the nervous tissue of the marine mollusc *Mytilus edulis. Brain Res.*, **181**, 445–50.

Stefano, G. B., Zukin, R. S. and Kream, R. M. (1982). Evidence for the presynaptic localization of a high affinity opiate binding site on dopamine neurones in the pedal ganglia of *Mytilus edulis. J. Pharmacol. Exp. Ther.*, **222**, 759–64.

Stefano, G. B., Leung, M. K., Zhao, X. and Scharrer, B. (1989a) Evidence for the involvement of opioid neuropeptides in the adherence and migration of immunocompetent invertebrate hemocytes. *Proc. Natl. Acad. Sci. USA*, **85**, 626–30.

Stefano, G. B., Cadet, P. and Scharrer, B. (1989b). Stimulatory effects of opioid neuropeptides on locomotory activity and conformational changes in invertebrate and human immunocytes: evidence for a subtype of delta receptor. *Proc. Natl. Acad. Sci. USA*, **86**, 6307–11.

Van Noorden, S., Fritsch, H. A. R., Grillo, T. A. I., Polak, J. M. and Pearse, A. G. E. (1980). Immunocytochemical staining for vertebrate peptides in the nervous system of a gastropod mollusc. *Gen. Comp. Endocrinol.*,, **40**, 375–87.

Voigt, K.-H. and Martin, R. (1986). Neuropeptides with cardioexcitatory and opioid activity in Octopus nerves. In: Stefano, G. B. (ed.), *Handbook of Comparative Opioid and Related Neuropeptide Mechanisms*, vol. 1. CRC Press, Boca Raton, FL, pp. 127–38.
Wang, X. C., Burbach, P. H., Verhoef, J. and De Weid, D. (1983). Characterization of cleavage sites by peptidases in synaptic membranes and formation of peptide fragments. *J. Biol. Chem.*, **258**, 7942–7.
Zipser, B. (1980). Identification of specific leech neurones immunoreactive to enkephalin. *Nature (Lond.)*, **283**, 857–8.

3 *Tetsuya Ikeda, Ichiro Kubota, Yasuo Kitajima and Yojiro Muneoka*

Structures and actions of neuropeptides isolated from an echiuroid worm, *Urechis unicinctus*

3.1. Introduction

In contrast to the considerable physiological, pharmacological and immunohistochemical studies that have been reported on neurotransmitters in annelids, only a few are concerned with echiuroids, an annelid-related phyla (for review see Gardner and Walker, 1982). Muneoka *et al.* (1981) and Muneoka and Kamura (1982) examined the pharmacological properties of mechanical responses of the body-wall muscles of an echiuroid worm, *Urechis unicinctus*, and suggested that acetylcholine (ACh) may be an excitatory transmitter while noradrenaline may be an inhibitory transmitter at the neuromuscular junctions in the body wall.

Recently, Ikeda *et al.* (1990) examined the biological activities of acid-water extracts of the ventral nerve cords of *U. unicinctus* on the isolated inner circular body-wall muscle of the worm, and suggested that there exist various bioactive substances in the nerve cords. Among the bioactive substances, hydrophilic ones were suggested to include at least four amine-like substances that have modulatory effects on twitch contraction of the body-wall muscle, one ACh-like and three catecholamine-like substances. Although the former substance showed cholinergic pharmacological properties in some other muscles, it seemed not to be ACh itself. In the inner circular body-wall muscle of *U. unicinctus*, this substance did not show contractile action but exhibited a potent twitch-contraction-inhibiting action. The bioactive substances retained by C-18 cartridges, on the other hand, were suggested to include peptidergic substances. That is, the retained material showed a potent contractile activity on the inner circular body-wall muscle and this activity was completely destroyed by treating the material with the peptidase subtilisin.

We attempted to purify the bioactive peptidergic substances in the retained material by using a HPLC system. Ten purified peptides that have contractile or twitch-contraction-inhibiting effect on the inner circular body-wall muscle were obtained. The structures of two of them were determined. In addition to these 10 peptides, at least 15 other species of bioactive peptidergic substances were suggested to be present in the retained material. They showed contractile, twitch-potentiating or twitch-inhibiting activity on the body-wall muscle. Some of these peptidergic substances also showed bioactivities on muscles of other animals, such as the anterior byssus retractor muscle (ABRM) of the mussel *Mytilus edulis* and the heart of the clam *Meretrix lusoria*. We report here the actions of the peptides and the structures determined on two of them.

3.2. **Extraction and separation of the bioactive peptides**

The ventral nerve cords of *U. unicinctus* were excised from 1500 specimens and immediately frozen on dry ice. The frozen nerve cords were steeped in 4% acetic acid, homogenized with a Polytron at 0°C, and centrifuged at 28,000*g* for 40 min at 4°C. The supernatant was applied to C-18 cartridges (Sep-Pak, Waters), and the retained material in the cartridges was eluted with methanol. The effluent was evaporated. The residue was taken up in 0·1 M acetic acid (0·5 ml), applied to a column (2·6 × 40 cm) of Sephadex G-15 and eluted with the same solvent. Thus, gel-filtrated fractions of 4 ml each were collected.

The biological activities of the fractions were examined on the isolated inner circular body-wall muscle of *U. unicinctus* mounted in an experimental chamber (2 ml). The tested dose of each fraction was 3 units (1 unit corresponds to one nerve cord/ml ASW). As shown in Fig. 3.1,

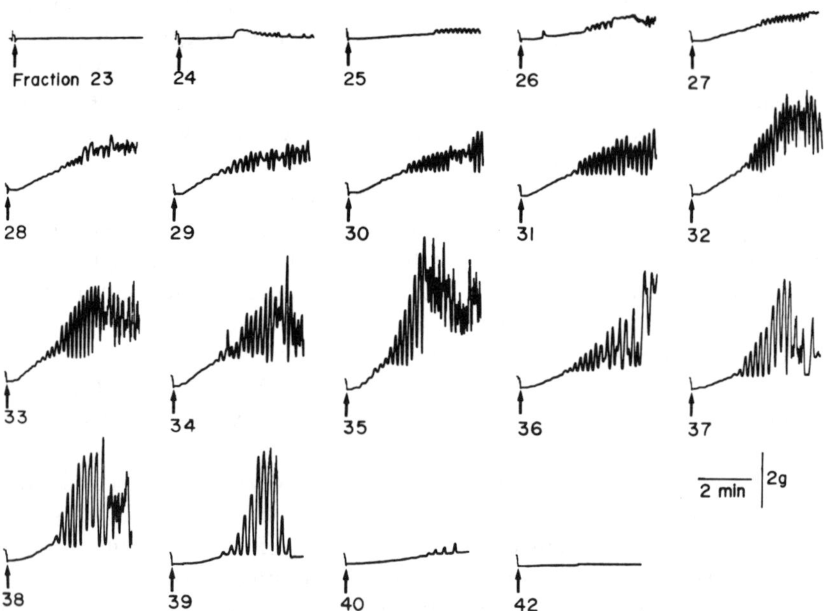

Fig. 3.1. Contractile effects of the gel-filtrated fractions 23–42 (each 3 units) on the inner circular body-wall muscle.

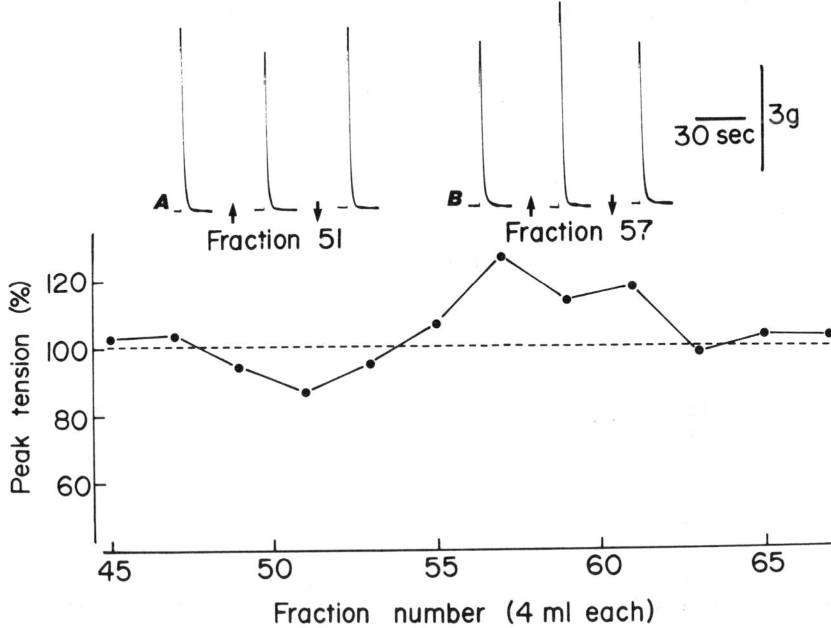

Fig. 3.2. Modulatory effects of the gel-filtrated fractions 45–67 (each 3 units) on twitch-contraction of the inner circular body-wall muscle. The twitch-contraction was elicited by stimulating the muscle with electrical pulse (20 V, 3 ms) at 10 min intervals. Each fraction was applied 8 min prior to the stimulation. Inserts A and B show inhibitory effect of the fraction 51 and potentiating effect of the fraction 57, respectively.

fractions 24–40 showed contractile activity on the body-wall muscle. The other fractions which did not show contractile activity were tested on twitch-contraction of the muscle. The twitch-contraction was elicited by applying an electrical pulse (20 V, 3 ms) of stimulation to the muscle at 10 min intervals. In these experiments we found two twitch-contraction-modulating peaks – one is of inhibition and the other is of potentiation. The maximal activities of the inhibitory and potentiating peaks were found at fractions 51 and 57, respectively (Fig. 3.2). However, the two peaks seemed to be eluted partially overlapping with each other. Therefore, we divided all the active fractions into two groups, group A (fractions 24–40) and group B (fractions 49–62). The former was the group having contractile activity and the latter was the group having modulatory activities. The active fractions of each group were pooled and then subjected to a HPLC system (Japan Spectroscopic Co. Ltd, Jas.co, TRI ROTAR-VI) to purify the bioactive substances in it. The columns used for the purification were two kinds of reversed-phase columns (C-8 and

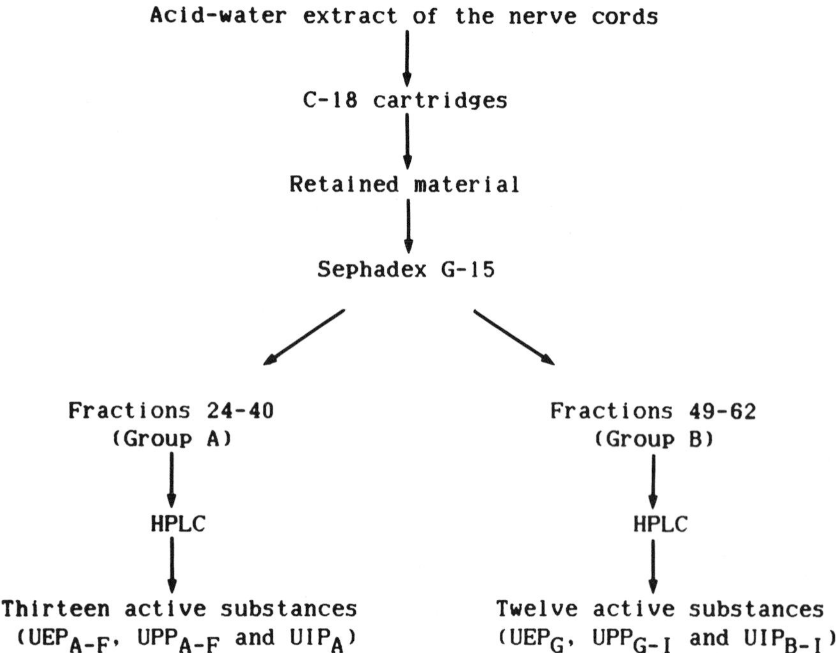

Fig. 3.3. The isolation procedures for bioactive peptides in the ventral nerve cords.

C-18 columns), a cation-exchange column and an anion-exchange column. The methods of bioassay in these purification experiments were the same as those used for the gel-filtrated fractions.

We separated 13 bioactive substances from group A (Fig. 3.3). The substances were confirmed to be peptides by treating them with subtilisin or aminopeptidase M. The substances were six contractile peptides (*Urechis* excitatory peptide A-F, UEP$_{A-F}$), six twitch-contraction-potentiating peptides (*Urechis* potentiating peptide A-F, UPP$_{A-F}$) and one twitch-contraction-inhibiting peptide (*Urechis* inhibitory peptide A, UIP$_A$). Four of the contractile peptides (UEP$_A$, UEP$_C$, UEP$_E$ and UEP$_F$) and the twitch-inhibiting peptide (UIP$_A$) were purified. The structures of UEP$_A$ and UEP$_C$ were determined. The structures of UEP$_E$, UEP$_F$ and UIP$_A$ are not yet determined; determination experiments are now in progress. The purification of the remaining eight peptidergic substances is incomplete, and experiments are in progress.

We separated 12 bioactive substances from group B (Fig. 3.3). The substances were confirmed to be peptides by enzyme-treatment experiments. The substances are one contractile peptide (UEP$_G$), three twitch-potentiating peptides (UPP$_{G-I}$) and eight twitch-inhibiting peptides

($UIP_{B\text{-}I}$). UPP_G and UPP_I also showed an inhibitory action on some preparations. It has been reported that endogenous peptides of *Mytilus edulis* show complex actions on ABRM (Hirata *et al.*, 1987; Fujisawa *et al.*, 1990). UEP_G, UIP_D, UIP_G, UIP_H and UIP_I were purified; structure-determination experiments are now in progress. The remaining peptidergic substances are not yet completely purified; purification experiments are in progress.

3.3. **Purification and structure determination of UEP_A and UEP_C**

3.3.1. *UEP_A*

The contractile peptide UEP_A was purified from the gel-filtrated fractions 24–40 through four HPLC-separation steps. At the first step a C-18 reversed-phase column (Finepak SIL $C_{18}S$, 4·6 × 150 mm, Jas.co) was used. The column was eluted with a 60-min linear gradient of 0–60% acetonitrile in 0·1% TFA (pH 2·2). We obtained three peaks of contractile activity. The fractions of the first peak which was eluted at around 20% acetonitrile were subjected to the next separation with a cation-exchange column (TSK gel SP-5PW, 7·5 × 75 mm, Tosoh). The column was eluted with a 60-min linear gradient of 0·1–0·4 M NaCl in 10 mM phosphate buffer (pH 6·8). An active peak was obtained at around 0·26 M NaCl. At the third step we used another C-18 reversed-phase column (TSKgel ODS-80T_M, 4·6 × 150 mm, Tosoh). The column was eluted with a 50-min linear gradient of 15–25% acetonitrile in 0·1% TFA. We obtained an absorbance peak (220 nm) which corresponded to the peak of contractile activity (Fig. 3.4A). Final purification was performed by applying the active material again to the reversed-phase column used at the first step and eluting isocratically with 19·5% acetonitrile in 0·1% TFA (Fig. 3.4B).

The purified active substance was subjected to amino acid analysis, amino acid sequence analysis by automated Edman degradation with a gas-phase sequencer (Applied Biosystems 470A) coupled with a PTH-amino acid analyser (Applied Biosystems 120A) and FAB-MS analysis (JEOL JMS HX-100).

Quantitative amino acid analysis showed the following amino acid composition normalizing on Lys = 2: $Asp_{2.1}$ $Ser_{1.9}$ $Gly_{2.2}$ $Ala_{4.0}$ $Leu_{0.8}$ $Tyr_{0.9}$ $Lys_{2.0}$. The determined sequence and detected amount (picomoles) of each amino acid were as follows: Ala(578)-Lys(145)-ND-Ser(126)-Gly(250)-Lys(129)-Trp(96)-Ala(343)-Asn(177)-Ser(91)-Tyr(148)-Leu(164)-ND-Ala(181)-Gly(104)-Ala(164)-Asn(58), in which ND means ‘not detected’. Asn is detected as Asp in the amino acid analysis. Trp is not usually detected in the amino acid analysis, because it is easily oxidized during analysis. Cys is not detected in either the amino acid or

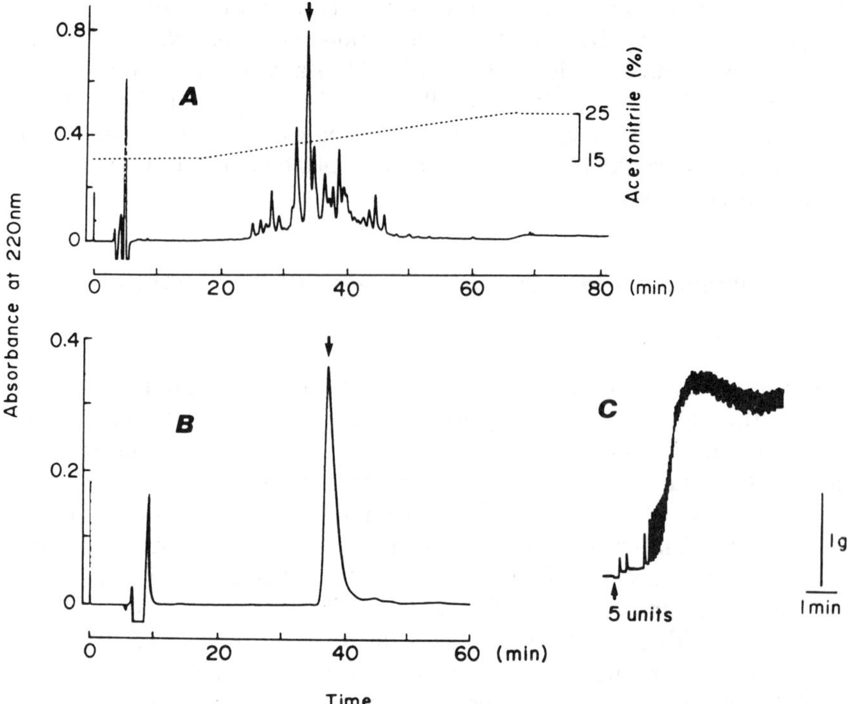

Fig. 3.4. HPLC profiles at the third step (A) and the fourth (last) step (B) of purification of UEP_A and the contractile effect of the purified UEP_A on the inner circular body-wall muscle (C). The absorbance peaks indicated by arrows corresponded to activity peaks.

amino acid sequence analysis; therefore there is no inconsistency between the result of the amino acid analysis and that of the amino acid sequence analysis. The amino acid residues which were not detected in the amino acid sequence analysis are assumed to be Cys. Supporting this assumption, 1·6 of $CysO_3H$ (normalized on Lys = 2) was detected when the amino acid analysis was performed on the active substance treated with performic acid.

A molecular ion peak in the FAB-MS spectrum of the active substance was at 1741·734 *m/z* $(M+H)^+$, suggesting that it is cyclized with a S–S bond between the two cysteine residues and that its C-terminus is free, not amidated. Thus, we propose that the structure of UEP_A is as follows:

H-Ala-Lys-Cys-Ser-Gly-Lys-Trp-Ala-Asn-Ser-Tyr-Leu-Cys-Ala-Gly-Ala-Asn-OH

Although we did not synthesize the peptide having the above structure, and therefore did not compare the pharmacological and chemical natures of native UEP_A with those of synthetic one, the proposed structure is probably correct.

3.3.2. *UEP_C*

As mentioned above, three peaks of contractile activity were observed at the initial step of purification of UEP_A, and the peptide was obtained from the first peak. The contractile peptide UEP_C was obtained from the second peak which was eluted at around 25% acetonitrile. At the next

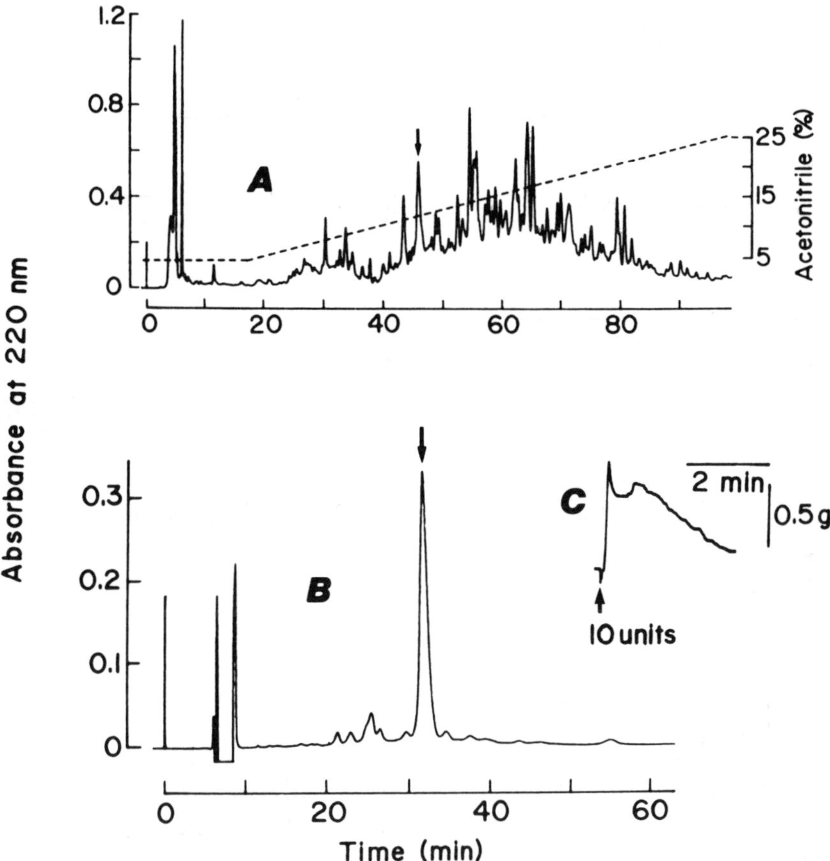

Fig. 3.5. HPLC profiles at the third step (A) and the fourth (last) step (B) of purification of UEP_C and the contractile effect of the purified UEP_C on the inner circular body-wall muscle (C). The absorbance peaks indicated by arrows corresponded to activity peaks.

step of purification of UEP_C the second peak was applied to an anion-exchange column (TSKgel DEAE-5PW, 7·5 × 75 mm, Tosoh) and eluted with a 60-min linear gradient of 0–0·3 M NaCl in 10 mM Tris buffer (pH 9·5). Contractile activity was observed in two groups of fractions, flowthrough fractions and fractions eluted at around 0·1 M NaCl. The active material in the latter group of fractions was then applied to a reversed-phase column (TSKgel ODS-$80T_M$) and eluted with an 80-min linear gradient of 5–25% acetonitrile in 5 mM phosphate buffer (pH 6·8). An active peak which corresponded to an absorbance peak (220 nm) was obtained (Fig. 3.5A). The active peak was subjected to the final HPLC separation. The column (Finepak SIL $C_{18}S$) was eluted with an isocratic system consisting of 22% acetonitrile and 0·1% TFA (Fig. 3.5B).

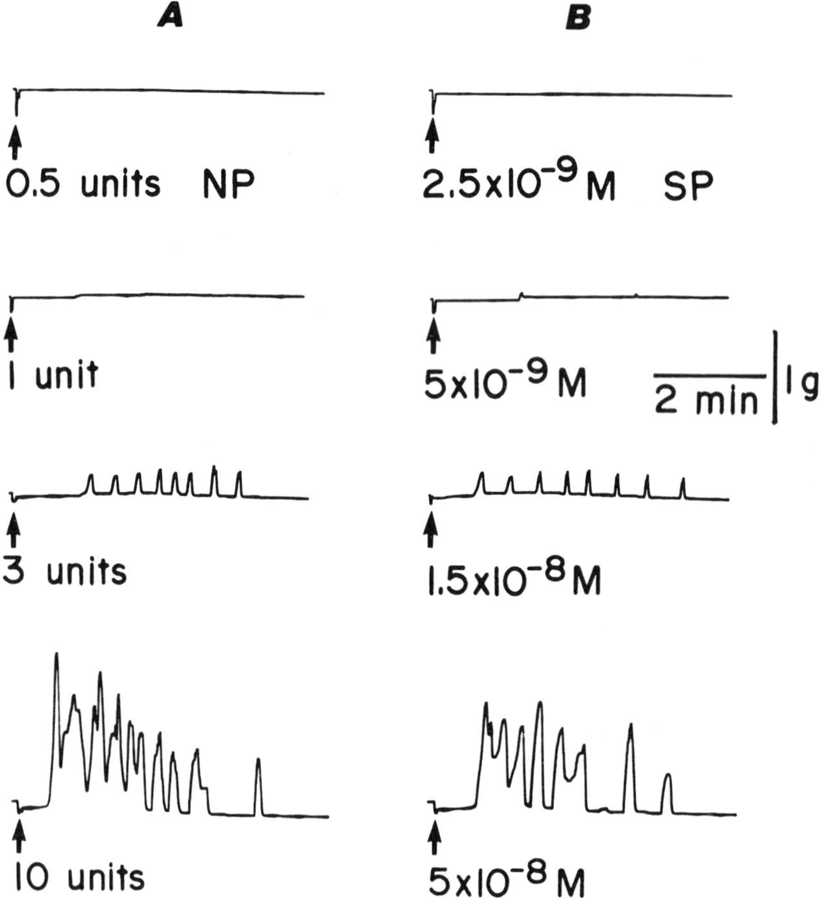

Fig. 3.6. Effects of native UEP_C (NP) and synthetic UEP_C (SP) on the inner circular body-wall muscle.

Structure analyses of UEP_C were carried out by the same methods used for UEP_A. Quantitative amino acid analyses of normal and oxidized (performic-acid-treated) UEP_C showed the following amino acid compositions, respectively (normalized on Leu = 1): $Asp_{2.2}$ $Thr_{1.7}$ $Ile_{0.9}$ $Leu_{1.0}$ $Phe_{1.3}$ and $Asp_{2.2}$ $Thr_{1.7}$ $Ile_{1.0}$ $Leu_{1.0}$ $Phe_{1.2}$ $Cys_{2.0}$. The determined sequence and detected amount (picomoles) of each amino acid in the amino acid sequence analysis were as follows: Thr(35)-Phe(98)-ND-Thr(21)-Ile(52)-Asp(18)-Leu(24)-Asn(54). From these results the sequence of UEP_C is suspected to be Thr-Phe-Cys-Thr-Ile-Asp-Leu-Asn-Cys. That is, the peptide has Cys at the third and last (ninth) positions.

A molecular ion peak in the FAB-MS spectrum of UEP_C was at 1027·404 *m/z* $(M+H)^+$, suggesting that it is cyclized with a S–S bond between the two cysteine residues and that its C-terminus is free. Thus, we propose that the structure of UEP_C is as follows:

H-Thr-Phe-Cys-Thr-Ile-Asp-Leu-Asn-Cys-OH (with an S–S bridge linking the two Cys residues)

The peptide having above structure was then synthesized by a solid-phase peptide synthesizer (Applied Biosystems 430A) followed by HF cleavage and HPLC purification. The structure of the synthesized peptide was confirmed to be correct by amino acid analyses, amino acid sequence analysis and FAB-MS analysis. The synthesized peptide showed a potent contractile activity on the inner circular body-wall muscle of *U. unicinc-*

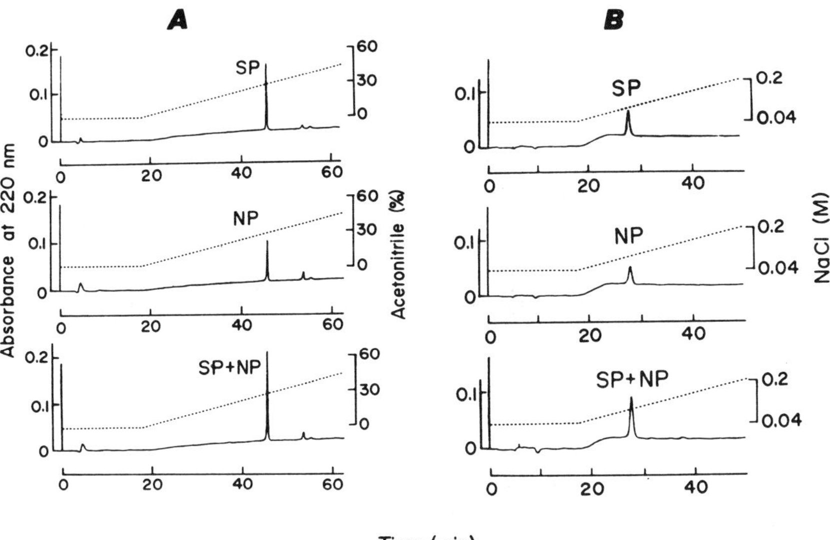

Fig. 3.7. Reversed-phase (A) and anion-exchange (B) chromatograms of synthetic UEP_C (SP), native UEP_C (NP) and their mixture (SP+NP).

tus. The dose–response relation of the peptide was found to be similar to that of native UEP_C (Fig. 3.6). As shown in Fig. 3.7, a mixture of the synthetic peptide and the native peptide showed a single absorbance peak when applied to the reversed-phase column (Finepak SIL $C_{18}S$) and the anion-exchange column (TSKgel DEAE-5PW).

3.4. **Actions of the peptides**

UEP_C showed contractile action on the inner circular body-wall muscle of *U. unicinctus* at 5×10^{-9} M or higher concentrations. However, the

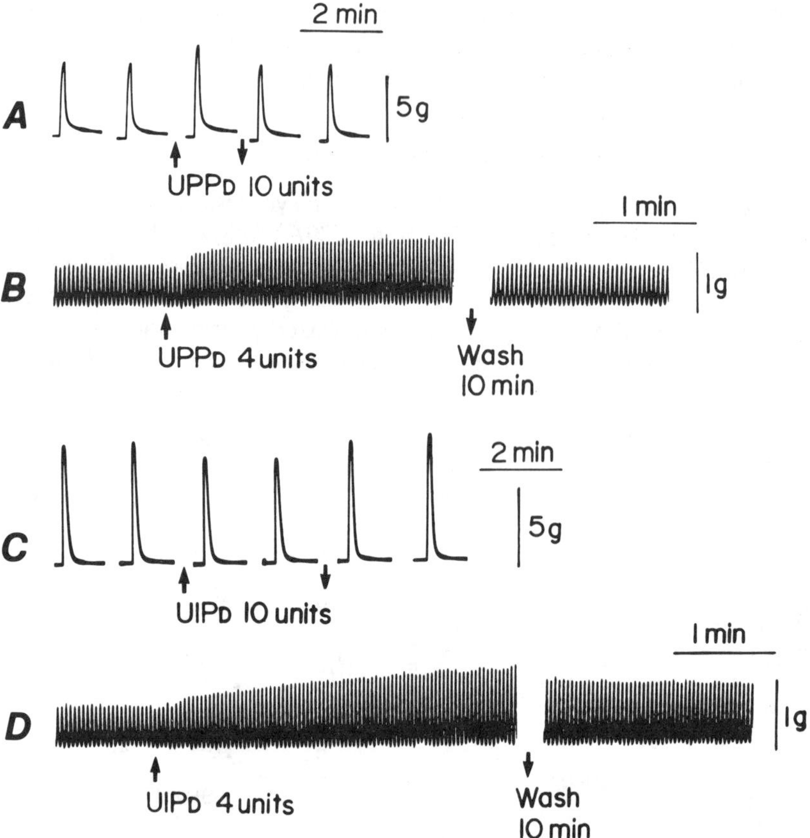

Fig. 3.8. Effects of UPP_D and UIP_D on molluscan muscles. A: Potentiating effect of UPP_D on phasic contraction of the ABRM of *Mytilus edulis*; B: potentiating effect of UPP_D on cardiac activity of *Meretrix lusoria*; C: inhibitory effect of UIP_D on phasic contraction of the ABRM of *M. edulis*; D: potentiating effect of UIP_D on cardiac activity of *M. lusoria*. The phasic contraction was elicited by stimulating the ABRM with repetitive electrical pulses (15 V, 3 ms, 10 Hz, for 5 s) at 10 min intervals, and the doses were applied 8 min prior to the stimulation.

peptide, even at 10^{-5} M, did not show any contractile or contraction modulating action on the ABRM of *Mytilus*, the radula retractor muscle of *Fusinus ferrugineus* (prosobranchia), the heart of *Meretrix lusoria* (Bivalvia), the longitudinal body-wall muscle of *Marphysa sanguinea* (Polychaeta) or the hind gut of *Gryllus bimaculatus* (Insecta). Using native UEP_A, we also examined its effect on the above muscles of the non-echiuroid animals but could not find any effect, though the tested dose was as low as 10 units, which was estimated to be approximately 5×10^{-8} M from the result of amino acid analysis.

In contrast to UEP_A and UEP_C, some other purified peptides showed biological activities on some non-echiuroid muscles. UPP_D, which was obtained from the third contractile peak observed at the initial step of HPLC purification of active substances in the gel-filtrated fractions 24–40, potentiated phasic contraction of the ABRM of *M. edulis* in response to repetitive electrical pulses of stimulation. The peptide also showed potentiating action on the spontaneous cardiac activity of *M. lusoria* (Fig. 3.8A). UPP_E of which behaviour on HPLC was similar to that of UPP_D showed almost identical actions on both of the muscles. UIP_D which was obtained from the gel-filtrated fractions 49–62 inhibited not only twitch-contraction of the *Urechis* muscle but also phasic contraction of the *Mytilus* muscle. However, it did not inhibit, but potentiated, the cardiac activity of *M. lusoria* (Fig. 3.8B).

3.5. **Conclusion**

By using the inner circular body-wall muscle of *U. unicinctus* as a bioassay system, we showed in the present study that there exist at least 25 species of bioactive peptides in the ventral nerve cord of the animal. They were found to have a contractile, twitch-potentiating or twitch-inhibiting effect on the muscle. Fujisawa *et al.* (1990) proposed that if we use appropriate muscles of an animal as bioassay systems, we may be able to detect most of the neuropeptides in the nervous system of the animal. The present experimental results support this notion. If we use a muscle as a bioassay system, further, we can easily carry out bioassay experiments and can examine the bioactivities of dozens of fractions in 1 or 2 days. Muscles of an animal seem to be good systems for bioassay of neurotransmitters and neuromodulators in the nervous system of the animal.

We determined the structures of two peptides obtained from the nerve cords of *U. unicinctus*; they are UEP_A and UEP_C. The amino acid sequences of both of the peptides are not homologous to any other vertebrate and invertebrate peptides. The peptides do not appear to be members of any other previously identified peptide family. However,

UEP_C seems to be bear some resemblance to crustacean cardioactive peptide (CCAP), as shown below:

UEP_C T F C T I D L N C-OH
CCAP P F C N A F T G C-NH_2

Both of the peptides are nonapeptides and have a disulphide bridge between Cys^3 and Cys^9. Further, the second amino acid residue is Phe in both of them. The most different point between them is that the C-terminus of CCAP is amidated but that of UEP_C is not.

CCAP was first isolated from the pericardial organs of a shore crab, *Carcinus maenas* (Stangier *et al.*, 1987). Later, this peptide was also found in the nervous system of an insect, *Locusta migratoria* (Stangier *et al.*, 1989). It can therefore be suspected that CCAP may be widely distributed in arthropods. The Echiuroidea is closely related to the Annelida, and the Annelida is a phylum phylogenically near the Arthropoda. Annelid worms might have a peptide related to both UEP_C and CCAP. Although UEP_C does not show any effect on the body-wall muscle of the annelid *M. sanguinea* or on the hind gut of the insect *G. bimaculatus*, it may be interesting to examine more precisely the effects of UEP_C, CCAP and their analogues, especially C-terminus-free CCAP and C-terminus-amidated UEP_C, on the muscles of echiuroids, annelids and arthropods.

It has been suggested that ACh may be an excitatory neurotransmitter in the body-wall muscle of *U. unicinctus* (Muneoka *et al.*, 1981). In the present study it was shown that UEP_A and UEP_C have a potent contractile action on the inner circular body-wall muscle. These peptides might also be excitatory transmitters in the muscle. ACh might be the principal excitatory transmitter and the peptides might be excitatory co-transmitters.

Some of the peptidergic substances, UPP_E, UPP_D and UIP_D, were found to have biological activities not only on the inner circular body-wall muscle of *U. unicinctus* but also on the ABRM of *M. edulis* and the heart of *M. lusoria*. We have tested more than 20 peptides of non-echiuroid animals on the body-wall muscle. None of them, except the molluscan neuropeptide CARP (catch-relaxing peptide), has been found to have any effect on the muscle. The above *Urechis* peptides seem not to be CARP-like peptides, because they did not show any catch-relaxing action on the ABRM. There might exist a novel family or families of peptides which include the above *Urechis* peptides and are interphyletically distributed.

Acknowledgement

A part of this work was supported by the Grant-in-Aid for General Scientific Research from the Ministry of Education, Science and Culture of Japan.

References

Fujisawa, Y., Kubota, I., Kanda, T., Kuroki, Y. and Muneoka, Y. (1990). Neuropeptides isolated from *Mytilus edulis* (Bivalvia) and *Fusinus ferrugineus* (Prosobranchia). This book, chapter 8.

Gardner, C. R. and Walker, R. J. (1982). The roles of putative neurotransmitters and neuromodulators in annelids and related invertebrates. *Prog. Neurboiol.*, **18**, 81–120.

Hirata, T., Kubota, I., Takabatake, I., Kawahara, A., Shimamoto, N. and Muneoka, Y. (1987). Catch-relaxing peptide isolated from *Mytilus* pedal ganglia. *Brain Res.*, **422**, 374–6.

Ikeda, T., Suematsu, K., Sakata, S. and Muneoka, Y. (1990). Bioactive substances in nerve-cord extracts from an echiuroid, *Urechis unicinctus* I. Amine-like substances. *Comp. Biochem. Physiol.*, **94C**, 603–12.

Muneoka, Y. and Kamura, M. (1982). Actions of noradrenaline and some other biogenic amines on the body-wall muscles of an echiuroid, *Urechis unicinctus*. *Comp. Biochem. Physiol.*, **72C**, 281–7.

Muneoka, Y., Ichimura, Y., Shiba, Y. and Kanno, Y. (1981). Mechanical responses of the body wall strips of an echiuroid worm, *Urechis unicinctus*, to electrical stimulation, cholinergic agents and amino acids. *Comp. Biochem. Physiol.*, **69C**, 171–7.

Stangier, J., Hilbich, C., Beyreuther, K. and Keller, R. (1987). Unusual cardioactive peptide (CCAP) from pericardial organs of the shore crab *Carcinus maenus*. *Proc. Natl. Acad. Sci. USA*, **84**, 575–9.

Stangier, J., Hilbich, C. and Keller, R. (1989). Occurrence of crustacean cardioactive peptide (CCAP) in the nervous system of an insect, *Locusta migratoria*. *J. Comp. Physiol. B*, **159**, 5–11.

Part II

Molluscan studies

4 *Hiroshi Takeuchi, Anchalee Yongsiri, Kah Hwi Kim, Yoshimi Kamatani, Hiroyuki Minakata, Kyosuke Nomoto, Ichiro Kubota and Yojiro Muneoka*

Peptide pharmacology on giant neurons of an African giant snail (*Achatina fulica* Férussac)

4.1. **Introduction**

An African giant snail, *Achatina fulica* Férussac, is easily obtainable in Japan due to its wide abundance in tropical and subtropical countries in Asia. We have imported this snail by air from Manila, Philippines, for about 20 years, in order to use its identifiable giant neurons for physiological and pharmacological experiments.

In the present report we summarize our results on the peptide pharmacology using *Achatina* giant neurons. First, we shall describe the structure of the snail's central nervous system and the identification of the giant neurons. Then the effects of biologically active peptides, which have been proposed as neurotransmitters in mammalian and invertebrate nervous systems, on the *Achatina* neurons are presented.

Further, the effects of *Mytilus* inhibitory peptides (MIPs), which have been isolated from the ganglia of a mussel, *Mytilus edulis*, on an Achatina neuron are reported. Finally, the effects of achatin-I, a neuroexcitatory tetrapeptide having a D-phenylalanine residue, which we have recently isolated from the *Achatina* ganglia, are described.

The formula of the physiological solution used throughout these experiments has been described previously (Takeuchi *et al.*, 1973b).

4.2. **Identification of giant neurons**

The central nervous system of *Achatina fulica* Férussac is composed of three ganglia – namely, suboesophageal, cerebral and buccal ganglia, which are identical to those of the other snail species (Fig. 4.1). To date, 34 giant neurons have been identified in these ganglia, based on their localization, their mode of spontaneous spike discharges, their sensitivities to putative neurotransmitters, and their axonal pathways as traced by intracellular injection of a dye, Lucifer Yellow CH (Takeuchi *et al.*, 1973a, 1975, 1976; Takeuchi and Yamamoto, 1982; Ku and Takeuchi, 1983a,b, 1984; Ku *et al.*, 1985; Boyles and Takeuchi, 1985; Matsuoka *et al.*, 1985,

A Buccal ganglia

(a) dorsal surface

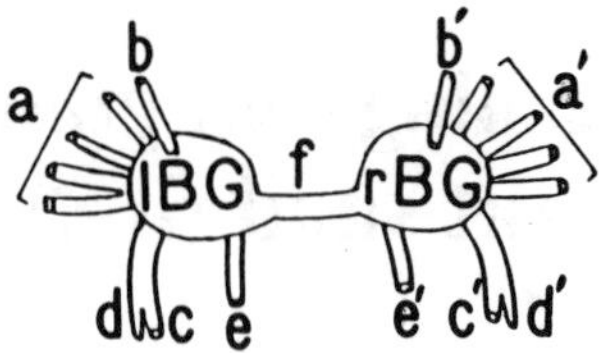

B Cerebral ganglia

(a) dorsal surface (b) ventral surface

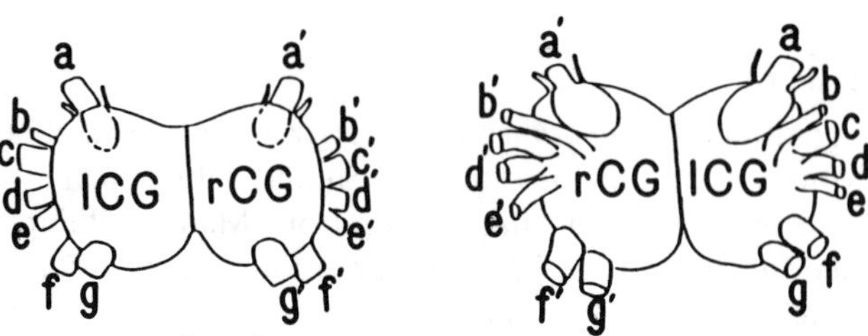

C Suboesophageal ganglia

(a) dorsal surface (b) ventral surface

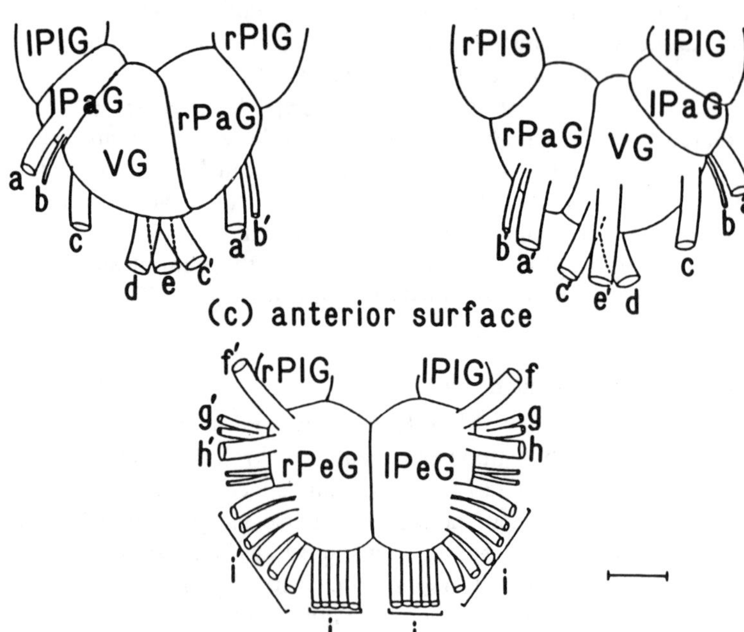

1986, 1987; Goto *et al.*, 1986; Yongsiri *et al.*, 1986a,b; Kim *et al.*, 1989). The localization of these neurons is illustrated in Fig. 4.2.

In Table 4.1 the effects of putative neurotransmitters and their derivatives – namely catecholamines, indoleamines, amino acids, imidazolamines and cholines – are presented as the number of identifiable neurons sensitive to these compounds. To obtain these results for screening trials, a glass microelectrode, filled with 2 M potassium acetate, was implanted into the soma of an identifiable neuron, and its intracellular potential was recorded under current clamp with a pen-writing galvanometer. The sensitivities of these individual neurons are reported in detail in our previous papers mentioned above. From these results, dopamine, 5-hydroxytryptamine, GABA, β-hydroxy-L-glutamic acid (L-BHGA), histamine and acetylcholine are considered to be putative neurotransmitters among the compounds listed in Table 4.1.

In particular L-BHGA, but not L-glutamic acid, was effective on these neurons. We intended to classify the L-BHGA receptors into several subtypes according to the sensitivities of the neurons to BHGA stereoisomers and their related compounds (Watanabe *et al.*, 1985; Nakajima *et al.*, 1985; Novales-Li *et al.*, 1987; Takeuchi *et al.*, 1989).

4.3. Peptides proposed as neurotransmitters in mammals, invertebrates, etc.

Biologically active peptides, which are proposed as mammalian and invertebrate neurotransmitters and are natural venoms, were examined on 30 identifiable giant neurons of *Achatina fulica* Férussac (Ku *et al.*, 1986;

←

Fig. 4.1. The three ganglia of *Achatina fulica* Férussac. **A**: Buccal ganglia (dorsal surface): lBG, left buccal ganglion; rBG, right buccal ganglion; a,a′, left and right lateral buccal nerves; b,b′, left and right dorsal buccal nerves; c,c′, left and right cerebral buccal connectives; d,d′, left and right accessory connective buccal nerves; e,e′, left and right medial buccal nerves; f, buccal commissure. **B**: Cerebral ganglia: (a), dorsal surface; (b), ventral surface; lCG, left cerebral ganglion; rCG, right cerebral ganglion; a,a′, left and right optic nerves; b,b′, left and right anterior tentacular nerves; c,c′, left and right supralabial nerves; d,d′, left and right sublabial nerves; e,e′, left and right cerebro-buccal connectives; f,f′, left and right cerebro-pedal connectives; g,g′, cerebro-pleural connectives. **C**: Suboesophageal ganglia: (a), dorsal surface; (b), ventral surface; (c), anterior surface; lPlG, left pleural ganglion; lPaG, left parietal ganglion; VG, visceral ganglion; rPaG, right parietal ganglion; rPlG, right pleural ganglion; rPeG, right pedal ganglion; lPeG, left pedal ganglion; a,a′, left and right anterior pallial nerves; b,b′, left and right anterior pallial accessory nerves; c,c′, left and right posterior pallial nerves; d, intestinal nerve; e, anal nerve; f,f′, left and right anterior parietal nerves; g,g′, left and right middle parietal nerves; h,h′, left and right posterior parietal nerves; i,i′, left and right anterior pedal nerves; j,j′, left and right posterior parietal nerves. Horizontal bar, scale (500 μm) (section 4.2).

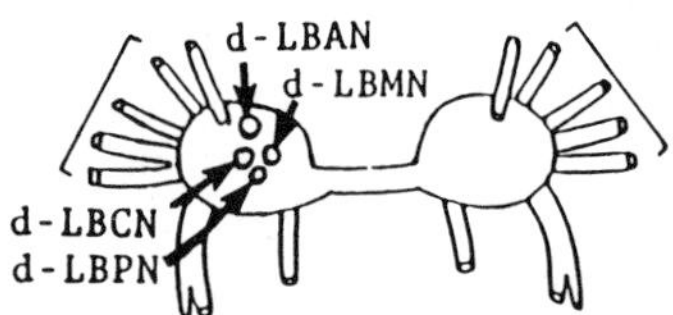
d-LBAN
d-LBMN
d-LBCN
d-LBPN

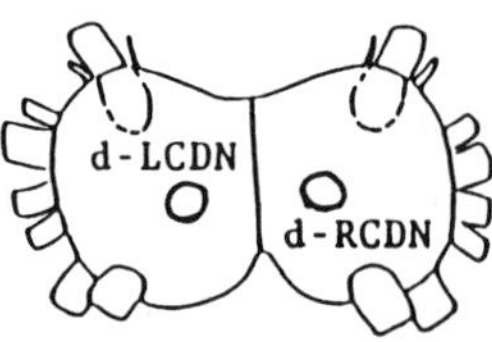
d-LCDN
d-RCDN

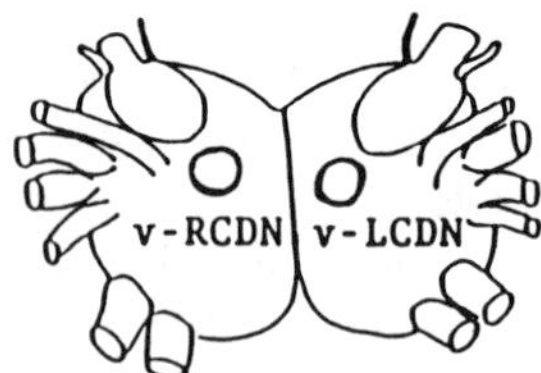
v-RCDN
v-LCDN

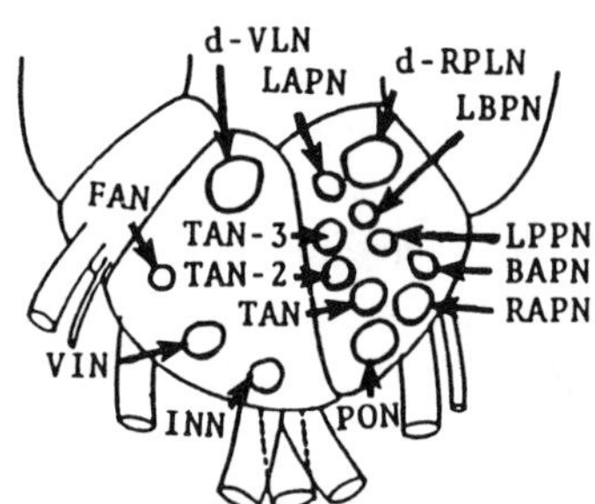
d-VLN
LAPN
d-RPLN
LBPN
FAN
TAN-3
TAN-2
TAN
LPPN
BAPN
RAPN
VIN
INN
PON

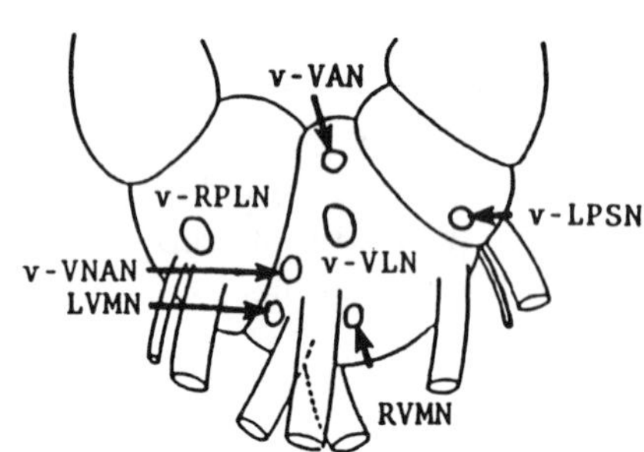
v-VAN
v-RPLN
v-LPSN
v-VNAN
v-VLN
LVMN
RVMN

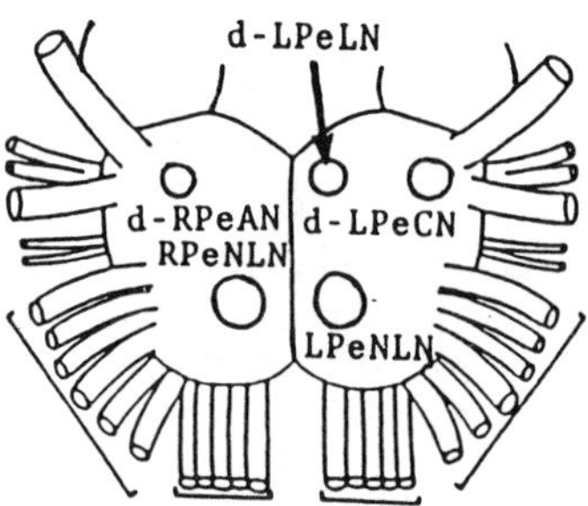
d-LPeLN
d-RPeAN
d-LPeCN
RPeNLN
LPeNLN

Yongsiri *et al.*, 1987; Kim *et al.*, 1987a,b; Matsuoka *et al.*, 1988). The amino acid sequences of these peptides are described in Table 4.2. The *Achatina* neurons tested were those used in experiments mentioned earlier and are shown in Fig. 4.2 (34 neurons), except for BAPN, LPPN, LBPN and LAPN (four neurons). The experimental protocol employed was basically similar to that mentioned above.

The results obtained are summarized in Table 4.3. Of the peptides tested, only three were effective on the *Achatina* neurons. Of the three, proctolin showed excitatory effects only on one neuron. Effects of oxytocin on these neurons were mainly excitatory, whereas those of FMRFamide were inhibitory. We consider that oxytocin and FMRFamide are excitatory and inhibitory putative neurotransmitters, respectively, on the *Achatina* neurons. The sensitivities of individual *Achatina* identifiable neurons to these peptides are described in detail in our previous papers mentioned above.

The peptides listed in Table 4.2, including Met-enkephalin, substance P, neurotensin, etc., other than the three effective peptides mentioned, were completely ineffective at screening trials of 10^{-4} M.

4.4. **Effects of *Mytilus* inhibitory peptides (MIPs)**

Hirata *et al.* (1988) isolated two congeneric hexapeptides, i.e. Ser^2-*Mytilus* inhibitory peptide (Ser^2-MIP) and Ala^2-MIP, from the pedal

←

Fig. 4.2. Identification of giant neurons. **A**: Buccal ganglia (dorsal surface): d-LBAN, dorsal-left buccal anterior neuron; d-LBMN, dorsal-left buccal medial neuron; d-LBCN, dorsal-left buccal central neuron; d-LBPN, dorsal-left buccal posterior neuron. **B**: Cerebral ganglia: (a), dorsal surface; d-LCDN, dorsal-left cerebral distinct neuron; d-RCDN, dorsal-right cerebral distinct neuron; (b), ventral surface; v-LCDN, ventral-left cerebral distinct neuron; v-RCDN, ventral-right cerebral distinct neuron. **C**: Suboesophageal ganglia: (**a**), dorsal surface: d-VLN, dorsal-visceral large neuron; FAN, frequently autoactive neuron; VIN, visceral intermittently firing neuron; INN, intestinal nerve neuron; PON, periodically autoactive neuron; TAN, tonically autoactive neuron; TAN-2; TAN-3; RAPN, right anterior pallial nerve neuron; d-RPLN, dorsal-right parietal large neuron; LAPN, left anterior pallial nerve neuron; LBPN, left bifurcate pallial nerve neuron; LPPN, left posterior pallial nerve neuron; BAPN, bilateral anterior pallial nerve neuron; (**b**), ventral surface: v-LPSN, ventral-left parietal silent neuron; v-VLN, ventral-visceral large neuron; v-VAN, ventral-visceral anterior neuron, old name: v-1-VOrN; LVMN, left visceral multiple spike neuron, old name: 1-VMN; RVMN, right visceral multiple spike neuron, old name: r-VMN; v-VNAN, ventral-visceral noisy autoactive neuron; v-RPLN, ventral right parietal large neuron; (**c**), anterior surface; d-LPeLN, dorsal-left pedal large neuron; d-LPeCN, dorsal-left pedal constantly firing neuron; LPeNLN, left pedal nerve large neuron; d-RPeAN, dorsal-right pedal autoactive neuron; RPeNLN, right pedal nerve large neuron. Horizontal bar, scale (500 μm) (section 4.2).

Table 4.1. Number of identifiable *Achatina* giant neurons (total 34 neurons) sensitive to putative neurotransmitters and their derivatives (screening tests at 10^{-3} M) (section 4.2)

No. Compound	*E*	*SE*	*I*	*SI*	*V*	*(−)*
Catecholamines						
1. Dopamine	5		24	1	2	2
2. L-Norepinephrine	5	6	11	2	2	8
3. L-Epinephrine	2	3	13	2	3	11
4. DL-Octopamine	6	5	9	1		13
5. L-Phenylalanine		1				33
6. L-Tyrosine		1				33
7. L-DOPA				2		32
Indoleamines						
8. L-Tryptophan		1				33
9. 5-Hydroxy-L-tryptophan						34
10. 5-Hydroxytryptamine	20		8	4		2
Amino acids						
11. Glycine		3				31
12. β-Alanine						34
13. γ-hydroxybutyric acid (GABA)	7	3	15	1		8
14. 1-β-hydroxy GABA (1-GABOB)	2	4	3	5		20
15. d-GABOB	3	2	2	2		25
16. L-Glutamic acid				1		33
17. L-Aspartic acid						34
18. L-Homocysteic acid	6	3	8	2		15
19. L-Homocysteine sulfonic acid	2	1	4	5		22
20. L-Methionine		1				33
21. Taurine						34
22. Erythro-L-β-hydroxyglutamic acid	7	3	9	3		12
Imidazolamines						
23. Histamine	11	4	6	1	2	10
24. L-Histidine		1	1			32
Cholines						
25. Acetylcholine	7		17	2	5	3
26. Propionylcholine	2	4	17	3	5	3
27. Butyrylcholine	3		17	3	5	6

E, excitatory effects; SE, slight excitatory effects; I, inhibitory effects; SI, slight inhibitory effects; V, varied effects; (−), no effect.

ganglia of *Mytilus edulis*. We examined the effects of these peptides and their fragments on *Achatina* giant neurons (Yongsiri *et al*., 1989).

Among several of the *Achatina* neurons tested, only RAPN (right anterior pallial nerve neuron), situated on the dorsal surface of the right parietal ganglion in the suboesophageal ganglia (Fig. 4.2), was sensitive to the two MIPs. Amino acid sequences and effects on RAPN of the two

Table 4.2. Amino acid sequences of the biologically active peptides tested (Ku *et al.*, 1986) (section 4.2)

Peptides proposed as neurotransmitters in mammals

1. Met-enkephalin
   H-Tyr-Gly-Gly-Phe-Met-OH
2. Substance P
   H-Arg-Pro-Lys-Pro-Gln-Gln-Phe-Phe-Gly-Leu-Met-NH_2
3. Neurotensin
   pGlu-Leu-Tyr-Glu-Asn-Lys-Pro-Arg-Arg-Pro-Tyr-Ile-Leu-OH
4. Luteinizing hormone-releasing hormone (LH-RH)
   pGlu-His-Trp-Ser-Tyr-Gly-Leu-Arg-Pro-Gly-NH_2
5. [Gln^{8}]-LH-RH
   pGlu-His-Trp-Ser-Tyr-Gly-Leu-Gln-Pro-Gly-NH_2
6. Oxytocin
   H-Cys-Tyr-Ile-Gln-Asn-Cys-Pro-Leu-Gly-NH_2
7. Arg-vasopressin
   H-Cys-Tyr-Phe-Gln-Asn-Cys-Pro-Arg-Gly-NH_2

Peptides proposed as neurotransmitters in invertebrates

8. Proctolin
   H-Arg-Tyr-Leu-Pro-Thr-OH
9. Molluscan cardioexcitatory neuropeptides (FMRFamide)
   H-Phe-Met-Arg-Phe-NH_2

Natural venom peptides

10. Bombesin
    pGlu-Gln-Arg-Leu-Gly-Asn-Gln-Trp-Ala-Val-Gly-His-Leu-Met-NH_2
11. Rabatensin C
    pGlu-Thr-Pro-Gln-Trp-Ala-Val-Gly-His-Phe-Met-NH_2
12. Vespakinin X
    H-Ala-Arg-Pro-Pro-Gly-Phe-Ser-Pro-Phe-Arg-Ile-Val-OH
13. Vespakinin M
    H-Gly-Arg-Pro-Hyp-Gly-Phe-Ser-Pro-Phe-Arg-Ile-Asp-OH

MIPs and their fragments tested in this work are shown in Table 4.4. The two MIPs and their two fragments, MIP (3–6) and MIP (4–6), showed inhibitory effects on this neuron, whereas the other two fragments, Ala^{2}, Val^{5}, Phe^{6}-MIP-NH_2 and Ala^{2}-MIP(1–4)-NH_2, were ineffective at screening trials of 10^{-3} M. Ser^{2}-MIP was also more potent than Ala^{2}-MIP on this neuron.

Pneumatic pressure ejection of Ser^{2}-MIP (1 kg/cm^2, 250 ms in duration and 10^{-3} M) with 0·5% fast green to RAPN, produced a marked hyperpolarization of the membrane potential of the neuron tested under current clamp. Thereafter, we adopted the conventional voltage clamp technique (Okamoto *et al.*, 1976) using two intracellular microelectrodes, filled with 2 M potassium acetate, to analyse precisely the effects of MIPs. Under voltage clamp at a holding voltage (V_h) of −50 mV, pressure

Table 4.3. Number of identifiable *Achatina* giant neurons (total 30 neurons) sensitive to biologically active peptides tested (screening tests at 10^{-4} M) (section 4.3)

No. Substance	*E*	*SE*	*I*	*SI*	*V*	*(−)*
1. Oxytocin	5	3	1	1		20
2. Proctolin	1					29
3. FMRFamide			6		1	23

E, excitatory effects; SE, slight excitatory effects; I, inhibitory effects; SI, slight inhibitory effects; V, varied effects; (−), no effect. The other peptides tested had no effect on these neurons.

Table 4.4. Amino acid sequences and effects on RAPN (right anterior pallial nerve neuron) of *Mytilus* inhibitory peptides (MIPs) and their fragments (screening tests at 10^{-3} M) (section 4.4)

No. Peptide	*Effects on RAPN*
1. Ser^2-MIP H-Gly-Ser-Pro-Met-Phe-Val-NH_2	I
2. Ala^2-MIP H-Gly-Ala-Pro-Met-Phe-Val-NH_2	I
3. MIP (3–6) H-Pro-Met-Phe-Val-NH_2	SI
4. MIP (4–6) H-Met-Phe-Val-NH_2	SI
5. Ala^2, Val^5, Phe^6-MIP-NH_2	(−)
6. Ala^2-MIP(1–4)-NH_2 H-Gly-Ala-Pro-Met-NH_2	(−)

I, inhibitory effects; SI, slightly inhibitory effects; (−), no effect.

ejection of the peptide produced an outward current with the same duration as that of the hyperpolarization obtained under current clamp. When repetitive hyperpolarizing pulses of 5 mV (500 ms in duration and 0·5 c/s) were superimposed on V_h to measure membrane conductance, the same application of the peptide produced an increase in conductance during the occurrence of the outward current (Fig. 4.3).

Pressure ejection of Ser^2-MIP in extracellular Ca^{2+}-free and Mg^{2+}-rich state also showed the outward current, indicating that the peptide had direct effects (not via synaptic influences) on the neuron tested.

The two MIPs, applied by bath, induced the slow outward current showing hardly any desensitization. The current was maintained for at least 30 min almost without decrement in the presence of the peptides.

Structure–activity relations of the two MIPs and their two effective fragments for producing both the outward current and the conductance increase of the RAPN membrane were studied by consecutive bath ap-

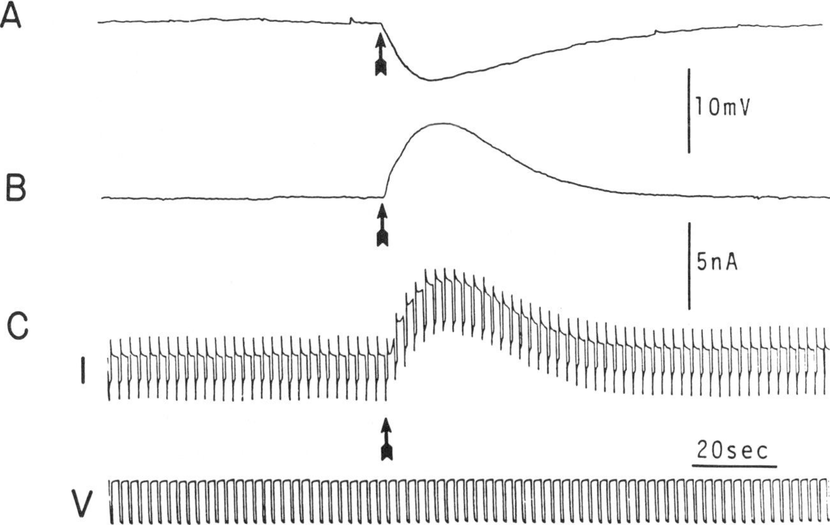

Fig. 4.3. Effects of Ser²-*Mytilus* inhibitory peptide (Ser²-MIP) on RAPN (right anterior pallial neuron). **A**: Hyperpolarization caused by the peptide under current clamp. **B**: Outward current produced by the peptide under voltage clamp. **C**: Outward current with an increase in conductance. Repetitive hyperpolarizing square pulses (5 mV, 500 ms, 0·5 c/s) were superimposed on a holding voltage (V_h) of −50 mV. I, membrane current; V, membrane voltage. In A, B and C, arrows indicate pneumatic pressure ejection of Ser²-MIP (1 kg/cm², 250 ms in duration and 10^{-3} M). Vertical bar, calibration (10 mV for A, 5 nA for B and C). Horizontal bar, time course (20 s) (Yongsiri *et al.*, 1990) (section 4.4).

plication of each peptide from 10^{-6} to 10^{-3} M. Their dose–response curves indicate that these four peptides produced both the outward current and conductance increase in a dose-dependent manner. The order of potency is: Ser²-MIP > Ala²-MIP > MIP(3–6) > MIP(4–6) (Fig. 4.4).

Based on these results we conclude that MIP(4–6), H-Met-Phe-Val-NH_2, is the minimum structure to produce the effects. Likewise, the presence of Pro³ potentiates the effects, and Gly¹-Ser² or Gly¹-Ala² further potentiates them.

The relations between the outward current (nA) and conductance increase (μS) of the RAPN membrane caused by these four peptides were linear, which were fitted to $Y = 0{\cdot}034X - 0{\cdot}011$ ($r = 0{\cdot}987$). This suggests that a unique ionic mechanism was involved in the inhibitory effects of these peptides on RAPN.

The equilibrium potential of Ser²-MIP ($E_{\text{S-MIP}}$) was determined by measuring the current–voltage relations (I–V curve) of the RAPN membrane in the absence and presence of Ser²-MIP at 3×10^{-5} M. The I–V curve was obtained by superimposing the hyperpolarizing ramp voltage

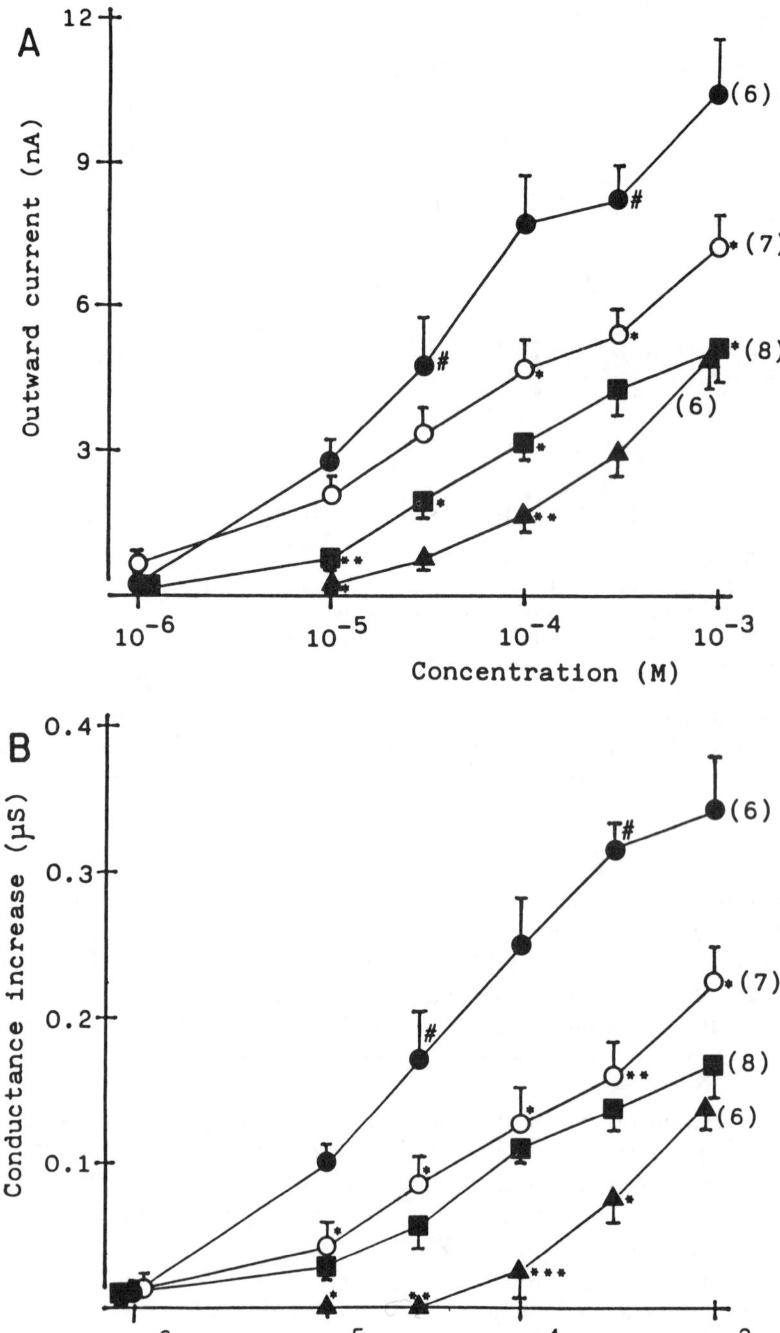
A
12
9
6
3
Outward current (nA)
(6)
#
(7)
#
(8)
(6)
10^{-6}
10^{-5}
10^{-4}
10^{-3}
Concentration (M)
B
0.4
0.3
0.2
0.1
Conductance increase (μS)
(6)
#
#
(7)
(8)
(6)
10^{-6}
10^{-5}
10^{-4}
10^{-3}
Concentration (M)

(with a slope of 70 mV/60 s) on V_h of −50 mV. $E_{S\text{-}MIP}$ was measured to be −87·6 ± 1·4 mV (mean ± SEM n = 11) in the normal state ($[K^+]_0$: 3·3 mM). $E_{S\text{-}MIP}$ values under varied $[K^+]_0$ were: −99·3 ± 3·2 mV (n = 8) for 1·7 mM of $[K^+]_0$ (× 0·5), −72·3 ± 1·8 mV (n = 8) for 6·6 mM (× 2) and −61·3 ± 1·9 mV (n = 5) for 9·9 mM (× 3). These values are fitted to the Nernst equation for E_K. On the other hand, $E_{S\text{-}MIP}$ was not affected by varied $[Na^+]_0$ and $[Cl^-]_0$. With these results we conclude that the inhibition of RAPN caused by Ser^2-MIP was due to the membrane permeability increase to potassium ions.

We consider that the MIPs, especially Ser^2-MIP, or their related peptides will be neurotransmitters on RAPN, though the presence of these peptides is not yet demonstrated in the ganglia of this snail species.

4.5. **Achatin-I, isolated from *Achatina* ganglia**

In collaboration with the Suntory Institute, Osaka, and School of Medicine, Gifu University, in Japan, and University of Santo Tomas, Manila, in the Philippines, we recently isolated (Kamatani *et al.*, 1989) a neuroactive tetrapeptide, termed achatin-I, peculiarly having a D-phenylalanine residue, from the *Achatina* ganglia. Bioassay for this purpose was performed mainly using an *Achatina* giant neuron, PON (periodically oscillating neuron), situated on the dorsal surface of the right parietal ganglion in the suboesophageal ganglia (Fig. 4.2).

From the suboesophageal and cerebral ganglia of 30,000 animals we finally obtained 50 μg of achatin-I (H-Gly-D-Phe-L-Ala-L-Asp), which was markedly effective on PON, and 17 μg of its stereoisomer, achatin-II (H-Gly-L-Phe-L-Ala-L-Asp), which was ineffective at the concentration of the screening tests, 3×10^{-4} M. Amino acid sequences and the effects on PON of achatin-I, its stereoisomers and related peptides examined in this study, are shown in Table 4.5. The isolation process and determination of the achatin-I structure have been described in detail in the original paper mentioned above. The synthetic achatin-I was confirmed to be identical to the natural peptide isolated from the *Achatina* ganglia, in ^{1}H-NMR, SIMS, CD, HPLC and biological activity.

←
Fig. 4.4. Dose–response relations of *Mytilus* inhibitory peptides (MIPs) and their effective fragments, applied by bath, for producing the outward current (**A**) and the membrane conductance increase (**B**) of RAPN at V_h of −50 mV. In A and B, ●, Ser^2-MIP; ○, Ala^2-MIP; ■, MIP(3–6); ▲, MIP(4–6). Bar shows SEM. Number of experimental trials to obtain each line is shown in parentheses except for # (n = 3). Student's t-tests (unpaired) were performed on Ala^2-MIP against Ser^2-MIP, MIP(3–6) against Ala^2-MIP, and MIP(4–6) against MIP(3–6) at each concentration (more than 10^{-5} M). *, $p < 0.05$; **, $p < 0.01$; ***, $p < 0.001$ (Yongsiri *et al.*, 1990) (section 4.4).

As described in Table 4.5, achatin-I and [D-Ala^3]-achatin-I (H-Gly-D-Phe-D-Ala-L-Asp) excited PON. The former was much more potent than the latter. Pneumatic pressure ejection of achatin-I (3 kg/cm^2, 300 ms in duration and 10^{-3} M) to PON produced a slow inward current (excitation) on the membrane of the neuron tested in both normal and Ca^{2+}-free and Mg^{2+}-rich states. This indicates that the peptide affected directly the neuron tested (Fig. 4.5A).

The dose–response curves of achatin-I and [D-Ala^3]-achatin-I were determined by consecutive bath application of each peptide up to 3×10^{-4} M using PON. These curves were analysed by the probit method (Litchfield and Wilcoxon, 1949) as follows: ED_{50}: $0{\cdot}23 \times 10^{-5}$ M (95% confidence limit: $0{\cdot}17$–$0{\cdot}31 \times 10^{-5}$ M), tangent slope in the flexion point (slope): 0·346, coefficient for ideal sigmoidal curve (r_1): 0·994, Hill coefficient: 0·583, coefficient for linear Hill plot (r_2): 0·991, and maximal effects (E_{max}, mean ± SEM): 5·53 ± 0·28 nA for achatin-I ($n = 6$); ED_{50}: $15{\cdot}46 \times 10^{-5}$ M (95% confidence limits: $11{\cdot}09$–$21{\cdot}56 \times 10^{-5}$ M), slope: 0·422, r_1: 0·994, Hill coefficient: 0·710, r_2: 0·991, E_{max}: 1·98 ± 0·17 nA for [D-Ala^3]-achatin-I ($n = 6$). Consequently, the ED_{50} of achatin-I is significantly lower than that of [D-Ala^3]-achatin-I, and the E_{max} of the former peptide is significantly greater than that of the latter ($p < 0{\cdot}001$). Hill coefficient values for the two peptides were far from 1·0, suggesting that these peptides may cause some allosteric changes in receptors, to affect their binding to the receptors (Fig. 4.5B).

Table 4.5. Amino acid sequences and effects on PON (periodically oscillating neuron) of achatin-I, its stereoisomers and related peptides (screening tests at 3×10^{-4} M (section 4.5)

No.	*Amino acid sequence*	*Effects on PON*
1.	H-Gly-L-Phe-L-Ala-L-Asp (achatin-II)	(−)
2.	H-Gly-D-Phe-L-Ala-L-Asp (achatin-I)	E
3.	H-Gly-D-Phe-L-Ala-L-Asp-NH_2 (achatin-Iamide)	(−)
4.	H-Gly-D-Phe-D-Ala-L-Asp ([D-Ala^3]-achatin-I)	SE
5.	H-Gly-D-Phe-D-Ala-D-Asp	(−)
6.	H-Gly-D-Phe-L-Ala-D-Asp	(−)
7.	H-Gly-L-Phe-D-Ala-L-Asp	(−)
8.	H-Gly-L-Phe-D-Ala-D-Asp	(−)
9.	H-Gly-L-Phe-L-Ala-D-Asp	(−)
10.	H-Gly-L-Phe-L-Ala-L-Asp-NH_2	(−)
11.	H-Gly-L-Phe-L-Ala-L-Asn	(−)
12.	H-Gly-L-Phe-L-Ala-L-Asn-NH_2	(−)

E, excitatory effects; SE, slightly excitatory effects; (−), no effect.

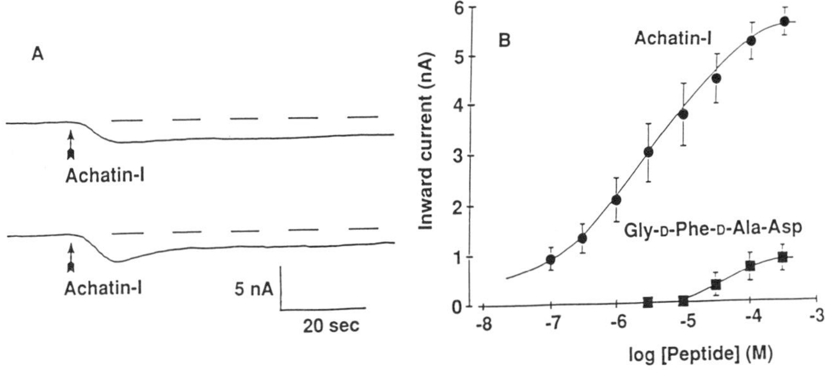

Fig. 4.5. Effects of achatin-I and [D-Ala[3]]-achatin-I on PON (periodically oscillating neuron). **A:** Pneumatic pressure ejection of achatin-I (3 kg/cm^2, 300 ms in duration and 10^{-3} M) in the physiological (upper trace) and the Ca^{2+}-free and Mg^{2+}-rich (lower trace) states. Vertical bar, calibration (5 nA); horizontal bar, time course (20 s). **B:** Dose–response relations of achatin-I (n = 6) and [D-Ala[3]]-achatin-I (n = 3), applied by bath. Ordinate, inward current (nA) (bar: SEM); abscissa, concentration of peptides in log scale (M). In A and B, V_h was maintained at −50 mV (Kamatani *et al.*, 1989) (section 4.5).

The current–voltage relations (I–V curves) of PON were measured in the absence and the presence of achatin-I applied by bath at 10^{-5} M. These I–V curves were measured by superimposing a depolarizing ramp voltage (with a slope of 100 mV/100 s) on V_h of −100 mV. This peptide caused a marked inward current in a wide range of the I–V curve in the physiological ($[Na^+]_0$: 65·6 mM) state. However, the inward current was hardly produced by the peptide at the same concentration in the Na^+-lacking (× 0·1, $[Na^+]_0$: 6·6 mM) state. With these results it is concluded that the PON excitation caused by the peptide was due to the membrane permeability increase to sodium ions (Fig. 4.6).

Achatin-I is proposed to be an excitatory neurotransmitter on PON, since this peptide is endogenous in the *Achatina* ganglia. We are at present studying in our laboratories the mapping of the *Achatina* giant neurons sensitive to this peptide, structure–activity relations of the peptide and related peptides, and the pharmacological characteristics in detail of the peptide on the *Achatina* neurons.

4.6. **Conclusion**

The peptide pharmacology of identifiable giant neurons of *Achatina fulica* Férussac, are summarized in this report. As the introductory presentation of the *Achatina* giant neurons, their identification in the ganglia and their sensitivities to putative neurotransmitters and their related compounds

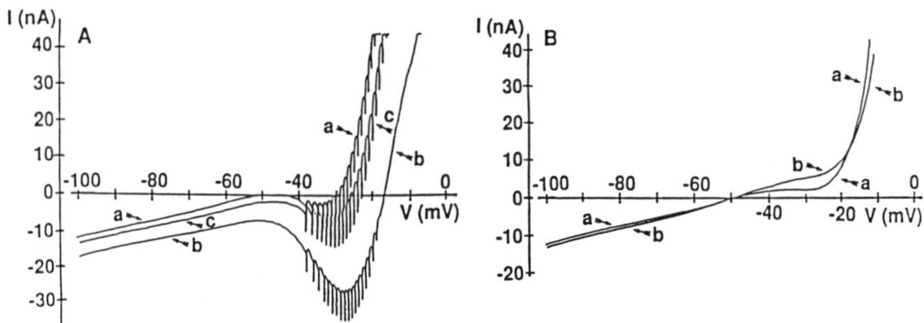

Fig. 4.6. Effects of achatin-I, applied by bath, at 10^{-5} M on the current–voltage relations (I–V curve) of PON membrane. **A:** Physiological state ($[Na^+]_0$: 65·6 mM). **B:** Na^+-lacking state ($[Na^+]_0$: 6·6 mM (0·1 ×); replaced with $Tris^+$). In A and B: a, control; b, achatin-I; c, after washing. I–V curves, shown in A and B, were obtained from the same PON (Kamatani *et al.*, 1989) (section 4.5).

(catecholamines, indoleamines, amino acids, etc.) are described. Of the peptides proposed as neurotransmitters in mammals and invertebrates, oxytocin showed mainly excitatory effects on the *Achatina* neurons, while FMRFamide had inhibitory effects. The two peptides are considered to be the putative neurotransmitters on these neurons.

The two *Mytilus* inhibitory peptides, Ser^2-MIP and Ala^2-MIP, of which the former is more potent than the latter, produced K^+-dependent inhibition (an outward current in voltage clamp) with an increase in the membrane conductance of an *Achatina* neuron, RAPN. These peptides also are considered to be putative neurotransmitters on the neuron, though their presence is not yet demonstrated.

A neuroactive tetrapeptide having a D-phenylalanine, termed achatin-I (H-Gly-D-Phe-L-Ala-L-Asp), was isolated from the *Achatina* ganglia. This peptide caused Na^+-dependent excitation (an inward current under voltage clamp) on another neuron, PON. This is proposed to be an excitatory neurotransmitter on PON, since this is endogenous in the *Achatina* ganglia.

We feel that the molluscan excitable tissues are really 'treasure islands' to isolate many novel neuroactive peptides. The possibility that some peptides, derived from molluscan tissues may exist and play important roles in the mammalian brain exists. We hope that future results demonstrate this claim.

Acknowledgements

This work was partly supported by Chiyoda Mutual Life Foundation Aid, Grant-in-Aid for Cooperative Research (No. 61304006) and Grant-in-Aid for Special Project to Gifu University in 1988 and 1989 from the Ministry of Education and Culture in Japan.

References

Boyles, H. P. and Takeuchi, H. (1985). Pharmacological characteristics of the three giant neurons, d-LPeLN, d-LPeCN and d-RPeAN, identified on the dorsal surface of the pedal ganglia of an African giant snail (*Achatina fulica* Férussac). *Comp. Biochem. Physiol.*, **81C**, 109–15.

Goto, T., Ku, B. S. and Takeuchi, H. (1986). Axonal pathways of giant neurones identified in the right parietal and visceral ganglia in the suboesophageal ganglia of an African giant snail (*Achatina fulica* Férussac). *Comp. Biochem. Physiol.*, **83A**, 93–104.

Hirata, T., Kubota, I., Iwasawa, N., Takabatake, I., Ikeda, T. and Muneoka, Y. (1988). Structure and actions of Mytilus inhibitory peptides. *Biochem. Biophys. Res. Commun.*, **152**, 1376–82.

Kamatani, Y., Minakata, H., Kenny, P. T. M., Iwashita, T., Watanabe, K., Funase, K., Sun, X. P., Yongsiri, A., Kim, K. H., Novales-Li, P., Novales, E. T., Kanapi, C. G., Takeuchi, H. and Nomoto, K. (1989). Achatin-I, an endogenous neuroexcitatory tetrapeptide from *Achatina fulica* Férussac containing a D-amino acid residue. *Biochem. Biophys. Res. Commun.*, **160**, 1015–20.

Kim, K. H., Yongsiri, A., Takeuchi, H., Yanaihara, N., Munekata, E. and Ariyoshi, Y. (1987a). Effects of synthetic peptides on giant neurones identified in the ganglia of an African giant snail (*Achatina fulica* Férussac). III. *Comp. Biochem. Physiol.*, **87C**, 59–61.

Kim, K. H., Yongsiri, A., Takeuchi, H., Yanaihara, N., Munekata, E. and Ariyoshi, Y. (1987b). Effects of synthetic peptides on giant neurones identified in the ganglia of an African giant snail (*Achatina fulica* Férussac). IV. *Comp. Biochem. Physiol.*, **88C**, 325–9.

Kim, K. H., Matsuoka, T., Takeuchi, H., Kubo, K. and Deura, S. (1989). Identification of four silent giant neurons on the dorsal surface of suboesophageal ganglia of an African giant snail (*Achatina fulica* Férussac). *Comp. Biochem. Physiol.*, **93C**, 61–5.

Ku, B. S. and Takeuchi, H. (1983a). Identification and pharmacological characteristics of the three peculiarly firing giant neurones in the visceral ganglion of an African giant snail (*Achatina fulica* Férussac). *Comp. Biochem. Physiol.*, **75C**, 103–110.

Ku, B. S. and Takeuchi, H. (1983b). Identification of three further giant neurones, r-APN, INN and FAN, in the caudal part on the dorsal surface of the suboesophageal ganglia of *Achatina fulica* Férussac. *Comp. Biochem. Physiol.*, **76C**, 99–106.

Ku, B. S. and Takeuchi, H. (1984). Identification and pharmacological characteristics of the two giant neurones, v-RPLN and v-VNAN, on the ventral surface in the suboesophageal ganglia of the African giant snail (*Achatina fulica* Férussac). *Comp. Biochem. Physiol.*, **77C**, 315–21.

Ku, B. S., Isobe, K. and Takeuchi, H. (1985). Pharmacological characteristics of four giant neurons identified in the cerebral ganglia of an African giant snail (*Achatina fulica* Férussac). *Comp. Biochem. Physiol.*, **80C**, 123–8.

Ku, B. S., Takeuchi, H., Yanaihara, N., Munekata, E. and Ariyoshi, Y. (1986). Effects of synthetic peptides on giant neurones identified in the ganglia of an

African giant snail (*Achatina fulica* Férussac). *Comp. Biochem. Physiol.*, **84C**, 391–6.

Litchfield, J. T. Jr and Wilcoxon, F. (1949). A simplified method of evaluating dose-effect experiments. *J. Pharmacol. Exp. Ther.*, **96**, 99–113.

Matsuoka, T., Watanabe, K. and Takeuchi, H. (1985) Pharmacological characteristics of three silent giant neurones, v-LPSN, v-1-VOrN and v-VLN, identified on the ventral surface in the left parietal and the visceral ganglia of the suboesophageal ganglia of an African giant snail (*Achatina fulica* Férussac). *Comp. Biochem. Physiol.*, **80C**, 331–6.

Matsuoka, T., Goto, T., Watanabe, K. and Takeuchi, H. (1986). Presence of TAN (tonically autoactive neurone) and its two analogous neurons, located in the right parietal ganglion of the suboesophageal ganglia of an African giant snail (*Achatina fulica* Férussac). Morphological and electrophysiological studies. *Comp. Biochem. Physiol.*, **83C**, 345–51.

Matsuoka, T., Yongsiri, A., Takeuchi, H. and Deura, S. (1987). Identification and pharmacological characteristics of giant neurones situated in the buccal ganglia of an African giant snail (*Achatina fulica* Férussac). *Comp. Biochem. Physiol.*, **88C**, 35–45.

Matsuoka, T., Takeuchi, H., Yanaihara, N., Munekata, E. and Ariyoshi, Y. (1988). Effects of synthetic biologically active peptides on giant neurones identified in the left buccal ganglion of an African giant snail (*Achatina fulica* Férussac). *Comp. Biochem. Physiol.*, **90C**, 347–50.

Nakajima, T., Nomoto, K., Ohfune, Y., Shiratori, Y., Takemoto, T., Takeuchi, H. and Watanabe, K. (1985) Effects of glutamic analogues on identifiable giant neurones, sensitive to β-hydroxy-L-glutamic acid, of an African giant snail (*Achatina fulica* Férussac). *Br. J. Pharmacol.*, **86**, 645–54.

Novales-Li, P., Watanabe, K., Takeuchi, H., Ohfune, Y., Kurokawa, N. and Kurono, M. (1987). Effects of compounds related to β-hydroxyglutamic acid (BHGA) on identifiable giant neurones of an African giant snail (*Achatina fulica* Férussac). *Eur. J. Pharmacol.*, **143**, 415–23.

Okamoto, H., Takahashi, K. and Yoshii, M. (1976). Membrane currents of the tunicate egg under the voltage-clamp condition. *J. Physiol. (Lond.)*, **254**, 607–38.

Takeuchi, H. and Yamamoto, N. (1982). Pharmacological characteristics of the two largest neurones symmetrically situated in the subesophageal ganglia of an African giant snail (*Achatina fulica* Férussac). *Comp. Biochem. Physiol.*, **73C**, 339–46.

Takeuchi, H., Mori, A. and Kohsaka, M. (1973a) Etude pharmacologique sur un neurone géant, identifié dans les ganglions sous-oesophagiens de l'Escargot géant africain (*Achatina fulica* Férussac), sensible à la 5-hydroxytryptamine et à la dopamine. *C.R. Séanc. Soc. Biol.*, **167**, 602–10.

Takeuchi, H., Morimasa, T., Kohsaka, M. and Morii, F. (1973b). Concentrations des ions inorganiques dans l'hémolymphe de l'Escargot géant africain (*Achatina fulica* Férussac) selon l'état de nutrition. *C.R. Séanc. Soc. Biol.*, **167**, 598–602.

Takeuchi, H., Yokoi, I., Mori, A. and Kohsaka, M. (1975). Effects of nucleic acid components and their relatives on the excitability of dopamine sensitive giant neurones, identified in subesophageal ganglia of the African giant snail (*Achatina fulica* Férussac). *Gen. Pharmacol.*, **6**, 77–85.

Takeuchi, H., Yokoi, I., Mori, A. and Ohmori, S. (1976). Effects of glutamic acid relatives on the electrical activity of an identified molluscan giant neurone (*Achatina fulica* Férussac). *Brain Res.*, **103**, 261–74.

Takeuchi, H., Watanabe, K., Novales-Li, P., Shimamoto, K. and Ohfune, Y. (1989). Classification of β-hydroxy-L-glutamic acid receptors using novel conformationally fixed analogues. *Comp. Biochem. Physiol*, **93C**, 263–8.

Watanabe, K., Takeuchi, H. and Kurono, M. (1985). Comparison of effects of four stereoisomers of β-hydroxyglutamic acid on identifiable giant neurones of an African giant snail (*Achatina fulica* Férussac). *Eur. J. Pharmacol.*, **111**, 57–63.

Yongsiri, A., Goto, T., Ku, B. S., Takeuchi, H. and Namba, M. (1986a). Histological and pharmacological investigations of the two symmetrically situated giant neurones, RPeNLN and LPeNLN, identified on the anterior surface of the pedal ganglia of an African giant snail (*Achatina fulica* Férussac). *Comp. Biochem. Physiol.*, **85C**, 233–8.

Yongsiri, A., Goto, T., Yamamoto, N., Araki, Y., Takeuchi, H. and Namba, M. (1986b). Axonal pathways of the four giant neurones identified in the cerebral ganglia of an African giant snail (*Achatina fulica* Férussac). *Comp. Biochem. Physiol.*, **85A**, 663–8.

Yongsiri, A., Kim, K. H., Takeuchi, H., Yanaihara, N., Munekata, E. and Ariyoshi, Y. (1987). Effects of synthetic peptides on giant neurones identified in the ganglia of an African giant snail (*Achatina fulica* Férussac). II. (*Comp. Biochem. Physiol.*, **86C**, 353–6.

Yongsiri, A., Takeuchi, H., Kubota, I. and Muneoka, Y. (1989). Effects of *Mytilus* inhibitory peptides on a giant neurone of *Achatina fulica* Férussac. *Eur. J. Pharmacol*, **171**, 159–65.

5 *Tibor Kiss*

Effects of the catch-relaxing peptide on identified neurons of the snail *Helix pomatia* L.

5.1. Introduction

Much interest has recently focused on the properties of neuropeptides that modulate the membrane conductances in different excitable cells. It was found that FMRFamide-like peptides, the small cardioactive peptides A and B (SCP_A, SCP_B) affect Ca-conductance in molluscan neurons (Cottrell *et al.*, 1984; Boyd and Walker, 1985; Colombaioni *et al.*, 1985; Acosta-Urquidi, 1988; Brezina *et al.*, 1987). It was also observed, however, that FMRFamide and some related peptides elicit multiple responses in identified neurons, depolarizing some and hyperpolarizing others (Cottrell *et al.*, 1984; Boyd and Walker, 1985).

For comparative purposes it was interesting to study the membrane effect of the catch-relaxing peptide (CARP), which was recently isolated from the pedal ganglia of *Mytilus edulis* by Muneoka and his co-workers (Hirata *et al.*, 1986). CARP showed a relaxing action on the catch tension in the anterior byssus retractor muscle (ABRM) of the *Mytilus* and modulatory effects on contractions in several other molluscan muscles (Hirata *et al.*, 1989a).

It has also been suggested that CARP exerts its effect by acting on postsynaptic sites in the muscle, and that the receptor site is different from the serotonin receptors (Hirata *et al.*, 1989b).

Although the possible site of action was supposed to be presynaptic or postsynaptic, Hirata *et al.* (1989a) found that CARP relaxes catch tension in the ABRM by acting directly on muscle fibres. Since CARP clearly affects the contractile properties of the muscle, we have suggested that it should influence membrane calcium conductance. We therefore investigated the membrane effect of the peptide in snail neurons. The nervous system of *Helix* has many advantages for the study of the basic membrane effects of different pharmacological agents because of the large size of its neurons and the ease with which the neurons can be identified. *Helix* therefore presents an excellent model system in which the basic principles of peptide modulation can be studied.

Neurons were voltage-clamped with a two-microelectrode system. The methods of current and potential recording and extracellular application of peptide were similar to those described in a previous paper (Kiss, 1988).

5.2. **Materials and methods**

The experiments were performed on the isolated and desheathed suboesophageal ganglionic mass of the land snail *Helix pomatia* L. After pronase treatment (0·5 mg/ml for 5–10 min) the connective tissue covering the dorsal surface of the ganglion was carefully removed and identified neurons (Fig. 5.1) were exposed.

Two microelectrode voltage-clamp techniques were used. Membrane current was measured using a virtual ground circuit connected to the bath

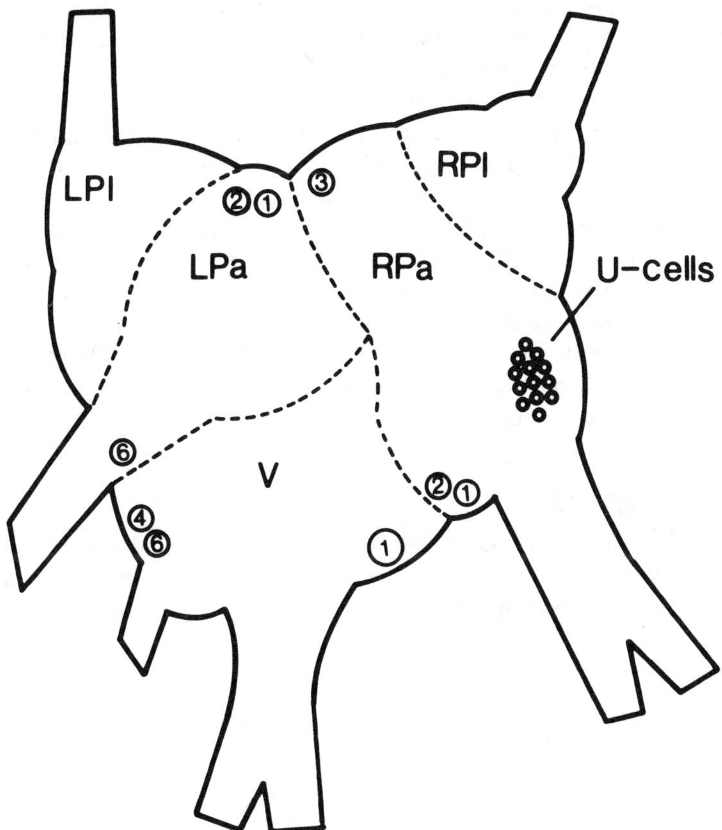

Fig. 5.1. The locations of the identified neurons on the dorsal surface of the suboesophageal ganglia. RPa: right parietal, LPa: left parietal, RPl: right pallial, LPl: left pallial, V: visceral ganglia.

reference Ag/AgCl electrode via an agar bridge. Microelectrodes were filled with 2·5 M KCl and had resistances of 2–5 MΩ.

An IBM minicomputer was used to acquire, store and analyse the experimental data (pCLAMP program set, Axon Instrument). Currents were filtered at 1 kHz and digitized at a sampling rate of 100–200 μs using a 12-bit analogue-to-digital converter. Currents were corrected by subtracting the linear components of capacitive and ionic currents, which give hyperpolarizing pulses, from the holding potential (HP). All records are averages of responses to three pulses identical in magnitude. Results were plotted on an *X–Y* HP plotter (70401). Depolarizing and hyperpolarizing voltage steps were generated by a computer.

The compositions of the solutions used are listed in Table 5.1. The pH of all solutions was adjusted by NaOH to 7·4–7·6, and 10 mM sucrose was added to the NR-r saline.

CARP (H-Ala-Met-Pro-Met-Leu-Arg-Leu-NH_2) was synthesized by I. Kubota at Suntory Institute (Osaka, Japan). The peptide was made up at 1 or 0·1 nM in normal physiological saline and was frozen. Small samples of this stock were added to different solutions before the experiment. Ganglia were continuously perfused (1 ml/min), and the appropriate concentration of the peptide was bath-applied. Experiments were carried out in winter and spring at room temperature (~20°C).

The inward currents of *Helix* neurons are carried by both Na and Ca ions, except in the U-cluster neurons, in which the inward current is suggested to be purely Ca^{2+}-dependent and the outward current is a Ca^{2+}-activated K-current (Lux and Hofmeier, 1982; Gola *et al.*, 1986). In the present experiments, therefore, the sodium current was eliminated by omitting Na ions from the perfusing saline, and outward potassium currents were blocked by using 4-aminopyridine (4AP) and/or TEACl.

5.2.1. *Effect of CARP on I_{Ca}*

Earlier we reported (Kiss, 1988) that, in neurons bathed in normal physiological saline, CARP increases the duration of action potentials and decreases the spike after hyperpolarization. The prolongation of AP

Table 5.1. Compositions of solutions used (in mM)

	NaCl	*KCl*	*$CaCl_2$*	*$MgCl_2$*	*TEACl*	*4AP*	*Tris*
NR-r	80	4	10	5	—	—	51
10Ca	—	4	10	5	—	—	120
27Ca	—	4	27	4	—	—	90
27CaTEA	—	4	27	4	75	—	10
27CaTEA+4AP	—	4	27	4	75	4	10

was observed in both normal physiological and Na^+-free 10CaTEA salines. It was therefore concluded that the prolongation of AP could be a result of either an increase of Ca inward or a decrease of Ca-dependent outward currents.

In voltage-clamp conditions currents elicited in CaTEA or CaTEA + 4AP salines are carried primarily by Ca ions. In these conditions the I_{Ca} receives little contamination from outward currents (Chad *et al.*, 1984), especially when depolarizations do not exceed 0 mV. In *Helix* CARP at 10^{-5}–10^{-6} M caused a transient suppression of the I_{Ca} in some neurons (Fig. 5.2). The holding potential was held at −50 mV and the peptide was added in CaTEA + 4AP saline. The suppressing effect of CARP was restored after washing out with the control saline. In some cases, independent of the type of neuron, CARP evoked an increase of inward current. The degree of peptide effect, however, depended on several factors, such as the concentration of peptide, extracellular Ca ion concentration, the HP and the length of time after the beginning of peptide application. The concentration and HP dependence of the CARP effect are shown in Fig. 5.3. Current measurements were made on the same neuron (RPa1) at −80 and −40 mV holding potentials using two different concentrations of the peptide. In Fig. 5.4 the time-dependence of the peptide effect is shown. In one case, following peptide application, there was a decrease for 10–12 min, which was followed by an increase in I_{Ca} amplitude. In the other case only an increase was observed. In both examples, however, a recovery was observed without washing out procedure in 30–60 min. Washing out with control saline accelerated recovery. Our measurements were usually made during the first 10 min after CARP administration, and in 80% of the neurons studied the I_{Ca} was decreased. In both cases, whether CARP increased or decreased the Ca inward current, the inactivation accelerated (Fig. 5.2). In some neurons CARP at a low concentration (10^{-6} M) increased the inward current, which was greater at HP = −80 than at HP = −40 mV (Fig. 5.3). Since even at a high peptide concentration (10^{-4} M) only a fraction of the I_{Ca} was suppressed, it was supposed that CARP specifically changed one type of inward Ca channel. In order to test this hypothesis, currents evoked at different holding potentials and different voltages were compared in control and peptide-containing salines. In Fig. 5.5 the neuron was depolarized from −50 mV HP to different command potentials. 10^{-6} M CARP decreased the peak I_{Ca} in a range of −50 and +40 mV. The difference in the current–voltage characteristic is also plotted, and it was concluded that the low-voltage activated channels were suppressed by the peptide. On the other hand, when an increase of I_{Ca} was obtained after peptide treatment, the increasing effect was connected to the enhancement of the very slowly inactivating current component. In Fig. 5.6 an example is shown of this effect at two different holding potentials. It can be seen that

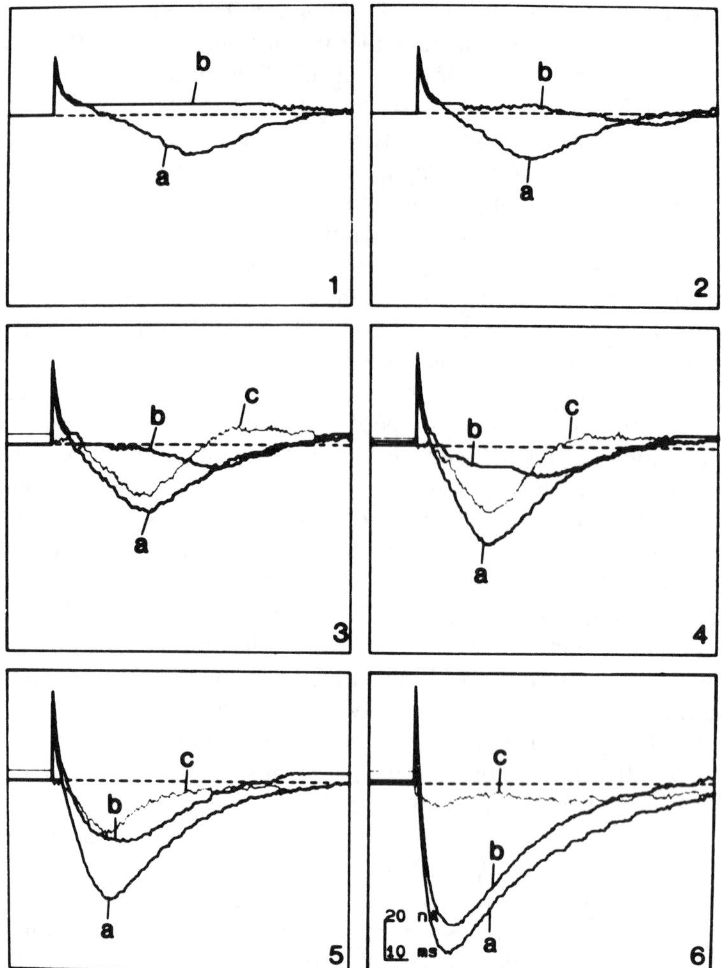

Fig. 5.2. Ca-inward currents recorded before (**a**) and after (**b**) 10^{-6} M CARP application at −20 (1), −15 (2), −10 (3), −5 (4), 0 (5) and +15 (6) mV. In few cases difference currents (c) are also plotted. Neuron V4, HP = −50 mV, CaTEA + 4AP saline. At low potentials an inactivating component was suppressed, while at high potentials a steady non-inactivating component was blocked.

at moderate depolarizations (below −10 mV) LVA current was suppressed, while at depolarizations higher than +10 mV a steady, almost non-inactivating, current component was developed.

When the peptide in CaTEA (without 4AP) saline was applied, the inactivating phase of I_{Ca} became faster. This enhancement was supposed to be connected with the activation of the outward current.

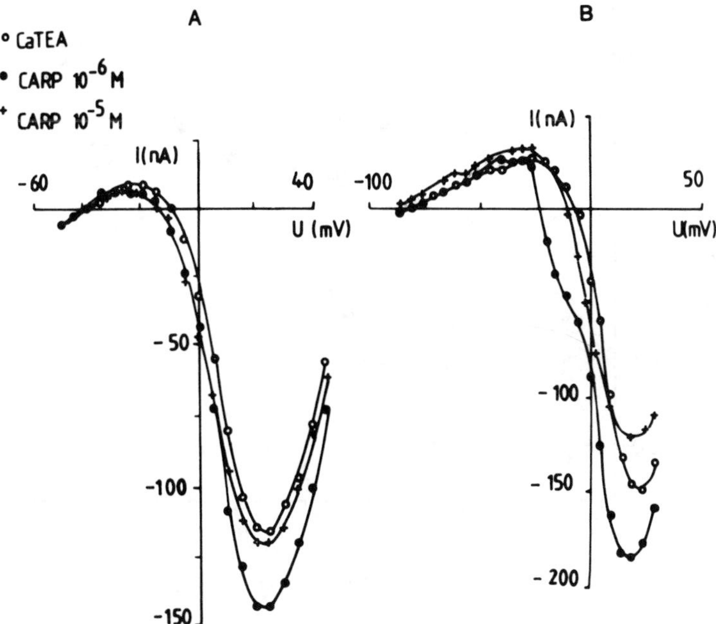

Fig. 5.3. The I–V characteristics of the inward Ca current at two different holding potentials (−40 (A) and −80 (B) mV) and in the presence of 10^{-6} and 10^{-5} M CARP. Neuron V8.

5.2.2. *Effect of CARP on Ca-outward current*

Earlier we described (Kiss, 1988) that the suppressing effect of CARP might be due to the increasing of the Ca-dependent K-conductance of the membrane. Since in the presence of potassium channel blockers application of peptide mostly decreased the peak amplitude of the I_{Ca}, the prediction was that Ca-activated K current will decrease. The experiment was therefore repeated in the absence of potassium channel blockers, and it was observed that outward current increased (Fig. 5.7). Previously it was found that CARP decreased the outward current in U-cluster neurons, in which the outward current is thought to be Ca-dependent. Our recent observations suggest that the peptide increases the potassium permeability in neurons other than those in U-clusters. As the peptide had no effect on the inactivation of the fast outward current, it was supposed that CARP affected the delayed rectifier.

5.3. **Conclusions**

The effect of the molluscan neuropeptide CARP on identified neurons in the suboesophageal ganglia of the snail *Helix pomatia* L. was investi-

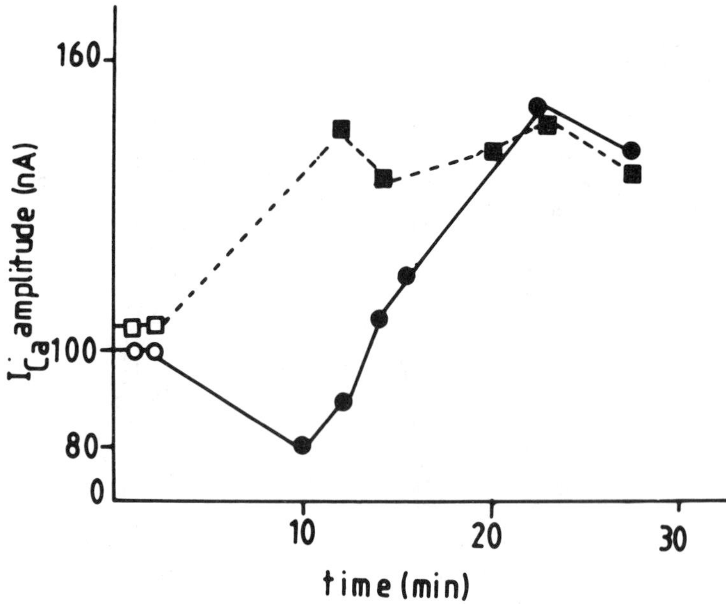

Fig. 5.4. The time-dependence of peptide effect.

gated. In many neurons the responses observed upon peptide application involve combinations of more than one type of action. On snail neurons CARP proved to be effective in much higher (order of 2) concentrations than on *Mytilus* muscle preparations. This can be explained either by the species specificity of the peptide or by the inappropriate selection of the neurons studied. The first possibility can be discarded, since it has been recently found that a number of cells in the pedal ganglia of *Helix* are labelled by anti-CARP (unpublished results). This means that closely related peptides, if not CARP itself, can be present in the snail brain. As the second possibility could be real, it would also be necessary to study the CARP effect on pedal ganglionic cells.

Nevertheless, some hint of the membrane effect of the catch-relaxing peptide can be obtained by using even higher concentrations of the peptide. We have found that CARP decreased the Ca inward current in certain neurons and increased it in others, depending on peptide concentration, potential and the neurons to be selected. When the amplitude of the Ca inward current was decreased we never observed a concentration-dependent full block of the current. Since only a part of the Ca current was blocked, we supposed that CARP had a specific effect on one type of Ca current. Mironov *et al.* (1985) and Haydon and Man-Son-Hing (1988) have reported the presence of two Ca currents (low- and high-voltage activated) in *Helix pomatia* and *Helisoma trivolvis*. As different types of

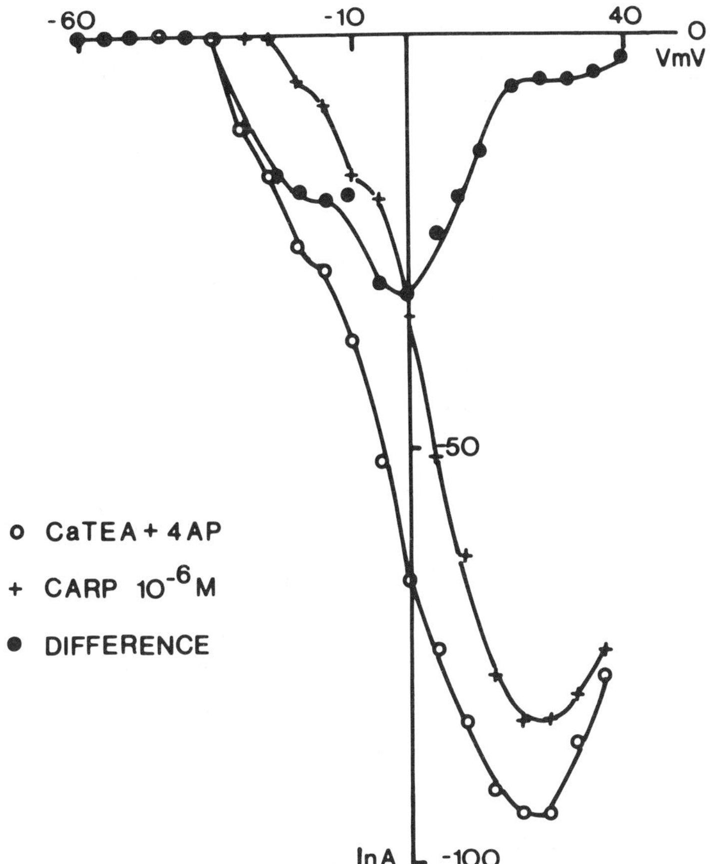

Fig. 5.5. Current–voltage relationship before (○) and following CARP treatment (+). The current component which was blocked is plotted by full circles.

Ca currents appear to serve different cellular functions, LVA and HVA Ca currents might have different distributions within the neurons. The soma, which does not release a neurotransmitter, contains both LVA and HVA Ca currents, while in the secretory soma the HVA current is dominant (Haydon and Man-Son-Hing, 1988). The presence of two Ca currents in *Helix* neurons, however, is not strongly evidenced. Experiments made by Mironov *et al.* (1985) could not be repeated using a two-microelectrode voltage clamp. Further experiments are needed to clarify the precise peptide effect on inward ionic conductances. The CARP effect differs from that of FMRFamide in spite of the fact that both peptides decreased the Ca conductance in *Helix* (Colombaioni *et al.*, 1985), since following CARP application the AP duration was increased,

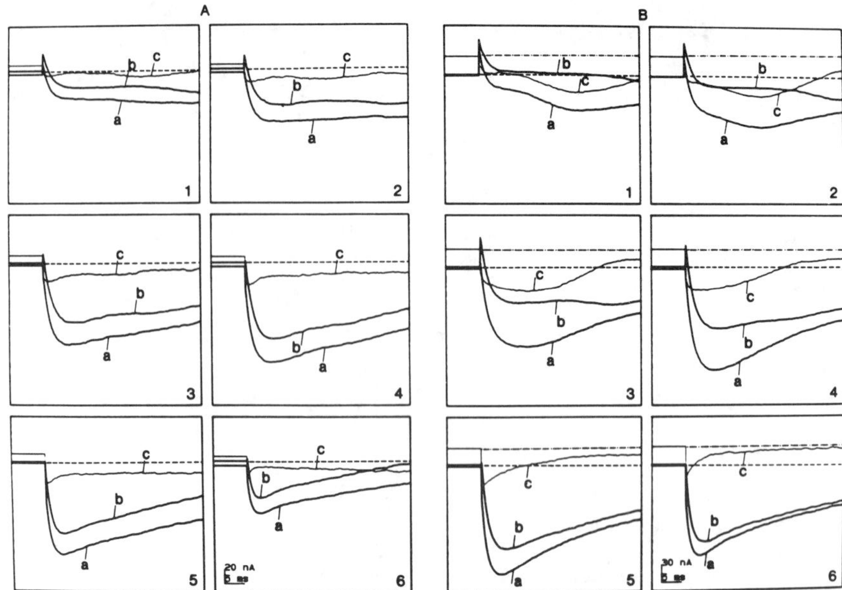

Fig. 5.6. CARP increased a steady non-inactivating component of Ca-current: a, CaTEA + 4AP saline; b, 10^{-6} M CARP; c, difference current. A and B, currents at two different holding potentials. The command levels from 1 to 6 were −5, 0, +5, +10, +20 and 30 mV respectively.

while in the presence of FMRFamide it was decreased. FMRFamide eliminates the fraction of Ca inward current without affecting the time course of the remaining current (Brezina *et al.*, 1987), while CARP speeds up the inactivation. This effect was also observed when CARP increased the Ca inward current (Fig. 5.6). Since the application of the peptide in the presence of K-channel blockers mostly decreased the I_{Ca}, it was predicted that the Ca-activated K current will also decrease. This was the case in U neurons, in which the outward current was supposed to be Ca-dependent (Kiss, 1988). Since both Ca-inward and Ca-activated currents decreased simultaneously in the presence of CARP so that the relationship between them was linear, it is suggested that depression of the K(Ca) current is merely the consequence of Ca inward current inhibition. In other neurons, however, a contrasting increase of outward current was observed. It is therefore supposed that CARP affected the delayed rectifier, but had no effect on inactivating fast potassium current (I_A).

It is known that FRMFamide suppressed both the K(S) current and the Ca inward current in several identified *Helix* neurons, while in others the Na current and different K conductances were suppressed (Colombaioni *et al.*, 1985; Cottrell *et al.*, 1984). Results obtained with FRMFamide and

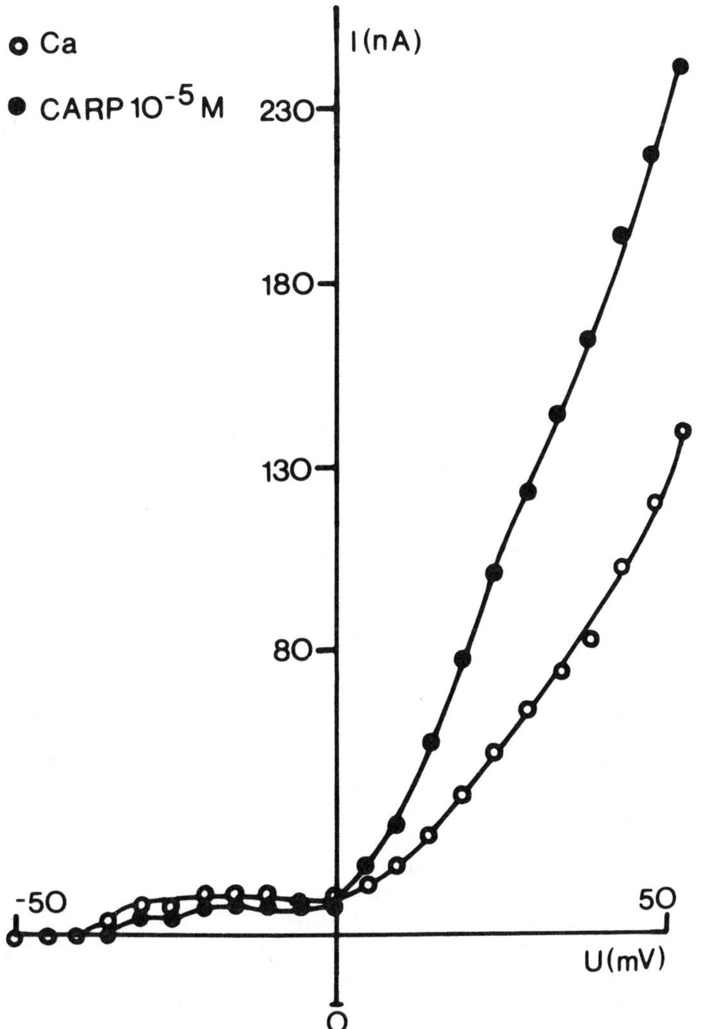

Fig. 5.7. Effect of CARP on the outward current: neuron RPa1, HP = −50 mV, 27b mM Ca saline.

related peptides, and also CARP, demonstrate that peptides isolated from molluscs can modulate different ion channels and produce different effects, depending mainly on the individual type of neuron investigated.

References

Acosta-Urquidi, J. (1988). Modulation of calcium current and diverse K^+-currents in identified *Hermissenda* neurons by small cardioactive peptide B. *J. Neurosci.*, **8**, 1694–703.

Boyd, P. J. and Walker, R. J. (1985). Actions of the molluscan neuro-peptide FRMF-amide on neurones in the suboesophageal ganglia of the snail *Helix aspersa. Comp. Biochem. Physiol.*, **81C**, 379–86.

Brezina, V., Eckert, R. and Erxleben, C. (1987). Suppression of calcium current by an endogenous neuropeptide in neurones of *Aplysia californica. J. Physiol.*, **388**, 565–95.

Chad, J., Eckert, R. and Ewald, D. (1984). Kinetics of calcium-dependent inactivation of calcium current in voltage-clamped neurones of *Aplysia californica. J. Physiol.*, **347**, 279–300.

Colombaioni, L., Paupardin-Tritsch, P., Vidal, P. P. and Gerschenfeld, H. M. (1985). The neuropeptide FRMF-amide decreases both the Ca^{2+} conductance and a cyclic 3′,5′-adenosine monophosphate-dependent K^+ conductance in identified molluscan neurons. *J. Neurosci.*, **5**, 2533–8.

Cottrell, G. A., Davies, N. W. and Green, K. A. (1984). Multiple actions of molluscan cardioexcitatory neuropeptide and related peptides on identified *Helix* neurons. J. Physiol., **356**, 315–33.

Gola, M., Hussy, N., Crest, M. and Ducreux, C. (1986). Time course of Ca and Ca-dependent K-currents during molluscan nerve cell action potentials. *Neurosci. Lett.*, **70**, 354–9.

Haydon, P. G. and Man-Son-Hing, H. (1988). Low- and high-voltage-activated currents: their relationship to the site of neurotransmitter release in an identified neuron of *Helisoma. Neuron*, **1**, 919–27.

Hirata, T., Kubota, I., Takabatake, I., Kawahara, A., Shimamoto, N. and Muneoka, Y. (1986). Catch-relaxing peptide isolated from *Mytilus* pedal ganglia. *Brain Res.*, **422**, 374–6.

Hirata, T., Kubota, I., Imada, M. and Muneoka, Y. (1989a). Pharmacology of relaxing response of Mytilus smooth muscle to the catch-relaxing peptide. *Comp. Biochem. Physiol.*, **92C**, 289–95.

Hirata, T., Kubota, I., Imada, M., Muneoka, Y. and Kobayashi, M. (1989b). Effects of the catch-relaxing peptide on molluscan muscles. *Comp. Biochem. Physiol.*, **92C**, 283–8.

Kiss, T. (1988). Catch-relaxing peptide (CARP) decreases the Ca-permeability of snail neuronal membrane. *Experientia*, **44**, 998–1000.

Lux, D. H. and Hofmeier, G. (1982). Activation characteristics of the calcium-dependent outward potassium current in *Helix. Pflugers Arch.*, **394**, 70–7.

Mironov, S. L., Tepikin, A. V. and Grishchenko, A. V. (1985). Two calcium currents in the somatic membrane of the mollusc neurons. *Neurophysiology*, **17**, 627–33 (in Russian).

6 *Katalin S.-Rózsa, David O. Carpenter, George B. Stefano and János Salánki*

Distinct responses to opiate peptides and FMRFamide on B-neurons of the *Aplysia* cerebral ganglia

6.1. Introduction

The presence of enkephalin-immunoreactive substance as well as specific binding of endogenous opiate ligands have been demonstrated in molluscs (Boer *et al.*, 1980, Stefano *et al.*, 1980). FMRFamide, the molluscan cardioactive peptide, has been localized by RIA and immunocytochemistry in a number of invertebrates and in the brain of several vertebrates (Greenberg and Price, 1979; Dockray *et al.*, 1983; Lehman and Greenberg, 1987), demonstrating that the same peptides can be found throughout the animal kingdom.

The opiate peptides were found to elicit both excitatory and inhibitory responses on the membrane of molluscan and vertebrate neurons (Zieglgänsberger *et al.*, 1979; Stefano *et al.*, 1980; Stone and Mayeri, 1981; Cottrell *et al.*, 1984; S.-Rozsa and Carpenter, 1985; Carpenter and Hall, 1986; McDonald and Werz, 1986). FMRFamide was also shown to depolarize or hyperpolarize individual *Aplysia* neurons (Stone and Mayeri, 1981; Ruben *et al.*, 1986; Brezina *et al.*, 1987; Belardetti *et al.*, 1987).

Although the biochemical characterization of opiate receptors has been described, little is known regarding the pharmacology of opiate analogues in molluscs. Effects of Leu-enkephalin and FMRFamide have been reported, but there are contradictory data as regards naloxone sensitivity of opiate responses in the nervous system of molluscs (Stefano *et al.*, 1980; Carpenter and Hall, 1986; Stefano, 1988).

The aim of the present investigation was to map the cerebral B-cells of *Aplysia californica* on the basis of their responses to opiate peptide analogues and FMRFamide, and to study the effect of these substances in comparison with the effect of the low molecular weight neurotransmitter ACh on these neurons.

6.2. Methods

Experiments were performed on *Aplysia californica* obtained from Pacific Biomarine Supply Co. (Venice, California) and maintained in an aquarium at 16°C. Cerebral ganglia were dissected from the animal pinned to a sylastic base (Sylgard 184, Dow Corning) and constantly perfused with artificial sea water (ASW) as described in Table 6.1.

The connective tissue sheath of the ganglia was removed by dissection above the cells of the B-cluster (Jahan-Parwar and Fredman, 1976). The cells of B-cluster were identified on the basis of their location and on the response to applied substances. For identification of the cells a camera lucida drawing of the dorsal surface of the cerebral ganglia was used (Jahan-Parwar, unpublished). The experiments were performed at room temperature (20°C).

In the course of the experiments neuronal somata were penetrated with two independent glass microelectrodes having a resistance of 1–5 MΩ and voltage-clamped with a Dagan Model 8500 amplifier. The current passing microelectrode was filled with 2 M potassium acetate at pH 7·0 and the potential-measuring microelectrode was filled with 3 M KCl. Electrodes and holders were wrapped in earthed aluminium foil to minimize cross-talk. Transmembrane current was recorded with a ground current monitor connected to the bath via an $Ag/AgCl_2$ junction.

Both current and voltage clamp modes of recording were used. Membrane resistance was monitored by applying bridge-balanced constant current pulses (0·5 nA) of 0·8 s duration every 4 s. Similar current pulses were used for determination of membrane rectification in the presence of DC current.

The peptides investigated were applied either by microperfusion or by pressure injection from a micropipette of 5 μm tip diameter, while the antagonist (naloxone) was added to the bath. The microejection electrode was positioned above the cell body so as to maximize the response of the cells. Fast green FCF (Aldrich) was added into the pressure pipette in order to visualize peptide ejection and removal.

Peptides were applied at 10–15 min intervals to avoid desensitization of cells. All the substances except dynorphin were dissolved in ASW. Dynorphin A was dissolved in a mixture of methanol and 0·1 N HCl (1:1) then diluted in ASW.

The following substances were used: (Leu^5) and (Met) enkephalins (acetate salt, Sigma), (D-Ala^2, D-Leu^5) enkephalin (acetate salt, Sigma), dynorphin A (Sigma), naloxone (hydrochloride, Sigma), 4-aminopyridine (Sigma), TEA chloride (ICN Pharmaceuticals).

Table 6.1. Composition of artificial sea water (ASW) (concentrations in mM)

	NaCl	*KCl*	*$CaCl_2$*	*$MgCl_2$*	*$MgSO_4$*	*Hepes*
ASW normal N-ASW	480	10	10	20	30	5
High Ca^{2+}, high Mg^{2+} H-ASW	212	10	66	165	—	10

Salines were adjusted to pH 7·8

Normal ASW or high Ca^{2+} + Mg^{2+} ASW with pH 7·8 was used for perfusion. Ion-free solutions were prepared as described by Pellmar and Carpenter (1980). The Na-free saline was made by adding mannitol or glucosamine in an equimolar concentration and the pH was adjusted with $Mg(OH)_2$ or NaOH. In K-free saline NaCl was substituted for K-ions, the Co^{2+} solution was made by adding 10–30 mM $CoCl_2$ to ASW, and solutions containing 5 mM 4-aminopyridine (4-AP) were also prepared in ASW. To avoid synaptic activity high Ca^{2+} and Mg^{2+} saline was used in a series of experiments prepared as shown in Table 6.1.

6.3. **Results**

6.3.1. *Identification of B-cluster neurons*

The two B-clusters of neurons are symmetrically situated on the dorsal surface of the cerebral ganglia at the midline of the two hemispheres. The designation of these cells as B-clusters was part of the cell identification carried out by Fredman and Jahan-Parwar in 1975. Each B-cluster contains about 35–40 neuronal somata, ranging in size from 45 to 200 μm diameter and having various degrees of pigmentation. The B-cells were silent or maintained a steady irregular firing. Their membrane potential varied between −35 and −60 mV and showed anomalous rectification beyond −60 mV. Many traditional neurotransmitters have receptors in the neuropil resulting in a large variety of effects (Gaillard and Carpenter, 1986). The B-cells have intensive synaptic input.

We have found it possible to identify some of the individual neurons within the dorsal surface B-cluster by a combination of size, pigmentation, location and physiological responses to transmitters and peptides. Because, as described below, the responses to peptides varied significantly from cell to cell in the B-cluster, it was very important to be able to study the same neuron in different preparations. Figure 6.1 illustrates a camera lucida drawing of the cerebral ganglion, and the areas encircled are the bilateral B-clusters on the dorsal surface. Individual neurons which were large and relatively constant in appearance and location were numbered and identified as Cerebral B right (CBR) or left (CBL) and by number. Using this method we have found that responses to opiates were predictable in neurons which could be identified.

6.3.2. *The effect of enkephalins and acetylcholine (ACh) on B-neurons*

B-cells showed a variety of types of responses to enkephalins and ACh. While all neurons showed responses to ACh, not all neurons responded to the peptides. However, the profile of responses to all of these substances was repeatable from preparation to preparation on neurons which

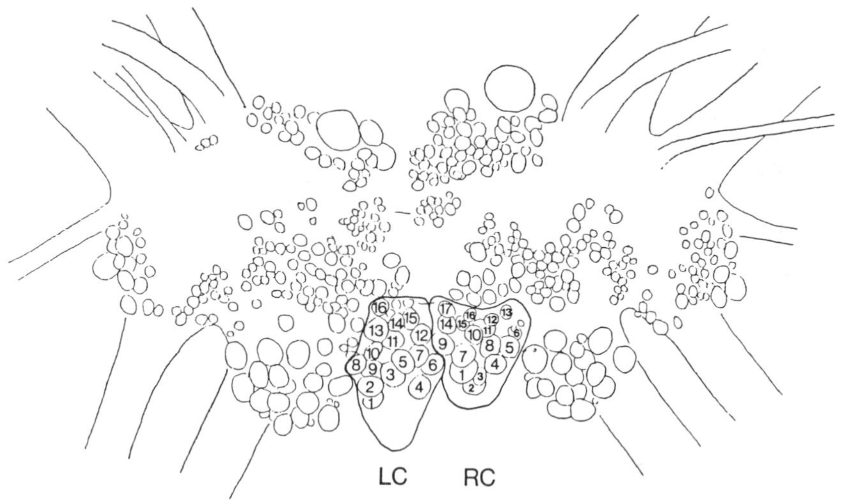

Fig. 6.1. Dorsal surface of the cerebral ganglia of *Aplysia californica* and location of the B-clusters. The numbering of the left (CBL) and right B-cluster neurons (CBR) is shown.

could be identified. Some neurons responded to the enkephalins with a monophasic depolarization or hyperpolarization. In other neurons, however, the responses to the peptides were complex and multiphasic, with depolarizing and hyperpolarizing potentials induced in sequences which varied from cell to cell. In most neurons ACh was more potent than were the opiates. Leu-enkephalin was the opiate used in most studies, except as indicated below. Met-enkephalin was also applied on some neurons, and on most neurons was found to elicit the same responses and to be about equal in potency to Leu-enkephalin.

In current-clamp experiments using Leu-enkephalin, applied by both pressure ejection and microperfusion, 24 neurons were found to respond (Table 6.2). Of these, 13% showed a monophasic depolorization and 39% a monophasic hyperpolarization. The remaining 39% showed bi- or polyphasic responses. All responses showed desensitization, although it appeared that the hyperpolarizing responses were more resistant to desensitization than were the depolarizing responses. As a consequence, responses which were initially biphasic would sometimes become a monophasic hyperpolarization if the agonist was applied without sufficient time between applications. While many of the responses were relatively brief (10–50 s), others were quite slow. Some responses, both depolarizing and hyperpolarizing, required up to 5 min of washing before returning to baseline.

Unlike responses to ACh, those to the enkephalins were usually not

Table 6.2. Type of responses of B-cluster cerebral neurons of *Aplysia californica* to enkephalins and FMRFamide

Responses to enkephalins at resting MP			*Response to FMRFamide at resting MP*		
D-type	*H-type*	*Biphasic*	*D-type*	*H-type*	*Biphasic*
CBL11	CBL1	CBL2	CBL1	CBL8	CBL2
CBR1	CBL3	CBL4	CBL4	CBL14	CBL5
CBR9	CBL5	CBL6	CBL11	CBR3	CBL7
	CBL8	CBL7	CBR5	CBR7	CBL15
	CBL13	CBL12	CBR8		CBR1
	CBR2	CBL14	CBR9		CBR7
	CBR4	CBR5			CBR11
	CBR8	CBR7			CBR13
	CBR13	CBR11			CBR17
	CBR14				
	CBR17				

associated with marked changes in membrane conductance. Figure 6.2 shows recording from a voltage-clamp experiment in which ACh, Leu- and Met-enkephalin were applied by pressure ejection. Membrane conductance was measured by applying brief voltage jumps from the holding potential of −50 mV to −70 and −30 mV. When agonist is applied a conductance increase is indicated by an increase in the current during the voltage step, while a conductance decrease response is indicated by a smaller current. As seen in the figure, there is a clear conductance increase in response to ACh, which on this neuron elicits an outward current. However, the responses to both Leu- and Met-enkephalin, which are slow inward currents of almost equal amplitudes, are not associated with a clear conductance change.

6.3.3. *Comparison of responses to FMRFamide and enkephalins*

Responses were also obtained from B neurons to FMRFamide. The responses were of the same general character as those to enkephalins, but they were distinct since in some neurons they were of opposite polarity or responses were not found for one or another substance. Furthermore, the identified neurons showed a consistent response to both FMRFamide and the enkephalins from one preparation to another. B-neuron responses to FMRFamide were also mono-, bi- and polyphasic, with depolarizations and hyperpolarizations in different sequences. In general, FMRFamide was more potent than the enkephalins.

Table 6.2 also lists the responses of the identified B-neurons to FMRFamide. Of 19 neurons studied 31% were depolarized, 21% hyperpolarized and 48% showed a biphasic response to FMRFamide. In seven

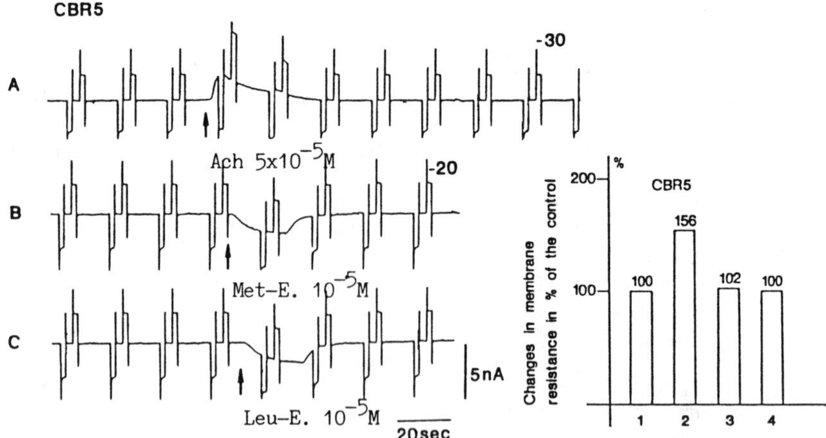

Fig. 6.2. Alteration in membrane conductance following pressure ejection of (**A**) acetylcholine (5×10^{-5} M), (**B**) Met-enkephalin (10^{-5} M) and (**C**) Leu-enkephalin (10^{-5} M) on cell CBR5. The neuron was voltage-clamped and conductance monitored by clamp pulses to −70 and −30 mV from the holding potential of −50 mV.

neurons the FMRFamide response was similar to that to Leu-enkephalin, and on two neurons it was opposite. Desensitization to FMRFamide was slow as compared to that to Leu-enkephalin.

Figure 6.3 illustrates FMRFamide and Leu-enkephalin responses recorded from cell CBL11. The upper trace of each pair is a current-clamp recording, while the lower trace is voltage-clamp. In this experiment both peptides were applied by pressure ejection at the concentrations indicated. In all such experiments the concentration of peptide with pressure ejection was much higher than that applied by microperfusion, indicative of the fact that pressure-applied peptide is diluted in the bath. Thus the concentrations applied are no indication of the threshold concentrations, although the relative potency of FMRFamide and enkephalin is significant. In this neuron both FMRFamide and Leu-enkephalin caused a slow depolarization. FMRFamide was associated with a clear conductance increase, while no conductance change was obvious to Leu-enkephalin.

Figure 6.4 shows recordings from a different cell in which both peptides caused hyperpolarizing responses, again with a clear conductance increase for FMRFamide but not Leu-enkephalin. In this cell the Leu-enkephalin response was both faster and more sensitive than that illustrated in Fig. 6.2. This figure also illustrates the pronounced desensitization to Leu-enkephalin. Part C shows a repeat application of Leu-enkephalin at a higher concentration 3 and 4 min after that shown in part B, and the response is much slower and smaller.

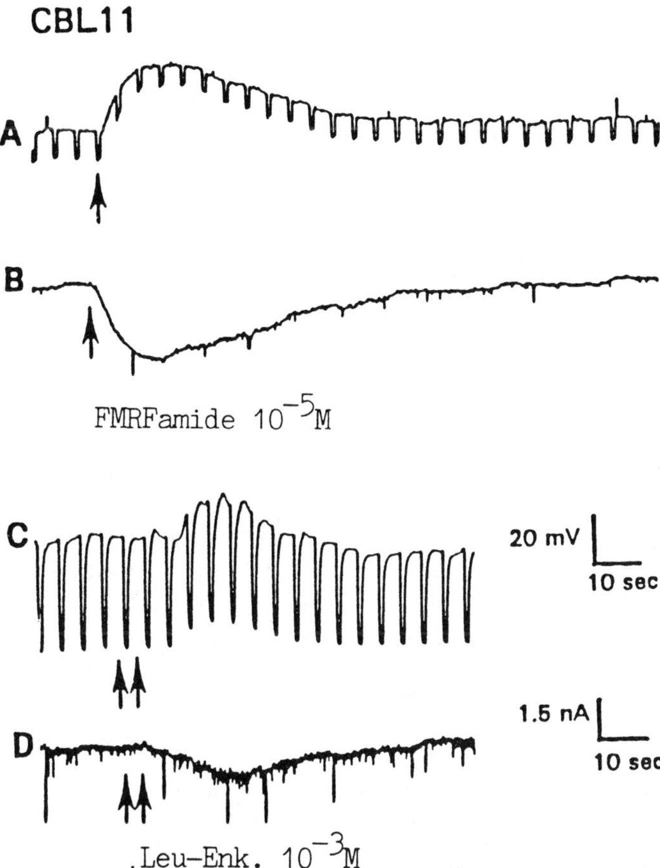

Fig. 6.3. Depolarizing response to (**A**) FMRFamide (10^{-5} M) and (**C**) Leu-enkephalin (10^{-3} M) applied by pressure ejection on cell CBL11 with corresponding inward currents generated at resting MP values (**B**,**D**). In the current clamp recording (**A**,**C**) the brief pulses are from application of 1 nA currents to monitor changes in membrane conductance.

6.3.4. *Effects of naloxone on responses to Leu-enkephalin and FMRFamide*

Naloxone is a relatively specific antagonist of opiate responses, and therefore its actions were determined on the enkephalin, FMRFamide and ACh responses of B-neurons. Responses were studied on neurons CBL2, CBL4, CBL7, CBL8, CBL11, CBL14, CBR4, CBR5, CBR7 and CBR8. In all but the last neuron responses to Met- and Leu-enkephalin were reduced or abolished following bath application of $1-10^{-5}$ to 5×10^{-3} M naloxone. The degree of blockade was greater in high Mg^{2+}, high Ca^{2+}

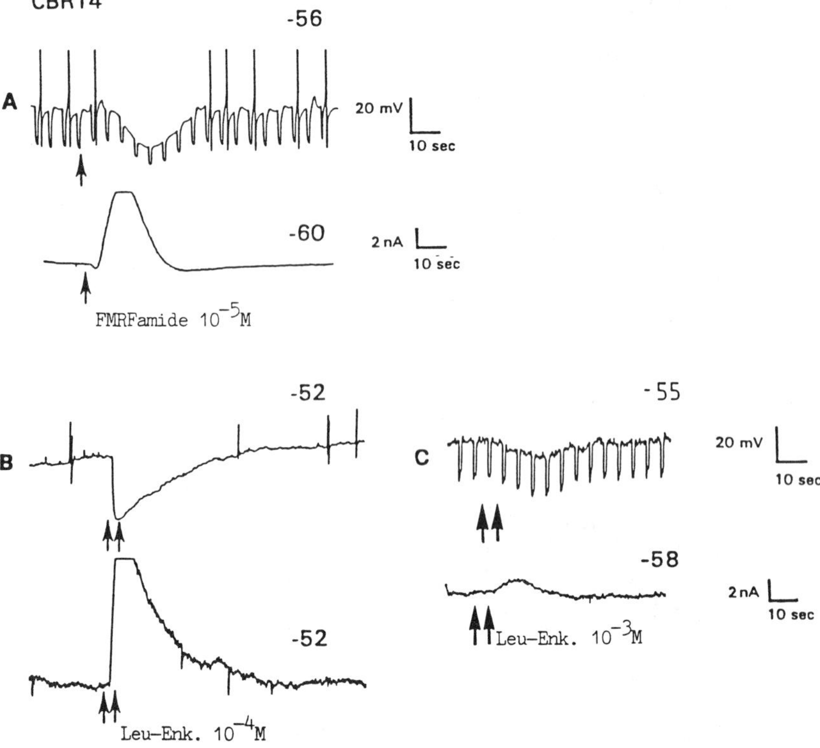

Fig. 6.4. Response of cell CBR14 to (**A**) FMRFamide (10^{-5} M) and (**B**) Leu-enkephalin (10^{-4} M) in current clamp (**A**,**C** upper) and voltage clamp (**B**,**D** lower); **C**, repeated application of Leu-enkephalin (10^{-3} M) shows a much smaller response, due to receptor desensitization.

sea water. In normal ASW naloxone frequently activated spontaneous synaptic activity which obscured the effects of naloxone.

Figure 6.5 illustrates results of application of naloxone on neuron CBR4, which in the control gave a slow outward current in response to Leu-enkephalin. After a 20 min application of 5×10^{-3} M naloxone the Leu-enkephalin response was totally abolished. The response had not recovered after 20 min of washing. In most experiments only incomplete recovery occurred after 40 or more minutes of wash. The inward current elicited by 10^{-3} M Leu-enkephalin in cell CBL11 was also naloxone-sensitive, being reduced to about 40% of control after a 15 min treatment with 5×10^{-3} M naloxone. No greater inhibition was seen after treating for a total of 30 min with naloxone. The degree of inhibition seen was often not more than 60–70%.

Since relatively high concentrations of naloxone were required to block the enkephalin responses, we investigated the effects of similar concentra-

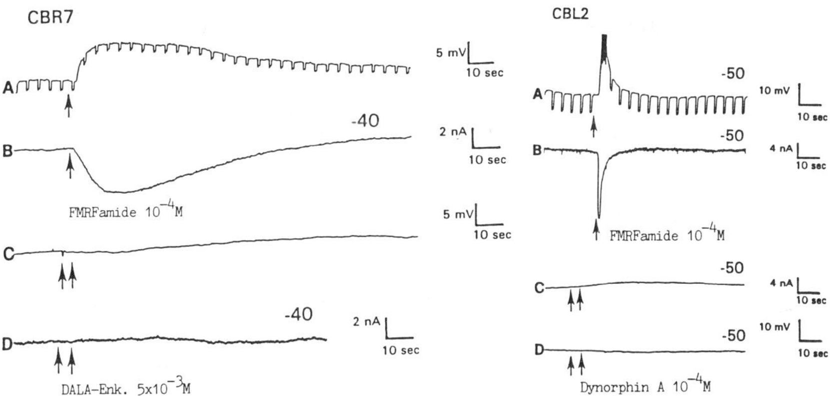

Fig. 6.5. Application of (D-Ala^2-, D-Leu^5) enkephalin and FMRFamide on cell CBR7, and dynorphin A and FMRFamide on cell CBL2. Recording is made in current-clamp (**A**,**C**) and voltage-clamp (**B**,**D**) mode at resting MP level. Conductance pulses are 1 nA currents.

tions on the responses to FMRFamide and ACh. At 5×10^{-3} M naloxone often had effects on responses to FMRFamide, but the effects were both facilitatory and inhibitory. Figure 6.6 shows results of one experiment testing the effects of naloxone on responses to Leu- and Met-enkephalin and FMRFamide. After 15 min of naloxone both enkephalin responses were blocked and the FMRFamide response was potentiated to 160% of control. However, with time in naloxone the FMRFamide response was depressed to about 50% of control. In some neurons only the facilitatory effects of naloxone were observed, while in others only the inhibitory actions were seen. In contrast, all enkephalin responses were depressed, albeit often not completely, but these very high concentrations of naloxone.

6.3.5. *Threshold and desensitization of opiate responses on B-neurons*

Even when using pressure application of the enkephalins on to B-neurons it was apparent that too-frequent application led to a reduced response, consistent with a process of receptor desensitization. Therefore, a series of experiments were done using microperfusion of peptides to both better determine the threshold concentration for actions and to study the process of desensitization.

Figure 6.7 illustrates responses to microperfusion of Met-enkephalin at concentrations between 10^{-8} and 10^{-4} M. The different concentrations were applied at intervals of 1 min. The threshold for effect was 10^{-8} M. With 10^{-7} M the response was larger. However, at 10^{-5} M the response amplitude was very much less, and there was little response at 10^{-4} M.

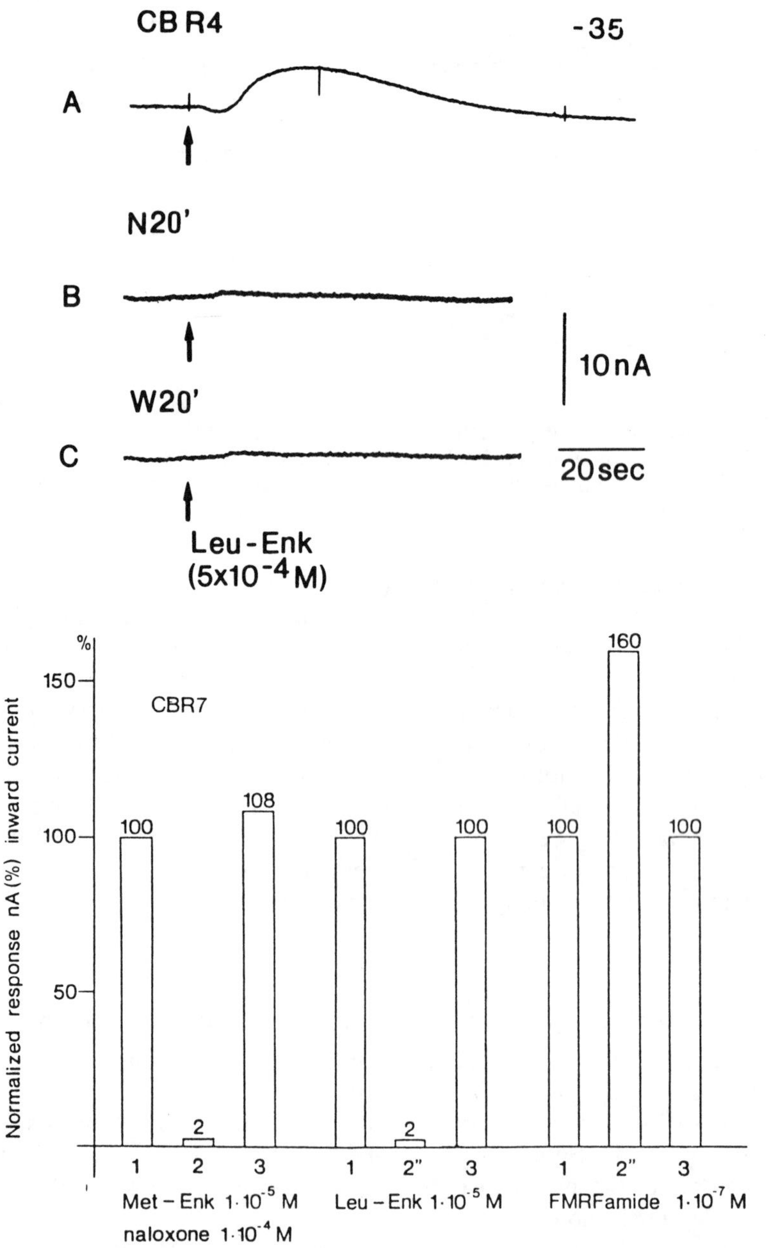

Fig. 6.6. Effect of naloxone on the Leu-enkephalin-induced current (5×10^{-4} M) at resting MP on cell CBR4. **A** is control **B** after 20 min in naloxone and **C** after 20 min wash. Lower part: effect of naloxone on the Met-enkephalin, Leu-enkephalin and FMRFamide-induced inward current on cell CBR7. 1 = Control, 2 = naloxone treatment, 3 = wash-out.

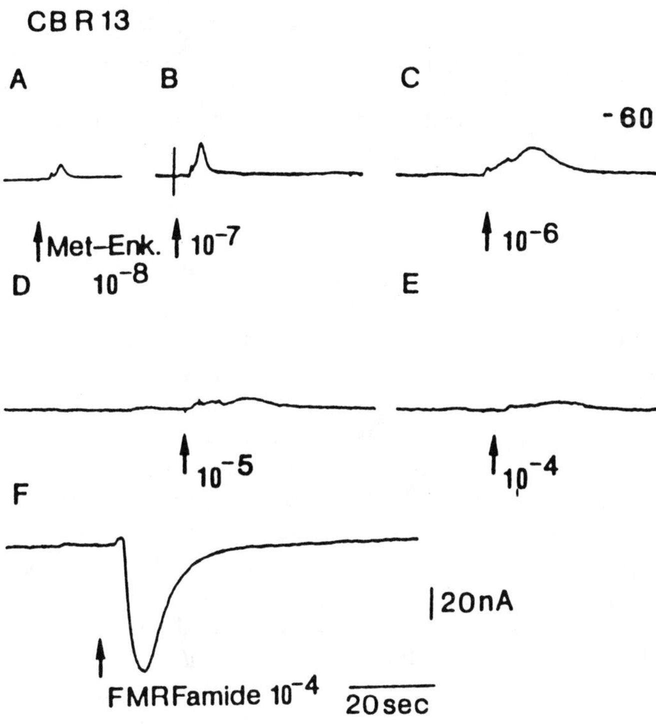

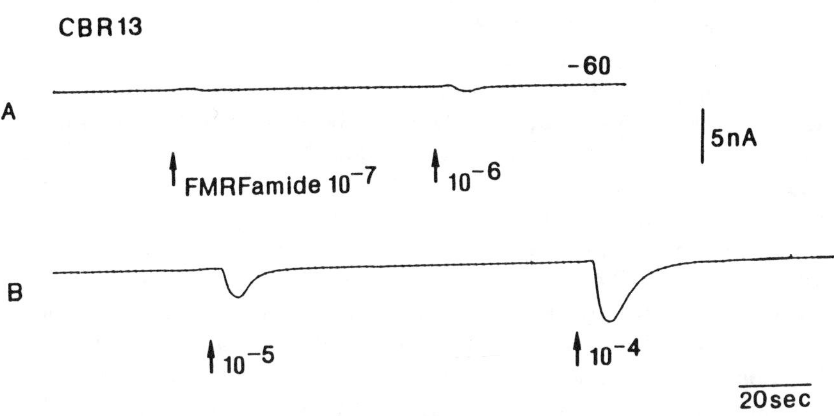

Fig. 6.7. Desensitization of cell CBR13 to Met-enkephalin using its increasing concentrations (**A**–**E**). The cell CBR13, desensitized to Met-enkephalin, responds to FMRFamide (**F**). Lower part: effect of increasing concentrations of FMRFamide on cell CBR13.

These observations suggest that the Met-enkephalin response shows marked desensitization. Similar rapid desensitization was apparent on all B-neurons studied with application of Leu-enkephalin. In contrast, responses to FMRFamide showed little or no desensitization with repeated and/or prolonged application (Fig. 6.8).

6.4. **Discussion**

The presence of stereospecific opiate receptors has been shown in invertebrates, including *Aplysia* (Kream *et al.*, 1980; Leung *et al.*, 1986). The effects of FMRFamide and Leu-enkephalin have also been studied on a number of gastropod species (Stefano *et al.*, 1980; Cottrell, 1982; Carpenter and Hall, 1986; Brezina *et al.*, 1987; Ichinose and McAdoo, 1989; Stone and Mayeri, 1981; Ruben *et al.*, 1986; Kramer *et al.*, 1988, etc.).

Our results show that opiate analogues and FMRFamide, like Leu-enkephalin (Carpenter and Hall, 1986), have complex responses on the neurons of the B-cluster in the cerebral ganglia of *Aplysia californica*. These peptides depolarize or hyperpolarize the individual B-neurons. On other *Aplysia* neurons FMRFamide has been shown to alter membrane excitability by several different mechanisms (Stone and Mayeri, 1981; Ruben *et al.*, 1986; Brezina *et al.*, 1987 Belardetti *et al.*, 1987). In the vertebrate CNS activation of opiate receptors also results in either inhibition (Williams *et al.*, 1982) or excitation (Corrigal, 1983; Madison and Nicoll, 1988) although the predominant action of opioids in mammals is to depress neuronal firing (McFadzean, 1988).

It is interesting that while the B-neurons appear to be a homogeneous population of neurons by other criteria (Jahan-Parwar and Fredman, 1976; Gaillard and Carpenter, 1986), they are not in terms of responses to peptides. The cerebral B-cells of *Aplysia* show a widespread sensitivity to opiate analogues, yet their transmitter role is not certain. The molluscan neuropeptide FMRFamide was shown to be more effective than any of the opiate analogues investigated including Met- and Leu-enkephalins. The effect of the low molecular weight neurotransmitter, ACh, was more consistent on the B-cells than that of the opiates. The alteration in membrane conductance evoked by opiates was very small in comparison with ACh effect. The observation that enkephalin actions are not associated with clear conductance changes indicates that opiate effects should not be considered to be the same as those of classical neurotransmitters. The opiates can be taken for modulators, acting through the intracellular mechanisms (McFadzean, 1988). The time course and duration of their membrane effects support this idea. Recently opiate peptides and FMRFamide were proposed to be endogenous modulators of morphine effects in vertebrates (Yang *et al.*, 1985) and to have a role in tolerance.

Although there is stereospecific binding of opiate peptides in the molluscan CNS (Stefano, 1982; Leung *et al.*, 1986) there are contradictions with regard to naloxone sensitivity of opiate actions in this group of animals (Stefano *et al.*, 1980; Carpenter and Hall, 1986). The data presented in this report suggest that naloxone does block actions of opiate peptides on the identified B-cells of *Aplysia*. However, high concentrations were needed and many opiate responses were not blocked completely. The FMRFamide and ACh responses were modulated by naloxone treatment but in a different way from that of the enkephalins.

References

Belardetti, F., Kandel, E. R. and Siegelbaum, S. A. (1987). Neuronal inhibition by the peptide FMRFamide involves opening of S K-channels. *Nature*, **325**, 153–6.

Boer, H. H., Schot, L. P. C., Veenstra, J. A. and Reichelt, D. (1980). Immunocytochemical identification of neural elements in the CNS of a snail, some insects, a fish and mammal with an antiserum to the molluscan cardioexcitatory tetrapeptide FMRFamide. *Cell Tiss. Res.*, **213**, 21–56.

Brezina, V., Eckert, R. and Erxleben, C. (1987). Modulation of potassium conductances by an endogenous neuropeptide in neurones of *Aplysia californica*. *J. Physiol.*, **382**, 267–90.

Carpenter, D. O. and Hall, A. F. (1986). Responses of *Aplysia* cerebral neurons to leucine enkephalin. In Stefano, G. B. (ed.), *CRC Handbook of Comparative Opioid and Related Neuropeptide Mechanisms*, Vol. II, CRC Press, Boca Raton, Florida, pp. 49–57.

Corrigal, W. A. (1983). Opiates and the hippocampus: a review of the functional and morphological evidence. *Pharmacol. Biochem. Behav.*, **18**, 255–62.

Cottrell, G. A. (1982). FMRFamide neuropeptides simultaneously increase and decrease K^+ currents in an identified neuron. *Nature (Lond.)*, **296**, 87–9.

Cottrell, G. A., Davies, N. W. and Green, K. A. (1984). Multiple actions of a molluscan cardioexcitatory neuropeptide and related peptides on identified *Helix* neurones. *J. Physiol. (Lond.)*, **356**, 315–33.

Dockray, G. J., Reeve, J. R., Shively, J., Gayton, R. J. and Barnard, C. S. (1983). A novel active pentapeptide from chicken brain identified by antibodies to FMRFamide. *Nature*, **305**, 328–30.

Fredman, S. M. and Jahan-Parwar, B. (1975). Synaptic connections in the cerebral ganglion of *Aplysia*. *Brain Res.*, **100**, 209–14.

Gaillard, W. D. and Carpenter, D. O. (1986). Spectra of neurotransmitter receptors and ionic responses on cerebral A and B neurons in *Aplysia californica*. *Brain Res.*, **373**, 303–10.

Greenberg, M. J. and Price, D. A. (1979). FMRFamide, a cardioexcitatory neuropeptide of molluscs: an agent in search of a mission. *Am. Zool.*, **19**, 163–74.

Ichinose, M. and McAdoo, D. (1989). The cyclic GMP-induced inward current in neuron R14 of *Aplysia californica*: Similarity to a FMRFamide-induced, inward current. *J. Neurobiol.*, **20**, 10–24.

Jahan-Parwar, B. and Fredman, S. M. (1976). Cerebral ganglia of *Aplysia*: Cellular organization and origin of nerves. *Comp. Biochem. Physiol.*, **54A**, 347–57.

Kramer, R. H., Levitan, E. S., Carrow, G. M. and Levitan, J. B. (1988).

Modulation of a subthreshold calcium current by the neuropeptide FMRFamide in *Aplysia* neuron R15. *J. Neurophysiol.*, **60**, 1728–38.

Kream, R. M., Zukin, R. S. and Stefano, G. B. (1980). Demonstration of two classes of opiate binding sites in the nervous tissues of the marine mollusc *Mytilus edulis*. *J. Biol. Chem.*, **55**, 9218–24.

Lehman, K. H. and Greenberg, M. J. (1987). The actions of FMRFamide-like peptides on visceral and somatic muscles of the snail *Helix aspersa*. *J. Exp. Biol.*, **131**, 55–68.

Leung, M. K., S-Rózsa, K., Hall, A., Kurovilla, S., Stefano, G. B. and Carpenter, D. O. (1986). Enkephalin-like substance in *Aplysia* nervous tissue and actions of leu-enkephalin on single neurons. *Life Sci.*, **38**, 1529–34.

McDonald, R. L. and Werz, M. A. (1986). Dynorphin A decreases voltage-dependent calcium conductance of mouse dorsal root ganglion neurones. *J. Physiol. (Lond.)*, **377**, 237–49.

McFadzean, J. (1988). The ionic mechanisms underlying opioid action. *Neuropeptides*, **11**, 173–80.

Madison, D. V. and Nicoll, R. A. (1988). Enkephalin hyperpolarizes interneurones in the rat hippocampus. *J. Physiol. (Lond.)*, **398**, 123–30.

Pellmar, T. C., and Carpenter, D. O. (1980). Serotonin induces a voltage-sensitive calcium current in neurons of *Aplysia californica*. *J. Neurophysiol.*, **44**, 423–39.

Ruben, P., Johnson, J. W. and Thompson, S. (1986). FMRFamide effects on *Aplysia* bursting neurons. *J. Neurosci.*, **6**, 252–9.

S.-Rózsa, K. and Carpenter, D. O. (1985). Distinct responses to leu-enkephalin and FMRFamide on identified neurons of the *Aplysia* cerebral ganglion. 15th Annual Meeting of Society for Neuroscience, Dallas, Texas, 20–25 October.

Stefano, G. B. (1982). Comparative aspects of opioid–dopamine interaction. *Cell. Mol. Neurobiol.*, **2**, 167–78.

Stefano, G. B. (1988). The evolvement of signal systems: conformational matching a determining force stabilizing families of signal molecules. *Comp. Biochem. Physiol.*, **90C**, 287–94.

Stefano, G. B., Vadász, I. and Salánki, J. (1980). Methionine enkephalin inhibits the bursting activity of the Br-type neuron in *Helix pomatia* L. *Experientia*, **36**, 666–7.

Stone, L. S. and Mayeri, E. (1981). Multiple actions of FMRFamide on identified neurons in the abdominal ganglion of *Aplysia*. *Soc. Neurosci. Lett.*, **27**, 25–30.

Williams, J. T., Egan, T. M. and North, R. A. (1982). Enkephalin opens potassium channels on mammalian central neurones. *Nature*, **299**, 74–7.

Yang, H. Y. T., Fratta, W., Majane, E. A. and Costa, E. (1985). Isolation, sequencing, synthesis and pharmacological characterization of two brain neuropeptides that modulate the action of morphine. *Proc. Natl. Acad. Sci. USA*, **82**, 7757–61.

Zieglgänsberger, W., French, E. D., Siggins, G. R. and Bloom, F. E. (1979). Opioid peptides may excite hippocampal pyramidal neurones by inhibiting adjacent inhibitory interneurones. *Science*, **205**, 415–17.

7 *Martin Kavaliers*

Opioid peptides, nociception and analgesia in molluscs

7.1. **Introduction**

In nature animals commonly face a spectrum of trivial to powerful aversive stimuli that influence their survival and ultimately fitness. In order to respond to these stimuli effectively, organisms require: (1) a mechanism for recognizing aversive stimuli, (2) a set of effectors which can react to the noxious stimulus, and (3) a system for producing co-ordinated and directed movements and behaviours in response to the stimuli. This has led to the development of defensive response systems that can be effectively triggered by the presence of 'painful' as well as 'non-painful' innate and learned danger-associated stimuli. In mammals, endogenous opioid peptide mediated stress-induced, or more appropriately, environmentally induced analgesia or antinociception is an important component of these defensive responses to aversive factors. This analgesia results in the facilitation of the animal's ability to perform (or conversely to inhibit) a motor response, the goal of which is the alleviation of the noxious quality of the aversive stimulus (Amit and Galina, 1986). This helps the animal to maintain its integrity in a potentially threatening situation.

Evidence is now accumulating for opioid modulation of nociception and analgesia in invertebrates as well as vertebrates (Kavaliers, 1988a, 1989b). This chapter briefly reviews evidence for endogenous opioid involvement in the modulation of nociception and analgesia in a number of species of molluscs.

7.2. **Nociception and pain**

The capability of animals to recognize and react to stimuli that can compromise their integrity is embodied in the term 'nociception' (Sherrington, 1906). Nociceptors are preferentially sensitive either to a noxious stimulus or to an aversive stimulus that would become noxious if prolonged, and code the intensity of the stimulus (Besson and Chaouch, 1987). In addition, the responses from the effectors are appropriate to the input from the receptors.

Nociception should not necessarily be considered synonymous with 'pain', which is defined in humans as 'an unpleasant sensory and emotional experience associated with either actual or potential tissue damage, or described in terms of such damage' (Merskey, 1983). In humans, nociceptors have been shown to respond before the applied stimulus is perceived as 'painful'. Pain is a very complex and subjective term which humans learn to recognize only through experience related to injury in early life. It should, however, be recognized that non-human animals do have levels of feeling, awareness (possibly including that of 'suffering'), anticipation and effect, and that there is a marked gradation and specialization in this among and between taxa. A more appropriate definition of pain for non-humans, and one that more effectively incorporates nociception, is 'an aversive sensory experience that elicits protective motor and vegetative reactions, results in learned avoidance and may modify species specific behavior, including social behavior' (Zimmerman, 1986).

As indicated, nociception can be used to provide an index of an animal's sensitivity to aversive environmental conditions, and thus can allow for the determination of the capacity to execute adaptive behaviour. Measurements of alterations in nociceptive related responses (decreases in sensitivity — antinociception or analgesia) are widely used to determine the behavioural and physiological status of animals following experimental procedures that involve exposure to aversive, or potentially aversive, stimuli. In rodents laboratory measures of nociception include: limb flexion or withdrawal (lifting a foot off an aversive, usually thermal, surface); active avoidance (flinch jump; jumping or moving from an aversive situation) and removal of the tail from a thermal stimulus (tail-flick) (Chapman *et al.*, 1985).

Parallel nociceptive responses have been observed in a variety of species of invertebrates. For example, the gastropod land snail, *Cepaea nemoralis*, when placed on a surface warmed at 40°C, displays within a few seconds the aversive nature of this temperature by lifting the anterior portion of its fully extended foot (Kavaliers *et al.*, 1983, 1985). This foot-lifting behaviour is not observed in hydrated snails exposed to temperatures normally present in their natural habitats, but becomes increasingly evident as the temperature is raised towards 40°C. This nociceptive response is comparable to the avoidance response occurring in rodents placed on a warmed surface where the animals quickly lift or lick their feet, or attempt to escape from the heated surface. Moreover, exposure to either actual or potentially aversive stimuli has been shown to increase the thermal nociceptive thresholds of *Cepaea*, as well as of the terrestrial slugs, *Arion ater* and *Limax maximus*, in a manner indicative of the induction of analgesia (Dalton and Widdowson, 1989; Kavaliers, 1987; Kavaliers and Hirst, 1986b). These increases in response latency are analogous to the antinociception or analgesia observed in rodents placed on a thermal surface after being exposed to noxious or potentially noxious stimuli.

A nociceptive function has also been indicated for specific mechanoafferent neurons innervating the tail, parapodia, and much of the foot and body wall of the marine mollusc *Aplysia californica* (Walters and Erickson, 1986). These cells display increasing discharges to progressively increasing pressure, with maximal responses occurring to stimuli that could potentially cause tissue damage (Walters, 1987). A similar graded pattern of response is used to define the activity of classical mammalian nociceptors (Besson and Chaouch, 1987). In addition, brief electrical or mechanical stimulation of the skin of *Aplysia* produces an enhancement of the defensive siphon-withdrawal reflex that is associated with long-lasting peripheral and central electrophysiological changes. These responses in *Aplysia* parallel the long-term increases in nociceptive responses, or hyperalgesia, observed in mammals.

7.3. **Comparative aspects of opioid systems**

In vertebrates endogenous opioid peptides coexist with diverse hormones in endocrine glands, and with classical or peptide transmitters in peripheral autonomic and sensory neurons. In addition, they are widely distributed in the central nervous system and themselves function as transmitters or neuromodulators, with the effects of shorter chain opioids being rapidly terminated by neuronal or extraneuronal enzymatic breakdown (Atweh and Kuhar, 1983).

There are three main gene families of opioid peptide precursors — pro-opiomelanocortin (POMC), proenkephalin A and proenkephalin B or prodynorphin. The three endogenous opioid precursors give rise to a number of fragments (peptides), among which those of POMC (e.g. β-endorphin) and proenkephalin A (e.g. methionine- and leucine-enkephalin) have affinities for mu- (and possibly mu_1 and mu_2 subtypes) and delta-opiate receptors; whereas those of prodynorpin (e.g. dynorphin) have affinities to kappa-opiate receptors (Martin, 1984; North, 1986).

It has now become apparent that opioid peptides have a much broader phylogenetic distribution and are not limited to vertebrates. Opioid peptides and opiate receptors have been identified from a variety of taxa of invertebrates, strongly suggesting a phylogenetic conservation of opioid peptide structure and function (Leung and Stefano, 1987; O'Neill *et al.*, 1988; Zisper *et al.*, 1988). However, to be certain of the structure of a peptide identified by immunocytochemistry in a heterologous species, even if it reacts with an antibody known to be 'specific' for a particular peptide, it is necessary to isolate and sequence it. In the limited number of species of invertebrates that this has been done, the opioid peptides identified (Leu-enkephalin) appear structurally identical to those present in mammals (Leung and Stefano, 1984; Leung *et al.*, 1986).

Results of behavioural, electrophysiological and pharmacological studies with molluscs have also shown that endogenous opioid peptides, and exogenous opiate agonists and antagonists, have behavioural and physiological actions resembling those induced in mammals (Kavaliers, 1988a; Leung and Stefano, 1984, 1987; Stefano, 1982). Binding studies have revealed that mu and kappa opiate binding sites are present in Mollusca and Crustacea (Kream *et al.*, 1980; Leung *et al.*, 1986). In addition, the presence of enkephalin-degrading enzymes in invertebrates is further evidence that endogenous opioids may be important in invertebrate physiology (Colette-Preverio *et al.*, 1985).

7.4. Opioid modulation and analgesia and nociception

There is substantial evidence that the opioid peptides are involved in the modulation of nociception and behavioural responses to aversive and stressful stimuli in mammals. Administration of small quantities of either endogenous opioid peptides such as Met-enkephalin, β-endorphin or dynorphin, or exogenous opiate agonists, such as the prototypic mu agonist, morphine, decrease pain sensitivity and have analgesic effects. Prototypic exogenous opiate antagonists, such as naloxone or naltrexone, suppress these anlagesic effects, and in certain cases reduce nociceptive responses and induce hyperalgesia (Martin, 1984).

Similar evidence for opioid involvement in the mediation of analgesia and nociception exists for molluscs. Morphine, as well as β-endorphin and Met-enkephalin, enhance in a dose-dependent fashion the nociceptive responses of *Cepaea* to a warmed surface in a manner analogous to that associated with the production of analgesia in mammals (Kavaliers *et al.*, 1983, 1985). The effect of morphine is also produced by the benzomorphan levorphanol, but not by the stereoisomer dextrophan, suggesting that the receptor that interacts with these opiates has stereospecific requirements (Hirst and Kavaliers, 1987). Naloxone suppresses and reverses the analgesic effects of morphine in *Cepaea*, as well as reducing the response times (hyperalgesia) of particular morphological types of control snails, further supporting opiate receptor involvement (Kavaliers *et al.*, 1983; Kavaliers, 1989b). As in mammals, after 5–7 days of daily administration of morphine, *Cepaea* develop tolerance to morphine-induced analgesia (Kavaliers and Hirst, 1983). Moreover, both associative (learning, environmental specificity) and non-associative (biochemical) mechanisms contribute to the development of tolerance in *Cepaea* (Kavaliers and Hirst, 1986a).

The specific mu and delta opioid agonists, (D-Ala^2, NMe-Phe^5, Glyol)-enkephalin (DAGO) and (D-Ala^2, D-Leu^4), enkephalin (DADLE), also have significant analgesic effects in *Cepaea* and *Arion*, suggesting the presence of mu and delta receptors (Dalton and Widdowson, 1989; Kavaliers *et al.*, 1985). In addition, the specific kappa opiate agonist, U-

50,488H, has significant antinociceptive effects in *Cepaea* (Kavaliers and Ossenkopp, 1988). As in mammals, the duration of the effect of U-50,488H is longer than that of morphine and there is a diminished sensitivity to naloxone. Taken together with the demonstrations of kappa opiate binding sites and the immunocytochemical localization of the endogenous kappa ligand, dynorphin, in invertebrates (Ford *et al.*, 1986), these analgesic effects suggest the presence of functional kappa opioid-mediated analgesic systems in *Cepaea*.

At a cellular level there is also evidence that the modulatory effects of opiates in molluscs and vertebrates may be associated with similar intermediary, second messenger systems. In vertebrates, activation of mu or delta opioid receptor types has been shown to increase potassium conductance and indirectly reduce calcium conductance, while activation of kappa receptors causes a reduction in voltage-dependent calcium conductance (North, 1986). In both cases the net result is a reduction in the rate of neuronal discharge and the amount of transmitter released. In rodents and *Cepaea* the dihydropyridine (DHP) and non-DHP calcium channel antagonists diltiazem, verapamil and nifedipine, differentially and significantly enhanced — while the DHP calcium channel agonist, BAY K8644, significantly reduced — exogenous opiate and stress-induced opioid analgesia (Kavaliers and Ossenkopp, 1987, 1988). This suggests similar roles for calcium channel related mechanisms in the mediation of opiate-induced analgesia in molluscs and vertebrates. In addition, pharmacological modifications of protein kinase C and G protein second messenger systems have similar effects on morphine-induced analgesia in *Cepaea* and rodents (in preparation). This suggests that a number of second messenger systems may be similarly involved in the mediation of opiate effects in *Cepaea* and rodents. Moreover, there are also data indicating that opiates have similar inhibitory effects on dopamine and possibly other monoamine systems in rodents and molluscs (Stefano, 1982). Taken together, these observations further support the suggestion that opiate effects may be similarly determined in molluscs and vertebrates.

7.5. **Day–night rhythms of nociception and analgesia**

There are significant day–night rhythms in the nociceptive responses and analgesic effects of morphine in *Cepaea* (Kavaliers *et al.*, 1989). *Cepaea* display heightened night-time levels of nociception and morphine-induced analgesia under both laboratory and natural lighting conditions (Kavaliers *et al.*, 1989; and in preparation). These rhythms are consistent with the nocturnal–crepuscular activity patterns of *Cepaea*. These diel rhythms in the latency of response to the aversive thermal stimulus may also be related to the day–night rhythms of behavioural thermoregulation and thermal preferences that have been recorded from molluscs (Ford and Cook, 1987; Kavaliers, 1980).

In nocturnally active rodents, day–night rhythms of nociception and analgesia have been related to the elevated night-time levels of opioid peptides and opiate receptor activity in the central nervous system (Kafka *et al.*, 1983; Naber *et al.*, 1981). Whether or not similar day–night variations exist in the levels of either opioid peptides or opiate receptors in *Cepaea* and other molluscs is at present unknown. These diel rhythms of nociception in *Cepaea* were, however, markedly disrupted by peripheral administration of β-funaltrexamine (B-FNA), an irreversible mu opioid receptor antagonist (in preparation). This suggests that mu opioid systems may be associated with the generation and/or expression of the day–night rhythm in this measure of nociception of *Cepaea*. Whether or not diel and possibly circadian rhythms of other behavioural measures are similarly affected by manipulations of opioid systems needs to be examined.

7.6. **Stress-induced opioid analgesia**

In rodents a variety of laboratory and more naturalistic, ecologically appropriate, stressful stimuli have been shown to increase endogenous opioid activity and induce a variety of integrated adaptive behavioural responses, including that of analgesia (Amit and Galina, 1986; Bodnar, 1986). This stress-induced analgesia can be reduced by exogenous opiate antagonist and prolonged by prevention of enzymatic degradation of opiates.

Similar environmentally induced analgesia is also evident in molluscs. Exposure to warm stress has been shown to increase the thermal nociceptive thresholds of *Cepaea*. The warm–stress-induced analgesia was blocked by naloxone and the delta opiate antagonist ICI 154,129, and was suppressed by a 24h pretreatment with B-FNA (Kavaliers, 1987). Brief body (tail) pinch stress of the slugs *Limax* and *Arion* also resulted in significant increases in their thermal response latencies indicative of the induction of analgesia (Dalton and Widdowson, 1989; Kavaliers and Hirst, 1986a). The analgesic response of *Limax* was blocked by naloxone, while that of *Arion* was reduced in a dose-dependent manner by naltrexone and the delta opiate antagonist, ICI 14864. Moreover, the duration of the stress-induced analgesia in *Arion* could be prolonged by injection of enkephalinase inhibitors. This further supports the involvement of endogenous opioid peptides in the mediation of stress-induced analgesia in molluscs. It should, however, be noted that although the present data are sufficient to indicate that these analgesic responses are opioid-mediated, it would be desirable to demonstrate cross-tolerance to exogenous opiate-induced analgesia, as well as to show changes in endogenous opioid peptide levels, receptor binding and expression of mRNAs for opioid peptides.

More naturalistic stimuli, such as new (novel) environments and chemostimuli that are suggestive of either actual or potential danger, have also been found to induce opioid-mediated analgesia in *Cepaea* (Kavaliers, 1988b; Kavaliers and Tepperman, 1988). This novelty-induced analgesia displays similar characteristics in *Cepaea* and rodents. In both cases the novelty-induced analgesia is of relatively short duration, evident only following the first exposure to the new environment stimulus, and is blocked by opiate antagonists such as naloxone.

Results of studies with rodents have also shown that, depending on the parameters of the aversive stimuli, the resulting stress-induced analgesia may be mediated by either endogenous opioids or by other non-opioid hormonal or neurochemical mechanisms (Bodnar, 1986). A similar divergence in analgesic responses is also evident in *Cepaea*. As indicated previously, warm stress-induced analgesia in *Cepaea* is blocked by opiate antagonists; however, cold stress-induced analgesia is relatively unaffected by these manipulations (Kavaliers, 1987). This suggests that, as in vertebrates, there are a number of different (opioid and non-opioid) neuromodulatory mechanisms involved in mediating the aversive environmental and nociceptive responses of *Cepaea*. The adaptive value and relations to survival of these nociceptive responses and their different neuromodulators remains, however, to be defined.

7.7. **Adaptive and genetic aspects of opioid–modulated nociception**

Evidence exists to suggest that opioid-mediated nociceptive and analgesic response of *Cepaea* are subject to genetic modulation and may be related to the fitness of the animal. *Cepaea* are genetically and morphologically polymorphic, displaying a variety of shell colour and banding types that have different thermal microhabitat preferences (Jones *et al.*, 1977). In laboratory studies it was observed that the effects of morphine and naloxone on the nociceptive responses of *Cepaea* varied with shell banding pattern and natural thermal microhabitat. In the three morphs that were examined in detail, the lightest shell type (yellow unbanded) with the highest basal response latency and morphine-induced analgesia was present at the warmest environmental temperatures, the shell type (yellow two-banded) with an intermediate basal response latency and morphine-induced analgesia was at field sites of intermediate temperature, and the darkest shell type (yellow five-banded) with the shortest response latency and lowest morphine sensitivity was present in the coolest microhabitats. This raises a possible environmental relatedness of this measure of nociception, and suggests that differences in opioid modulation of thermal responses may contribute to the polymorphic thermal responses of natural populations of *Cepaea*.

Heightened endogenous opioid activity (e.g. synthesis and/or release of opioid peptides; receptor binding; activation of second messenger systems) has been correlated with an increased tolerance to various environmental stimuli, including that of temperature (Amit and Galina, 1986; Clark, 1981). This suggests a possible physiological mechanism whereby snails that experience warmer thermal environments are more tolerant of them. Opioid systems may be involved in the adaptation of individuals to their thermal environments. Thus, differences in the activation and/or expression of endogenous opioid systems may, in part, contribute to the polymorphic thermal responses of *Cepaea*.

Although the heritability of the variation in opioid-mediated responses of *Cepaea* was not directly examined, similar inter-morph variations in opiate responses have been recorded across years (generations) and field sites. This consistency across sites and generations further supports the proposal for the presence of a genetically influenced polymorphism in opiate-mediated thermal nociceptive responses of *Cepaea*.

7.8. Conclusions

There is accumulating evidence for a phylogenetic continuity in the expression and regulation of fundamental behaviours of essential survival value. Results of behavioural studies with molluscs and rodents have shown that endogenous opioid systems are similarly involved in the modulation of nociception and analgesic responses that are activated by actual and potential danger-associated stimuli.

Acknowledgement

The studies described here were supported by a Natural Sciences and Engineering Research council of Canada grant to M. K.

References

Amit, Z. T. and Galina, Z. H. (1986). Stress-induced analgesia: adaptive pain suppression. *Physiol. Rev.*, **66**, 1091–1120.

Atweh, S. F. and Kuhar, M. J. (1983). Distribution and physiological significance of opioid receptors in the brain. *Br. Med. Bull.*, **39**, 47–52.

Besson, J.-M. and Chaouch, A. (1987). Peripheral and spinal mechanisms of nociception. *Physiol. Rev.*, **67**, 68–186.

Bodnar, R. J. (1986). Neuropharmacological and neuroendocrine substrates of stress-induced analgesia. *Ann. N.Y. Acad. Sci.*, **467**, 345–60.

Chapman, C. R., Casey, K. L., Dubner, R., Foley, K. M., Gracely, R. H. and Reading, A. E. (1985). Pain measurement: an overview. *Pain*, **22**, 1–31.

Clark, W. G. (1981). Effects of opioid peptides on thermoregulation. *Fed. Proc.*, **40**, 2754–9.

Colette-Preverio, M. A., Mattras, H., Zwilling, R. and Preverio, A. (1985). Enkephalin-degrading activity in arthropod hemolymph. *Neuropeptides*, **6**, 405–15.

Dalton, L. M. and Widdowson, M. (1989). The involvement of opioid peptides in stress-induced analgesia in the slug, *Arion ater*. *Peptides*, **10**, 9–13.

Ford, D. J. G. and Cook, A. (1987). The effects of temperature and light on the circadian activity of the pulmonate slug, *Limax pseudoflavus*. *Anim. Behav.*, **34**, 1754–6.

Ford, R., Jackson, D. M., Tetrault, L., Torres, J. C., Assanh, P., Harper, J., Leung, M. K. and Stefano, G. B. (1986). A behavioral role for enkephalins in regulating locomotor activity in the insect *Leucophaea maderae*: evidence for high affinity kappa-like opioid binding sites. *Comp. Biochem. Physiol.*, **85C**, 61–6.

Hirst, M. and Kavaliers, M. (1987). Levorphanol but not dextrophan suppresses the foot-lifting response to an aversive thermal stimulus in the terrestrial snail (*Cepaea nemoralis*). *Neuropharmacology*, **26**, 121–3.

Jones, J. S., Leith, B. H. and Rawlings, P. (1977). Polymorphism in *Cepaea*: a problem with too many solutions. *Ann. Rev. Ecol. Syst.*, **8**., 109–143.

Kafka, M. S., Wirz-Justice, A., Naber, D. and Moore, R. Y. (1983). Circadian rhythms in rat brain neurotransmitter receptors. *Fed. Proc.*, **42**, 2796–2801.

Kavaliers, M. (1980). A circadian rhythm of behavioural thermoregulation in a freshwater gastropod, *Helisoma trivolis*. *Can. J. Zool.*, **58**, 2152–5.

Kavaliers, M. (1987). Evidence for opioid and non-opioid forms of stress-induced analgesia in the snail, *Cepaea nemoralis*. *Brain Res.*, **410**, 111–15.

Kavaliers, M. (1988a). Evolutionary and comparative aspects of nociception. *Brain Res. Bull.*, **21**, 923–31.

Kavaliers, M. (1988b). Novelty-induced opioid analgesia in the terrestrial snail, *Cepaea nemoralis*. *Physiol. Behav.*, **42**, 29–32.

Kavaliers, M. (1989a). Evolutionary aspects of the neuromodulation of nociceptive behaviors. *Am. Zool.*, **29**, 1345–53.

Kavaliers, M. (1989b). Polymorphism in opioid modulation of the thermal responses of the land snail, *Cepaea nemoralis*. *Can. J. Zool.* (In press).

Kavaliers, M. and Hirst, M. (1983). Tolerance to morphine-induced thermal responses in the terrestrial snail, *Cepaea nemoralis*. *Neuropharmacology*, **22**, 1321–6.

Kavaliers, M. and Hirst, M. (1986a). Environmental specificity of tolerance to morphine-induced analgesia in a terrestrial snail: generalization of the behavioral model of tolerance. *Pharmacol. Biochem. Behav.*, **23**, 1201–6.

Kavaliers, M. and Hirst, M. (1986b). Naloxone-reversible stress-induced feeding and analgesia in the slug, *Limax maximus*. *Life Sci.*, **38**, 203–9.

Kavaliers, M. and Ossenkopp, K-P. (1987). Calcium channel involvement in magnetic field inhibition of morphine-induced analgesia. *Naunyn-Schmiedberg's Arch. Pharmacol.*, **336**, 308–15.

Kavaliers, M. and Ossenkopp, K-P. (1988). Magnetic fields inhibit "analgesic" behaviors of the terrestrial snail, *Cepaea nemoralis*. *J. Comp. Physiol.*, **161**, 551–8.

Kavaliers, M. and Tepperman, F. S. (1988). Exposure to novel odors induces opioid-mediated analgesia in the land snail, *Cepaea nemoralis*. *Behav. Neural Biol.*, **50**, 2855–299.

Kavaliers, M., Hirst, M. and Teskey, G. C. (1983). A functional role for an opiate system in snail thermal behavior. *Science*, **220**, 99–101.

Kavaliers, M., Hirst, M. and Teskey, G. C. (1985). The effects of opioid and FMRF-amide peptides on thermal behavior in the snail. *Neuropharmacology*, **24**, 621–6.

Kavaliers, M., Ossenkopp, K-P. and Lipa, S. M. (1990). Day–night rhythms in the inhibitory effects of 60 Hz magnetic fields on opiate-mediated 'analgesic' behaviors of the land snail, *Cepaea nemoralis. Brain Res.* **517**, 276–82.

Kream, R. M., Zukin, R. S. and Stefano, G. B. (1980). Demonstration of two classes of opiate binding sites in the nervous tissue of the marine mollusc *Mytilus edulis. J. Biol. Chem.*, **255**, 9218–24.

Leung, M. K. and Stefano, G. B. (1984). Isolation and identification of enkephalins in pedal ganglia of *Mytilus edulis* (Mollusca). *Proc. Natl. Acad. Sci.*, **81**, 955–8.

Leung, M. K. and Stefano, G. B. (1987). Comparative neurobiology of opioids in invertebrates with special attention to senescent alterations, *Prog. Neurobiol.*, **18**, 131–59.

Leung, M. K., Nixon, B., Kuruvilla, S. Boer, H. H. and Stefano, G. B. (1986). Biochemical evidence for opioids in ganglia of the snail *Lymnaea stagnalis. Soc. Neurosci. Abst.*, **12**, 408.

Martin, W. R. (1984). Pharmacology of opioids. *Pharmacol. Rev.*, **335**, 283–323.

Merskey, D. M. (1983). Classification of chronic pain. *Pain* (Suppl. 3), S217.

Naber, D., Wirz-Justice, A. and Kafka, A. (1981). Circadian rhythm in rat brain opiate receptor. *Neurosci. Lett.*, **21**, 45–50.

North, R. A. (1986). Opioid receptor types and membrane ion channels. *Trends Neurosci.*, **9**, 114–17.

O'Neill, J. B., Pert, C. B., Ruff, M. R., Smith, C. C., Higgins, W. J. and Zisper, B. (1988). Identification and characterization of the opiate receptor in the ciliated protozoan, *Tetrahymena. Brain Res.*, **450**, 303–315.

Sherrington, C. S. (1906). *The Integrative Action of the Nervous System.* Yale University Press, New Haven.

Stefano, G. B. (1982). Comparative aspects of opioid-dopamine interaction. *Cell. Mol. Neurobiol.*, **2**, 167–78.

Walters, E. T. (1987). Site-specific sensitization of defensive reflexes in *Aplysia*: a simple model of long-term hyperalgesia. *J. Neurosci.*, **7**, 400–7.

Walters, E. T. and Erickson, M. T. (1986). Directional control and the functional organization of defensive responses in *Aplysia. J. Comp. Physiol.*, **159**, 339–51.

Zimmerman, M. A. (1986). Behavioral investigations of pain in animals. In: Duncan, I. J. A. and Molony, V. (eds), *Assessing Pain in Farm Animals* Commission of the European Communities, Luxembourg, pp. 30–5.

Zisper, B., Ruff, M. R., O'Neill, J. B., Smith, C. C., Higgins, W. J. and Pert, C. B. (1988). The opiate receptor: a single 100 KDa recognition molecule appears to be conserved in *Tetrahymena*, leech and rat. *Brain Res.*, **463**, 296–304.

8 *Yuko Fujisawa, Ichiro Kubota, Tomoko Kanda, Yoshihiro Kuroki and Yojiro Muneoka*

Neuropeptides isolated from *Mytilus edulis* (Bivalvia) and *Fusinus ferrugineus* (Prosobranchia)

8.1. Introduction

Hirata *et al.* (1987, 1988) isolated three bioactive peptides from the pedal ganglia of the bivalve mollusc *Mytilus edulis*. One was a heptapeptide termed catch-relaxing peptide (CARP), that has a potent relaxing effect on catch tension of the anterior byssus retractor muscle (ABRM) of the animal. The others were two congeneric hexapeptides termed *Mytilus* inhibitory peptides (MIPs; Ser^2-MIP and Ala^2-MIP), that have a potent inhibitory effect on the ABRM but do not have any catch-relaxing effect. If these peptides, isolated from the pedal ganglia, are actually neurotransmitters or neuromodulators controlling the ABRM, they should exist in the muscle.

It has been shown that the molluscan neuropeptide FMRFamide, first isolated from the ganglia of the bivalve mollusc *Macrocallista nimbosa* (Price and Greenberg, 1977), and related peptides (FMRFamide-related peptides, FaRPs) are widely distributed in molluscs (for reviews, Price *et al.*, 1987; Cottrell, 1989). In the ABRM of *Mytilus*, FMRFamide causes a contraction (Painter, 1982) or a relaxation (Muneoka and Saitoh, 1986) depending on its concentration. Some synthetic FaRPs, such as Tyr-Gly-Gly-Phe-Met-Arg-Phe-NH_2 and acetyl-Phe-Nle-Arg-Phe-NH_2, show 10–30 times more potent contractile action on the ABRM but do not show any catch-relaxing action (Muneoka and Saitoh, 1986). Painter (1982) has reported that a considerable amount of FMRFamide-like immunoreactive substance is present in the ABRM. It is supposed from these facts that the ABRM may be also regulated by FaRP and that the peptide might be an N-terminal-extended analogue of FMRFamide.

For the foregoing reasons we attempted to isolate bioactive peptides from the ABRM of *Mytilus edulis*, and found many peptides including CARP, Ser^2-MIP, Ala^2-MIP, a new MIP-related peptide, FMRFamide and a new FaRP with an unusual structure. We report here the structures and actions of these peptides.

The structure of CARP is closely related to that of the neuropeptide myomodulin, which was first isolated from the opisthobranch mollusc

Aplysia californica by Cropper *et al.* (1987). Only two amino acid residues are different between them, indicating that CARP is an analogue of myomodulin. Therefore, there may exist a myomodulin-CARP family of peptides in molluscs. In fact there is evidence that myomodulin-CARP-related peptides (MCRPs) are widely distributed in molluscs. Immunohistochemical studies using anti-CARP have shown the presence of CARP-like substances in the nervous systems of various molluscs, including pulmonate, cephalopod and polyplacophoran molluscs (personal communications from Dr S. Moffett and from Dr K. Kuwasawa). Further, CARP has been shown to have biological effects on various molluscan muscles and neurons (Kiss, 1988; Hirata *et al.*, 1989b).

In the radula retractor and protractor muscles of *Fusinus perplexus*, CARP shows a potent inhibitory action on their contractions (Hirata *et al.*, 1989a). In contrast, FMRFamide potentiates the contractions of the muscles at low concentrations (10^{-9}–10^{-8} M), and at high concentrations (higher than 10^{-8} M) it induces a contraction by itself. Thus, it is supposed that MCRP and FaRP might be present in the nervous system of *Fusinus*. In fact, Kanda *et al.* (1989) fractionated acetone extract of the ganglia of *Fusinus ferrugineus* by using a gel-filtration column and obtained several peaks of biological activities, including CARP-like and FMRFamide-like activities.

For the reasons mentioned above, we attempted to isolate bioactive peptides from the ganglia of *Fusinus ferrugineus*, and found many peptides, such as myomodulin, a novel MCRP, FMRFamide, FLRFamide, a novel FaRP and two novel peptides functionally related to CARP but structurally related to FMRFamide. We also report here the structures and actions of these peptides.

8.2. **Identification of the peptides**

8.2.1. *Peptides in the ABRMs of* Mytilus

Acetone extract of the ABRMs excised from 10,000 specimens of *Mytilus edulis* was forced through C-18 cartridges (Sep-Pak, Waters) and the retained material was eluted with methanol. The material was then applied to a column (2.6 × 40 cm) of Sephadex G-15, and fractions of 4 ml each were collected. Bioactivities of the fractions were examined on phasic contraction and catch tension of the ABRM bundles mounted in an experimental chamber (2 ml). The phasic contraction was evoked by stimulating the ABRM with repetitive electrical pulses (15 V, 3 ms, 10 Hz, for 5 s) and the catch tension was induced by applying 10^{-4} M acetylcholine (ACh) to the muscle for 2 min.

The gel-filtrated fractions of the retained material showed many peaks of biological activities on the ABRM; three peaks of phasic-contraction-potentiating activity, two peaks of phasic-contraction-inhibiting activity and three peaks of catch-tension-relaxing activity. At high doses, fractions of the potentiating peaks were found to show contractile activity by themselves. All of the activities were found to be destroyed by treating the fractions with subtilisin or aminopeptidase M, suggesting that the active substances in the peaks are peptides.

Each of the foregoing bioactive peaks seemed to be eluted partially overlapping with one or two other peaks. Therefore, we divided the active fractions into two groups—group A (fractions 23–40) and group B (fractions 41–55). In group A, one potentiating peak, two inhibitory peaks and one catch-relaxing peak were included. In group B two other potentiating peaks and two other catch-relaxing peaks were included. Each group was then subjected to a HPLC system (Toyo Soda CCPM with a detector UV-8000) to purify the bioactive substances in it. The columns used for the purification were three kinds of reversed-phase columns (C-8, C-18 and phenyl columns), a cation-exchange column and an anion-exchange column.

We separated 10 bioactive substances from group A; two contractile, six contraction-inhibiting and two catch-relaxing substances (Fig. 8.1). Five of the 10 substances were purified and their structures were determined. They are one contractile peptide (*Mytilus* contractile peptide 1, MCP_1), three contraction-inhibiting peptides (*Mytilus* inhibitory peptide 1–3, MIP_{1-3}) and one catch-relaxing peptide (*Mytilus* relaxing peptide 1, MRP_1). The contractile peptide, MCP_1, also showed contraction-potentiating activity at low doses. The structures of the other five bioactive substances in group A are not yet determined, and determination experiments are in progress.

We separated seven bioactive substances from group B; one contractile, three contraction-potentiating and three contraction-inhibiting substances (Fig. 8.1). Two of the three contraction-potentiating substances showed an inhibitory action in some preparations. It has been reported that CARP, at low doses such as 5×10^{-9} M, shows a potentiating or inhibitory action, depending on muscle preparation (Hirata *et al.*, 1987). Unlike CARP, however, the above two substances did not show any catch-relaxing action. The contractile substance, which we temporarily termed *Mytilus* contractile peptide 2 (MCP_2), showed catch-relaxing and contraction-potentiating actions at slighty lower concentrations than the threshold for contractile action. It has been reported that FMRFamide has such actions (Muneoka and Matsuura, 1985). As described later, MCP_2 was purified and its structure was determined to be the same as that of FMRFamide. The structures of the other six bioactive substances in group B are not yet determined, and determination experiments are also in progress.

The purified active substances were used for amino acid analyses,

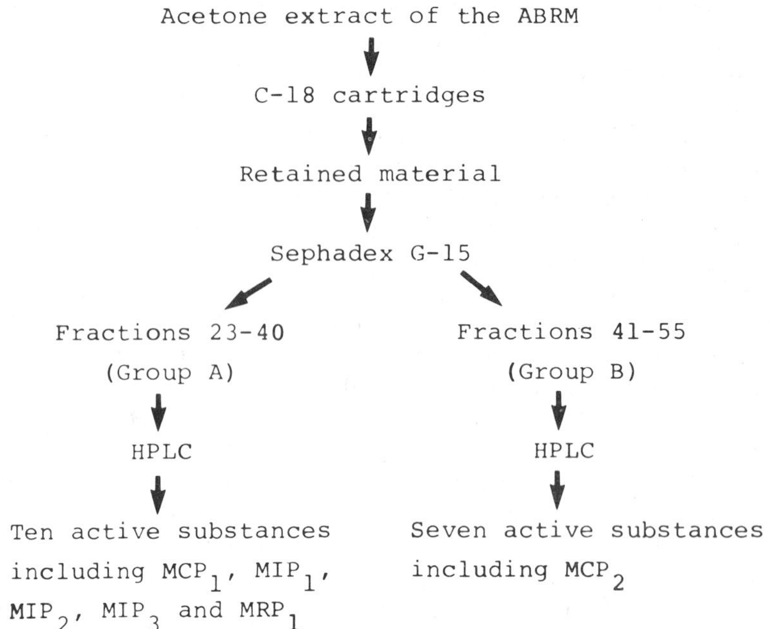

Fig. 8.1. The isolation procedures for bioactive peptides in the ABRM of *Mytilus*.

amino acid sequence analyses by automated Edman degradation with a gas-phase sequencer (Applied Biosystems 470A) coupled with a PTH-amino acid analyser (Applied Biosystems 120A) and FAB-MS analyses (JEOL JMS HX-100), and their probable structures were determined. The determined structures of the purified peptides (MCP_1, MCP_2, MIP_1, MIP_2, MIP_3 and MRP_1) are shown in Table 8.1.

MCP_1 was found to be a novel FMRFamide-like decapeptide and MCP_2 to be FMRFamide itself. MIP_1 and MIP_2 were found to be Ser^2-MIP and Ala^2-MIP, respectively; that is, the MIPs first isolated from the pedal ganglia (Hirata *et al.*, 1988) are also shown to be present in the ABRM. In addition to these MIPs, a novel MIP analogue, MIP_3, was

Table 8.1. The peptides isolated from the ABRM of *Mytilus*

MCP_1	H-Ala-Leu-Ala-Gly-Asp-His-Phe-Phe-Arg-Phe-NH_2
MCP_2	H-Phe-Met-Arg-Phe-NH_2 (FMRFamide)
MIP_1	H-Gly-Ser-Pro-Met-Phe-Val-NH_2 (Ser^2-MIP)
MIP_2	H-Gly-Ala-Pro-Met-Phe-Val-NH_2 (Ala^2-MIP)
MIP_3	H-Asp-Ser-Pro-Leu-Phe-Val-NH_2
MRP_1	H-Ala-Met-Pro-Met-Leu-Arg-Leu-NH_2 (CARP)

shown to be present in the muscle. MRP_1 was found to be CARP which was first isolated from the pedal ganglia (Hirata *et al.*, 1987).

MCP_1 was synthesized by a solid-phase peptide synthesizer (Applied Biosystems 430A) followed by HF cleavage and HPLC purification. The structure of the synthesized MCP_1 was confirmed to be correct by amino acid analysis, amino acid sequence analysis and FAB-MS analysis. HPLC profiles of native MCP_2, MIP_1, MIP_2 and MRP_1 were compared with those of the synthetic peptides by using a reversed-phase column and a cation-exchange column. Biological activity of native FMRFamide, Ser^2-MIP, Ala^2-MIP and CARP were also compared with that of the synthetic peptides. The native peptides were found to show identical HPLC profiles and identical dose–response relation with those of the synthetic peptides.

It has been reported that Ser^2-MIP and Ala^2-MIP show almost identical inhibitory action on contractions of the ABRM (Hirata *et al.*, 1988, 1989c). In the present experiments we found that native MIP_3 showed similar inhibitory action on the ABRM to those of synthetic Ser^2-MIP and Ala^2-MIP. Although we did not synthesize MIP_3, and hence could not compare the natures of native MIP_3 with those of the synthetic one, the proposed structure of the peptide seems to be correct.

In the foregoing experiments on the biological activities, the concentrations of the native peptides were estimated from the results of amino acid analyses.

8.2.2. *Peptides in the ganglia of* Fusinus

Acetone extract of the ganglion masses (cerebral, suboesophageal and buccal ganglia) excised from 1100 specimens of *Fusinus ferrugineus* was forced through C-18 cartridges, and the retained material was eluted. The retained material was then fractionated by gel filtration. Biological activities of the fractions were examined on a train of twitch contractions of the radula retractor muscle of the animal. The train twitch contractions (five twitches) were evoked by applying electrical pulses (15 V, 2 ms, 0.2 Hz, five pulses) of stimulation to the muscle. The procedures for the gel filtration and the bioassay experiments were basically the same as those used for the ABRM of *Mytilus*.

The gel-filtrated fractions of the retained material showed four peaks of biological activities on twitch contractions of the radula retractor; three peaks of contraction-potentiating activity and one peak of contraction-inhibiting activity. All activities were found to be destroyed by treating the active fractions with subtilisin, suggesting that the active substances in the peaks are peptides (Kanda *et al.*, 1989).

One of the contraction-potentiating peaks showed its maximum activity at fraction 22, and the contraction-inhibiting peak showed its maximum activity at fractions 28–30. However, these two peaks seemed to be eluted partially overlapping each other. The maximum activities of the other two contraction-potentiating peaks were observed at fractions 44 and 50, respectively. That is, the two peaks were also eluted partially overlapping each other. Therefore, we divided the gel-filtrated active fractions into two groups—group A (fractions 21–40) and group B (fractions 41–58). Each of the groups was then subjected to a HPLC system (Jasco Tri-Roter IV) to purify the bioactive substances in it. The columns used for the purification were the same as those used in experiments on the ABRM.

We separated 13 bioactive substances from group A; five contraction-potentiating and eight contraction-inhibiting substances (Fig. 8.2). Three of the 13 were purified and their structures determined. There was one potentiating peptide (*Fusinus* excitatory peptide 1, FEP_1) and two inhibitory peptides (*Fusinus* inhibitory peptide 1 and 2, $FIP_{1,2}$). The potentiating peptide, FEP_1, did not show contractile activity even at a concentration 100 times higher than the threshold for potentiation of twitch contraction. Structure-determination experiments on the other 10 bioactive substances in group A are now in progress.

We separated 10 bioactive substances from group B; seven contraction-potentiating and three contraction-inhibiting substances (Fig. 8.2). Four of the 10 were purified and their structures determined. There were two

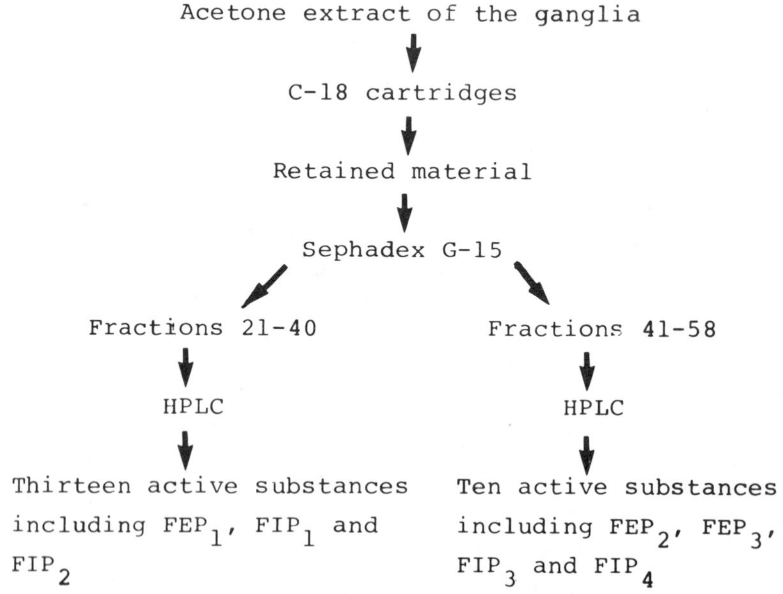

Fig. 8.2. The isolation procedures for bioactive peptides in the ganglia of *Fusinus*.

potentiating peptides (*Fusinus* excitatory peptide 2 and 3, $FEP_{2,3}$) and two inhibitory peptides (*Fusinus* inhibitory peptides 3 and 4, $FIP_{3,4}$). At high doses the two contraction-potentiating peptides, FEP_2 and FEP_3, showed a contractile activity by themselves. Structure-determination experiments on the other six bioactive substances in group B are also in progress.

The probable structures of the purified active substances were determined by the same methods used for the peptides in the ABRM. The determined structures of the substances are shown in Table 8.2.

FEP_1 was found to be a novel FMRFamide-like decapeptide closely related to MCP_1 which was found in the ABRM. FEP_2 and FEP_3 were found to be FMRFamide and FLRFamide, respectively. FIP_1 was myomodulin itself. That is, myomodulin is present not only in the opisthobranch mollusc *Aplysia* (Cropper *et al.*, 1987) but also in the prosobranch mollusc *Fusinus*. FIP_2 was a congener of FIP_1. Only one amino acid residue is different between them. That is, in addition to myomodulin and CARP, the third MCRP was found. FIP_3 and FIP_4 were also found to be congeneric peptides. These peptides, as well as MCRPs, show inhibitory action on the radula retractor muscle but their structures are related to FaRPs rather than to MCRPs.

FEP_1 was synthesized by the same methods used for the synthesis of MCP_1, and the structure of the synthesized peptide was confirmed to be correct by the foregoing analyses. HPLC profiles of native FEP_1 were then compared with those of the synthetic peptide by using a reversed-phase column and a cation-exchange column. Biological activity of the native peptide was also compared with that of the synthetic one. All the results obtained in these experiments indicated that both of the peptides have identical properties. Thus, we concluded that the proposed structure of FEP_1 is correct.

By the same procedures we synthesized FIP_3 and FIP_4 and compared the properties of each of the synthetic peptides with those of the native peptide. The results indicated that the proposed structures of FIP_3 and FIP_4 are correct. We also compared HPLC profiles and bioactivities of FEP_2, FEP_3 and FIP_1 with those of commercial FMRFamide, FLRFa-

Table 8.2. Peptides isolated from the ganglia of *Fusinus*

FEP_1	H-Ala-Leu-Thr-Asn-Asp-His-Phe-Leu-Arg-Phe-NH_2
FEP_2	H-Phe-Met-Arg-Phe-NH_2 (FMRFamide)
FEP_3	H-Phe-Leu-Arg-Phe-NH_2 (FLRFamide)
FIP_1	H-Pro-Met-Ser-Met-Leu-Arg-Leu-NH_2 (Myomodulin)
FIP_2	H-Pro-Met-Asn-Met-Leu-Arg-Leu-NH_2
FIP_3	H-Gly-Ser-Leu-Phe-Arg-Phe-NH_2
FIP_4	H-Ser-Ser-Leu-Phe-Arg-Phe-NH_2

mide and myomodulin, respectively. The results indicated that the proposed structures of the peptides are correct. We did not synthesize the new MCRP, FIP_2. This peptide showed almost identical biological action with myomodulin, though the HPLC profiles of the two peptides were not identical. The proposed structure of FIP_2 is probably correct.

8.3. **Aspects of structures and actions of the peptides**

8.3.1. *FMRFamide-related peptides*

Many species of FaRPs have been isolated from molluscs. As shown in Table 8.3, we can classify these FaRPs into three subgroups according to the number of amino acid residues of the peptides: tetrapeptide, pentapeptide and heptapeptide subgroups. According to the structures of C-terminal parts of the FaRPs, we can also classify them into two subgroups: FMRFamide and FLRFamide subgroups. In the present experiments we found two decapeptides, MCP_1 in the ABRM of *Mytilus* and FEP_1 in the ganglia of *Fusinus*. FEP_1 is regarded as a FaRP, because the structure of its C-terminal part is -Phe-Leu-Arg-Phe-NH_2. As we will mention later, the actions of FEP_1 on some molluscan muscles are qualitatively similar to those of FMRFamide and FLRFamide.

The structure of C-terminal part of MCP_1 is neither -Phe-Met-Arg-Phe-NH_2 nor -Phe-Leu-Arg-Phe-NH_2; it is -Phe-Phe-Arg-Phe-NH_2. As shown in Table 8.4, however, MCP_1 is apparently an analogue of FEP_1. Only three amino acid residues are different between them. The actions of MCP_1 on some molluscan muscles are qualitatively similar to those of FEP_1. MCP_1 can be regarded as a FaRP; thus, we may propose that there exists a decapeptide subgroup of FaRPs in molluscs. Furthermore, it is suspected that, in addition to FMRFamide and FLRFamide subgroups, there might be a FFRFamide subgroup of FaRPs in molluscs.

To investigate structure–activity relations of MCP_1 and FEP_1, we synthesized some fragments of these peptides (Table 8.4). They are

Table 8.3. The FMRFamide-related peptides in molluscs

Tetrapeptides	H-Phe-Met-Arg-Phe-NH_2
	H-Phe-Leu-Arg-Phe-NH_2
Pentapeptides	H-Ala-Phe-Leu-Arg-Phe-NH_2
	H-Thr-Phe-Leu-Arg-Phe-NH_2
Heptapeptides	pGlu-Asp-Pro-Phe-Leu-Arg-Phe-NH_2
	H-Asn-Asp-Pro-Phe-Leu-Arg-Phe-NH_2
	H-Ser-Asp-Pro-Phe-Leu-Arg-Phe-NH_2
	H-Gly-Asp-Pro-Phe-Leu-Arg-Phe-NH_2

For further details see Price *et al.* (1987) and Cottrell (1989).

Table 8.4. The FMRFamide-related decapeptides and synthetic analogues

Sequence	Name
H-Ala-Leu-Ala-Gly-Asp-His-Phe-Phe-Arg-Phe-NH_2	(MCP_1)
H-Ala-Leu-Thr-Asn-Asp-His-Phe-Leu-Arg-Phe-NH_2	(FEP_1)
H-Asp-His-Phe-Phe-Arg-Phe-NH_2	(MCP_1 5–10)
H-Asp-His-Phe-Leu-Arg-Phe-NH_2	(FEP_1 5–10)
H-Phe-Phe-Arg-Phe-NH_2	(MCP_1 7–10, FFRFamide)

MCP_1 5–10, FEP_1 5–10 and MCP_1 7–10 (FFRFamide). The actions of these fragment peptides were then examined on several molluscan muscles, and they were compared with those of MCP_1, FEP_1, FMRFamide and FLRFamide (FEP_1 7–10).

MCP_1 potentiated phasic contraction of the ABRM in response to repetitive electrical pulses of stimulation. The threshold concentration for the potentiation was found to be very low, being between 10^{-12} and 10^{-11} M. At higher than 10^{-8} M the peptide induced a contraction of the ABRM by itself (Fig. 8.3A). It has been shown that FMRFamide relaxes catch tension of the muscle at 10^{-8}–10^{-7} M (Muneoka and Matsuura, 1985). In contrast to FMRFamide, MCP_1 did not show any relaxing action. The actions of the MCP_1 fragments, MCP_1 5–10 and FFRFamide, on the ABRM were almost identical with those of MCP_1, though the threshold concentrations of these fragments for evoking contraction were about 10 times higher than that of MCP_1.

FEP_1 potentiated phasic contraction of the ABRM at higher than 10^{-9} M and induced a contraction at higher than 10^{-7} M (Fig. 8.3B). The peptide, as well as MCP_1 and its fragments, did not show catch-relaxing action. That is, FEP_1 and MCP_1 showed qualitatively identical actions on the ABRM, but the former is about 1000 times less potent than the latter in potentiating phasic contraction. The FEP_1 fragment, FEP_1 5–10, showed actions identical with those of FEP_1. FEP_1 7–10 (FLRFamide) showed catch-relaxing action at 10^{-8}–10^{-7} M. The other actions of the peptide were identical with those of FEP_1. FMRFamide showed potentiating action at higher than 10^{-10} M, evoked a contraction at higher than 10^{-7} M and relaxed catch at 10^{-8}–10^{-7} M. That is, FMRFamide and FLRFamide showed qualitatively identical actions, though the former is about 10 times more potent than the latter in potentiating phasic contraction.

The foregoing actions of the FaRPs on the ABRM are summarized in Table 8.5. As shown in the table, MCP_1 and its fragments (FFRFamide analogues) are the most potent peptides in potentiating phasic contraction, and the next is FMRFamide. Several FMRFamide analogues have been also shown to have a strong potentiating effect. They are Tyr-Gly-Gly-Phe-Met-Arg-Phe-NH_2 (Muneoka and Matsuura, 1985), Trp-Nle-Arg-Phe-NH_2 (Takemoto *et al.*, 1986), Tyr-Phe-Met-Arg-Phe-NH_2 and

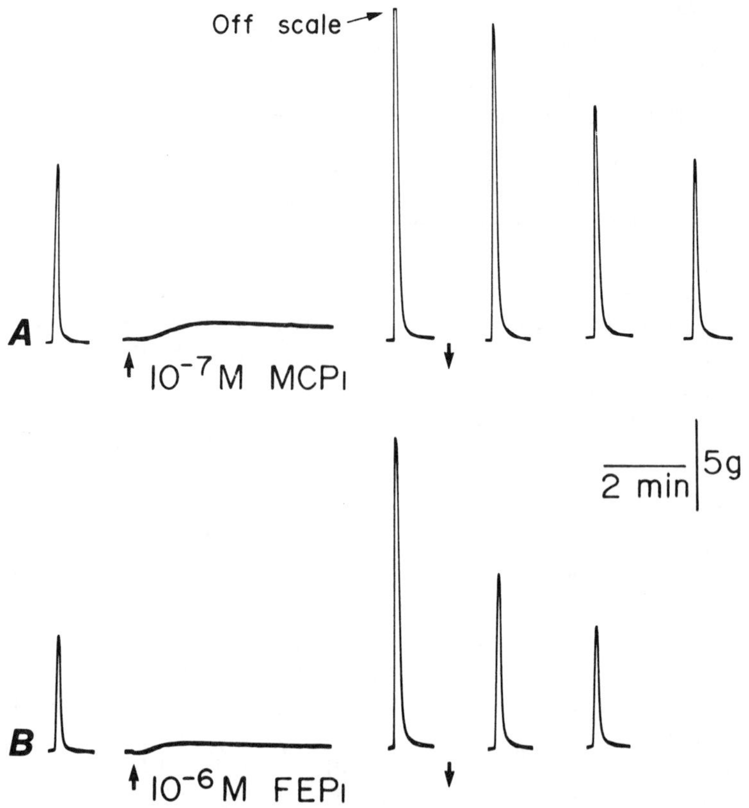

Fig. 8.3. Contractile and phasic-contraction-potentiating effects of MCP_1 and FEP_1 on the ABRM of *Mytilus*. The phasic contractions were elicited by stimulating the muscle with repetitive electrical pulses (15 V, 3 ms, 10 Hz, for 5 s) at 10 min intervals. Each peptide was applied to the muscle 8 min prior to the electrical stimulation.

Table 8.5. Effects of FMRFamide-related peptides on the ABRM of *Mytilus*

	Threshold for potentiation	*Threshold for contraction*	*Relaxation at* 10^{-8}–10^{-7} M
MCP_1	10^{-12}–10^{-11} M	10^{-8}–10^{-7} M	No
MCP_1 5–10	10^{-12}–10^{-11} M	10^{-7}–10^{-6} M	No
MCP_1 7–10 (FFRFamide)	10^{-12}–10^{-11} M	10^{-7}–10^{-6} M	No
FEP_1	10^{-9}–10^{-8} M	10^{-7}–10^{-6} M	No
FEP_1 5–10	10^{-9}–10^{-8} M	10^{-7}–10^{-6} M	No
FEP_1 7–10 (FLRFamide)	10^{-9}–10^{-8} M	10^{-7}–10^{-6} M	Yes
FMRFamide	10^{-10}–10^{-9} M	10^{-7}–10^{-6} M	Yes

acetyl-Phe-Nle-Arg-Phe-NH_2 (unpublished). FLRFamide and its analogues, such as FEP_1, are less potent than FFRFamide, FMRFamide and their analogues in potentiating phasic contraction. In the present experiments we isolated MCP_1 and FMRFamide from the ABRM, but did not isolate FLRFamide and its analogues. FLRFamide or its analogue might not be present in the ABRM. Thus it is supposed that MCP_1 and FMRFamide, but not FLRFamide, might be neuromodulators having a function of potentiating contraction of the muscle. FMRFamide also shows catch-relaxing action in the ABRM. However, it is not considered that the relaxing effect reflects the physiological role of the peptide, because the relaxing response is not elicited after the ABRM has been briefly treated with high concentrations (10^{-6} M or higher) of FMRFamide (Muneoka and Matsuura, 1985).

In the radula retractor muscle of *Fusinus*, FEP_1 is the most potent peptide in the FaRPs in potentiating twitch contraction. It potentiates the contraction at higher than 10^{-10} M. The other FaRPs are about 10 times less potent than FEP_1. At higher concentrations all of the FaRPs, except MCP_1 5–10, evoke contraction of the muscle by themselves. Among these peptides, FMRFamide and FLRFamide are the most potent contractile peptides. They exhibit the contractile action at higher than 10^{-8} M. The other tetrapeptide FFRFamide is about 10 times less potent than FMRFamide and FLRFamide. These actions of the FaRPs are summarized in Table 8.6.

At 10^{-5} M, MCP_1 and FEP_1 induce contraction of the radula retractor. However, twitch contraction of the muscle in 10^{-5} M MCP_1 is depressed, though the contraction in 10^{-5} M FEP_1 is markedly potentiated (Fig. 8.4A,B). It is very interesting that MCP_1 5–10 does not induce contraction of the muscle even at 10^{-5} M. At this concentration the peptide, as

Table 8.6. Effects of FMRFamide-related peptides on the radula retractor muscle of *Fusinus*

	Threshold for potentiation	*Threshold for contraction*
MCP_1	10^{-9}–10^{-8} M	10^{-6}–10^{-5} M
MCP_1 5–10	10^{-9}–10^{-8} M	—*
MCP_1 7–10 (FFRFamide)	10^{-9}–10^{-8} M	10^{-7}–10^{-6} M
FEP_1	10^{-10}–10^{-9} M	10^{-6}–10^{-5} M
FEP_1 5–10	10^{-9}–10^{-8} M	10^{-6}–10^{-5} M
FEP_1 7–10 (FLRFamide)	10^{-9}–10^{-8} M	10^{-8}–10^{-7} M
FMRFamide	10^{-9}–10^{-8} M	10^{-8}–10^{-7} M

* At 10^{-5} M of the peptide, no contraction was elicited and twitch contractions were inhibited.

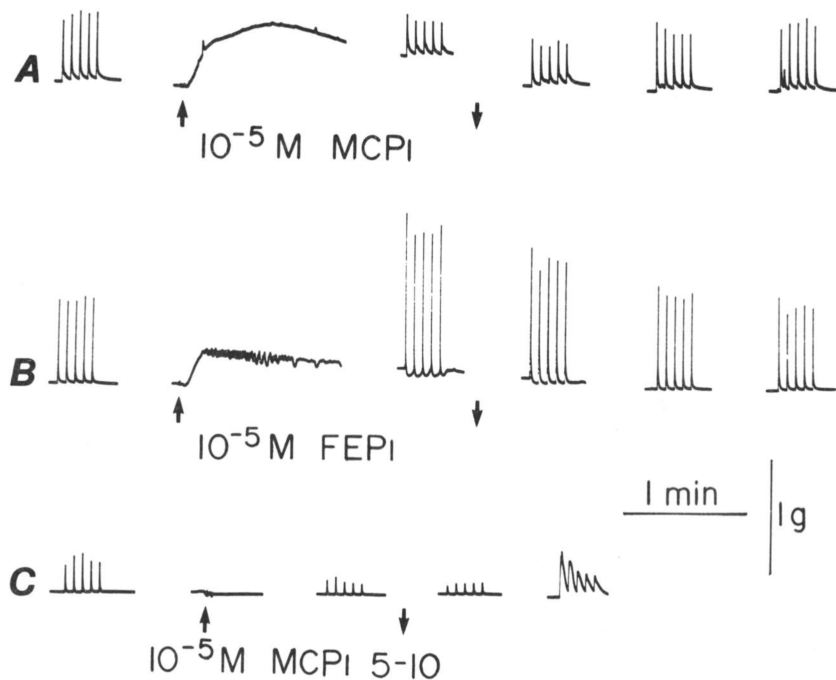

Fig. 8.4. Contractile and twitch-contraction-inhibiting effects of MCP_1, contractile and twitch-contraction-potentiating effects of FEP_1 and twitch-contraction-inhibiting effect of MCP_1 5–10 on the radula retractor muscle of *Fusinus*. The trains of twitch contractions were elicited by stimulating the muscle with train electrical pulses (15 V, 2 ms, 0·2 Hz, 5 pulses) at 10 min intervals. Each peptide was applied to the muscle 8 min prior to the electrical stimulation.

well as MCP_1, inhibits twitch contraction (Fig. 8.4C), though at 10^{-8}–10^{-7} M it potentiates the contraction.

In the present experiments we have shown that FEP_1, FMRFamide and FLRFamide are present in the ganglia of *Fusinus*. FEP_1 might be a neuromodulator having a function of potentiating contraction of the radula retractor. FMRFamide and FLRFamide might be excitatory neurotransmitters. It seems that none of the FFRFamide-group peptides is present in *Fusinus*. These peptides seem to act as agonists or antagonists of FMRFamide, FLRFamide or FEP_1, and thus exhibit their potentiating, contractile or inhibitory effect.

8.3.2. Mytilus *inhibitory peptides*

Three species of inhibitory hexapeptides, whose structures are closely related, were isolated from the ABRM. As shown in Table 8.1, they are

MIP_1 (Ser^2-MIP), MIP_2 (Ala^2-MIP) and MIP_3 (Asp^1, Ser^2, Leu^4-MIP). The inhibitory actions of these peptides on phasic contraction of the ABRM in response to repetitive electrical pulses of stimulation are similar to one another, though in some muscles of other molluscs their actions may not be identical (Hirata *et al.*, 1989c).

We synthesized some fragments of the MIPs and examined their actions on phasic contraction of the ABRM to compare them with those of the MIPs. They are Pro-Met-Phe-Val-NH_2 (MIP 3–6) and Met-Phe-Val-NH_2 (MIP 4–6). The results obtained are summarized in Table 8.7. MIP 3–6 was found to be 10–30 times less potent than MIPs in inhibiting phasic contraction. The effect of MIP 4–6 was found to be extremely weak; that is, the peptide was 3000–10,000 times less potent than MIPs (Fig. 8.5). These results suggest that Pro^3 of MIPs is very important to exhibit the inhibitory activity, though the residue is not essential. The critical structure in the MIPs for exhibition of the inhibitory action may be -Phe-Val-NH_2. Gly^1, Ser^2 and Met^4 of Ser^2-MIP can be substituted for Asp, Ala and Leu, respectively. Thus, it is supposed that the three MIPs act on the same receptors, and that the differences of structures among the peptides do not have important meanings.

8.3.3. *Myomodulin-CARP-related peptides and others*

In the present experiments we showed that CARP is present in the ABRM of *Mytilus*. This fact supports the notion that CARP is a co-relaxing neurotransmitter in the ABRM (Hirata *et al.*, 1989b). It is well known that the principal relaxing neurotransmitter in the muscle is serotonin (Twarog, 1954; Satchell and Twarog, 1978; for review see Muneoka and Twarog, 1983).

We also isolated two species of MCRPs from the ganglia of *Fusinus* in the present experiments. They are FIP_1 (myomodulin) and FIP_2 (Asn^3-myomodulin). These peptides can also relax catch tension of the ABRM. However, they are about 10 times less potent than CARP. That is, the

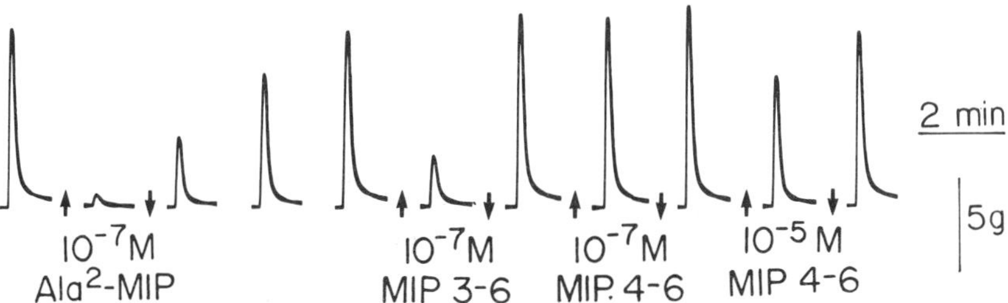

Fig. 8.5. Inhibitory effects of Ala^2-MIP and fragment peptides on phasic contraction of the ABRM of *Mytilus*. The procedures are the same as in Fig. 8.3.

Table 8.7. Equipotent molar ratios (EPMR) of *Mytilus* inhibitory peptides and synthetic analogues in inhibition of phasic contraction of the ABRM

		EPMR
MIP_1 (Ser^2-MIP)	H-Gly-Ser-Pro-Met-Phe-Val-NH_2	1
MIP_2 (Ala^2-MIP)	H-Gly-Ala-Pro-Met-Phe-Val-NH_2	1
MIP_3	H-Asp-Ser-Pro-Leu-Phe-Val-NH_2	1–10*
MIP 3–6	H-Pro-Met-Phe-Val-NH_2	10–30
MIP 4–6	H-Met-Phe-Val-NH_2	3000–10,000

* Native peptide (concentrations were estimated from the result of amino acid analysis).

threshold concentration of CARP for relaxation of catch is 3×10^{-10}–10^{-9} M (Hirata *et al.*, 1987), while those of myomodulin and Asn^3-myomodulin are 3×10^{-9}–10^{-8} M. The latter two peptides may relax catch by acting on CARP receptors of the ABRM as agonists.

In the radula retractor muscle of *Fusinus*, myomodulin and Asn^3-myomodulin showed equipotent inhibitory effect on twitch contraction (Fig. 8.6A). These peptides may be inhibitory neurotransmitters or neuromodulators which act on the same receptors of the muscle. CARP also shows an inhibitory action on twitch contraction of the muscle, but it is about 10 times less potent than myomodulin and Asn^3-myomodulin.

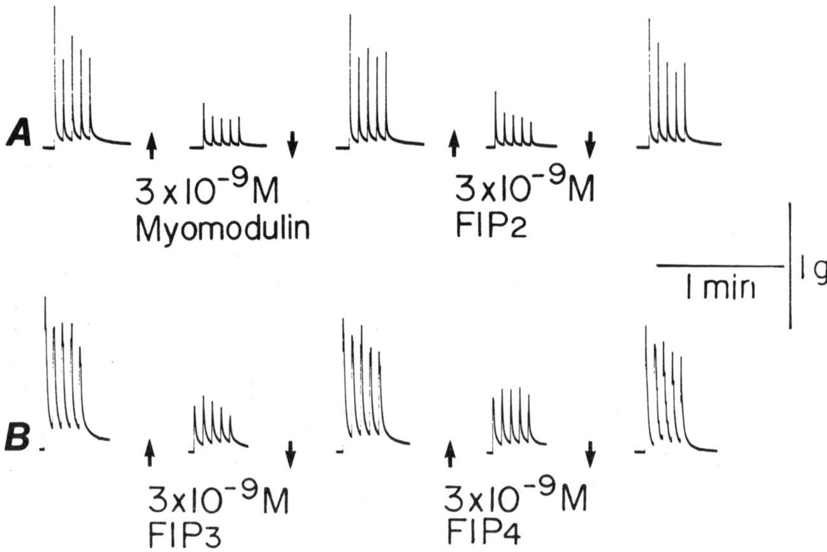

Fig. 8.6. Inhibitory effects of myomodulin, FIP_2, FIP_3 and FIP_4 on train twitch contractions of the radula retractor muscle of *Fusinus*. The procedures are the same as in Fig. 8.4.

That is, the threshold concentration of CARP for the inhibition is 10^{-9} – 5×10^{-9} M while those of myomodulin and Asn^3-myomodulin are 10^{-10} – 5×10^{-10}. The former may inhibit the contraction by action as an agonist of the latter two peptides.

It is probable that MCRPs are widely distributed in molluscs. Further, it is supposed that peptides resembling MCRPs might be distributed in non-molluscan animals including mammals, because CARP has been shown to have potentiating or inhibitory effect on several non-molluscan muscles, such as the coxal muscle of a cockroach and the aorta of guinea pig (Table 8.8).

FIP_3 and FIP_4, as well as MCRPs, inhibit twitch contraction of the radula retractor muscle of *Fusinus*. Their inhibitory potencies are almost identical (Fig. 8.6B). The threshold concentrations for the inhibition are between 3×10^{-10} and 10^{-9} M, being slightly higher than those of myomodulin and Asn^3-myomodulin. The structures of C-terminal parts of FIP_3 and FIP_4 resemble those of FaRPs. FIP_3 and FIP_4 have -Leu-Phe-Arg-Phe-NH_2, while FaRPs have -Phe-Met-Arg-Phe-NH_2, -Phe-Leu-Arg-Phe-NH_2 or -Phe-Phe-Arg-Phe-NH_2. However, FIP_3 and FIP_4 do not show FaRP-like excitatory action on the radula retractor muscle but do show MCRP-like inhibitory action.

In the ABRM of *Mytilus*, FIP_3 and FIP_4 show neither FaRP-like action nor MCRP-like action. They depress relaxation of catch tension in response to repetitive electrical pulses of stimulation (Fig. 8.7). The threshold concentrations for the depression are about 10^{-9} M in both cases. Relaxations of catch in response to serotonin and CARP are not affected by FIP_3 and FIP_4. The depression of relaxation of catch by repetitive electrical stimulation may therefore be brought about by an inhibitory action of the peptides on the intramuscular relaxing-nerve elements. It is well known that brief repetitive electrical pulses of stimula-

Table 8.8. Effects of CARP on muscle contractions in non-molluscan animals

Animals	*Contractions*	*Effects*	*Threshold*
Urechis unicinctus (Echiuloidea)	Twitch contraction of the body-wall muscle	Potentiation	10^{-8} M
Marphysa sanguinea (Polychaeta)	Twitch contraction of the body-wall muscle	Inhibition	10^{-7} M
Periplaneta americana (Insecta)	Twitch and tetanic contractions of the coxal muscle	Potentiation	10^{-7} M*
Guinea pig (Mammalia)	Noradrenaline and histamine contractions of the aorta	Inhibition	10^{-6}–10^{-5} M

* From Dr H. Washio

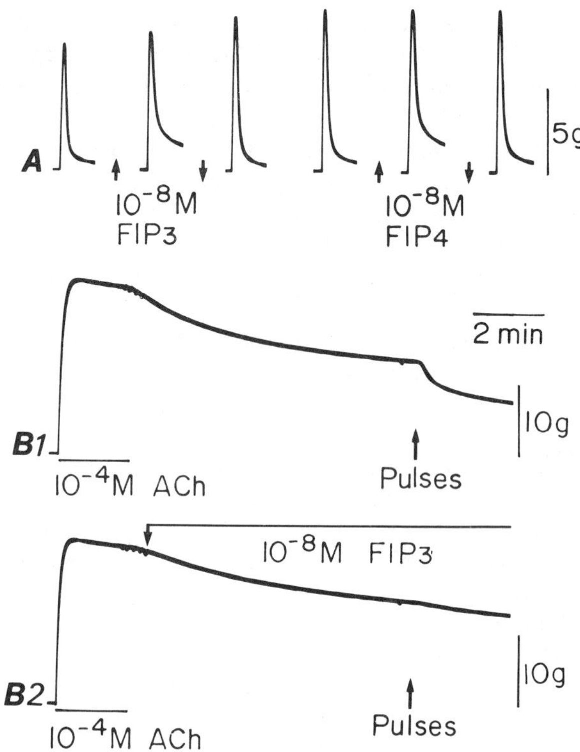

Fig. 8.7. Inhibitory effects of FIP_3 and FIP_4 on relaxations of the ABRM of *Mytilus*. **A**: Effects of FIP_3 and FIP_4 on relaxation of phasic contraction. The procedures are the same as in Fig. 8.3. **B**: Effect of FIP_3 on relaxation of ACh-induced catch tension in response to low frequency of repetitive electrical pulses (15 V, 3 ms, 1 Hz, 10 pulses) of stimulation.

tion elicit a contraction followed by a rapid relaxation (phasic contraction) in the ABRM by acting on both excitatory and relaxing nerve fibres in the muscle (for review see Muneoka *et al.*, 1990). As shown in Fig. 8.7A, FIP_3 and FIP_4 do not inhibit tension development of phasic contraction of the ABRM. Therefore, non-specific inhibitory action of the peptides on excitatory and relaxing nerve elements may be ruled out. The peptides might inhibit release of relaxing neurotransmitter and thus depress relaxation of catch tension in response to repetitive electrical pulses of stimulation.

8.4. **Conclusion**

Thirteen molluscan peptides were isolated in the present experiments — six from the ABRM of *Mytilus edulis* and seven from the ganglia of

Fusinus ferrugineus. The *Mytilus* peptides are as follows: MCP_1, MCP_2 (FMRFamide), MIP_1 (Ser^2-MIP), MIP_2 (Ala^2-MIP), MIP_3 and MRP_1 (CARP). The *Fusinus* peptides are as follows: FEP_1, FEP_2 (FMRFamide), FEP_3 (FLRFamide), FIP_1 (myomodulin), FIP_2, FIP_3 and FIP_4. Among these peptides, newly isolated ones are as follows: MCP_1, MIP_3, FEP_1, FIP_2, FIP_3 and FIP_4 (Tables 8.1 and 8.2).

In addition to the six peptides isolated from the ABRM, at least 11 other peptides were suggested to be present in the muscle. It has been proposed that four biogenic amines (ACh, serotonin, dopamine and octopamine) may be involved in the regulation of the ABRM (for review see Muneoka *et al.*, 1990). Although it may not be the case that all of the peptides present in the ABRM are involved in the regulation of the muscle, multiple mechanisms of regulation by multiple neurotransmitters and neuromodulators seem to exist in the muscle. Such multiplicity is considered to be general in invertebrate muscles, as well as vertebrate visceral muscles. Clarification of physiological meaning of the multiplicity may be one of the most important subjects in the field of neurobiology.

In addition to the seven peptides isolated from the ganglia of *Fusinus*, at least 16 other peptides, which showed biological activities on the radula retractor muscle of the animal, were suggested to exist in the ganglia. It is well known in vertebrates that most of the neuropeptides found in the central nervous system show bioactivities in the visceral organs, and vice-versa. This seems also to be the case in invertebrate muscles and central nervous systems. Of course, not all the neuropeptides present in an invertebrate nervous system show bioactivities on a muscle of the animal. For example, none of the three MCRPs found in the present experiments has any effect on the proboscis retractor muscle of *Fusinus*, though all the peptides show potent inhibitory effect on the radula retractor of the animal. However, if we use some suitable muscles of an invertebrate for bioassay, we may be able to detect most of the neuropeptides in its nervous system. The most important thing in such bioassay experiments using muscles is to know the precise pharmacological properties of the muscles before the experiments.

Acknowledgement

Part of this work was supported by the Grant-in-Aid for General Scientific Research from the Ministry of Education, Science and Culture of Japan.

References

Cottrell, G. A. (1989). The biology of the FMRFamide-series of peptides in molluscs with special reference to *Helix*. *Comp. Biochem. Physiol.*, **93A**, 41–5.

Cropper, E. C., Tenenbaum, R., Kolks, M. A. G., Kupfermann, I. & Weiss, K. R. (1987). Myomodulin: a bioactive neuropeptide present in an identified cholinergic buccal motor neuron of *Aplysia. Proc. Natl. Acad. Sci. USA*, **84**, 5483–6.

Hirata, T., Kubota, I., Takabatake, I., Kawahara, A., Shimamoto, N. and Muneoka, Y. (1987). Catch-relaxing peptide isolated from *Mytilus* pedal ganglia. *Brain Res.*, **422**, 374–6.

Hirata, T., Kubota, I., Iwasawa, N., Takabatake, I., Ikeda, T. and Muneoka, Y. (1988). Structures and actions of *Mytilus* inhibitory peptides. *Biochem. Biophys. Res. Commun.*, **152**, 1376–82.

Hirata, T., Kubota, I., Imada, M., Muneoka, Y. and Kobayashi, M. (1989a). Effects of the catch-relaxing peptide on molluscan muscles. *Comp. Biochem. Physiol.*, **92C**, 283–8.

Hirata, T., Kubota, I., Imada, M. and Muneoka, Y. (1989b). Pharmacology of relaxing response of *Mytilus* smooth muscle to the catch-relaxing peptide. *Comp. Biochem. Physiol.*, **92C**, 289–95.

Hirata, T., Kubota, I., Iwasawa, N., Fujisawa, Y., Muneoka, Y. and Kobayashi, M. (1989c). Effects of *Mytilus* inhibitory peptides on mechanical responses of various molluscan muscles. *Comp. Biochem. Physiol.*, **93C**, 381–8.

Kanda, T., Takabatake, I., Fujisawa, Y., Ikeda, T., Muneoka, Y. and Kobayashi, M. (1989). Biological activities of ganglion extracts from a prosobranch mollusc, *Fusinus ferrugineus. Hiroshima J. Med. Sci.*, **38**, 106–16.

Kiss, T. (1988). Catch-relaxing peptide (CARP) decreases the Ca-permeability of snail neuronal membrane. *Experientia*, **44**, 998–1000.

Muneoka, Y., Fijisawa, Y., Fujimoto, N. and Ikeda, T. (1990). The regulation and pharmacology of muscles in *Mytilus*. In: Stefano, G. B. (ed.), *Neurobiology of* Mytilus edulis. Manchester University Press, Manchester.

Muneoka, Y. and Matsuura, M. (1985). Effects of the molluscan neuropeptide FMRFamide and the related opioid peptide YGGFMRFamide on *Mytilus* muscle. *Comp. Biochem. Physiol.*, **81C**, 61–70.

Muneoka, Y. and Saitoh, H. (1986). Pharmacology of FMRFamide in *Mytilus* catch muscle. *Comp. Biochem. Physiol.*, **85C**, 207–14.

Muneoka, Y. and Twarog, B. M. (1983). Neuromuscular transmission and excitation-contraction coupling in molluscan muscle. In: Saleuddin, A. S. M. and Wilbur, K. M. (eds), *The Mollusca*, Vol. 4, Part 1, Academic Press, New York, pp. 35–76.

Painter, S. D. (1982). FMRFamide catch contractures of a molluscan smooth muscle: pharmacology, ionic dependence and cyclic nucleotides. *J. Comp. Physiol.*, **148**, 491–501.

Price, D. A. and Greenberg, M. J. (1977). Structure of a molluscan cardioexcitatory neuropeptide. *Science*, **197**, 670–1.

Price, D. A., Davies, N. W., Doble, K. E. and Greenberg, M. J. (1987). The variety and distribution of the FMRFamide-related peptides in molluscs. *Zool. Sci.*, **4**, 395–410.

Satchell, D. G. and Twarog, B. M. (1978). Identification of 5-hydroxytryptamine (serotonin) released from the anterior byssus retractor muscle of *Mytilus californianus* in response to nerve stimulation. *Comp. Biochem. Physiol.*, **59C**, 81–5.

Takemoto, M., Saitoh, H. and Muneoka, Y. (1986). Relaxing action of Trp-Nle-Arg-Phe-NH_2 on the anterior byssus retractor muscle of *Mytilus*. *Hiroshima J. Med. Sci.*, **35**, 381–8.

Twarog, B. M. (1954). Responses of a molluscan smooth muscle to acetylcholine and 5-hydroxytryptamine. *J. Cell. Comp. Physiol.*, **44**, 141–63.

9 *David J. Prior and Ian G. Welsford*

Neuropeptide modulation of central neuronal networks and peripheral targets in terrestrial slugs

9.1. Introduction

It is well established that multiple regulatory systems are involved in the integration of diverse physiological and behavioural responses. During long-term adaptation, and short-term changes in responsiveness, physiological systems respond in concert. In an effort to further explore the mechanisms underlying the simultaneous control of central neural networks and peripheral target organs, the control of feeding and cardiovascular function has been studied in the terrestrial slug, *Limax maximus*.

Due to their moist integument, terrestrial slugs are extremely susceptible to dehydration (Machin, 1964). In active slugs the water loss due to evaporation from the integument and the deposition of a dilute mucus trail can be as great as 16% of their initial body weight (% IBW) per hour (Dainton, 1954). Thus in drying conditions slugs can experience severe dehydration stress within a few hours. In these organisms there is little short-term regulation of the ionic and osmotic concentration of the haemolymph. In *Limax maximus*, dehydration to 70% IBW results in an increase in the osmolality of the haemolymph from 140 mOsm/kg H_2O to 200–220 mOsm/kg H_2O (Prior *et al.*, 1983). There is a concomitant increase in the concentrations of Na^+ (from 60 to 90 mM) and K^+ (from 2.5 to 5 mM; Hess, 1982).

Because the nervous system of these organisms is exposed to severe changes in osmolality and ionic concentration, the question is raised whether, as in some arthropod systems, there is a haemolymph/CNS barrier (see Treherne, 1980) or whether the neurons are uniquely tolerant to the stress. In *Limax* the connective tissue sheath surrounding the central ganglia appears to pose no barrier to diffusion of ions. In isolated CNS preparations the responses of central neurons to changes in osmolality of the perfusion saline occur immediately (Prior, 1981). Because these organisms can survive more than 100% increases in haemolymph osmolality, it appears that the central neurons in this system are uniquely tolerant to osmotic stress.

Slugs do not readily control water balance by physiological mechanisms such as alteration of urine concentration, thus they must rely upon modification of behavioural patterns.

9.2. **Behavioural regulation of body hydration**

Although slugs are unusually susceptible to dehydration, they possess an array of behaviours effective in minimizing the extent of the stress. These include contact-rehydration (Prior, 1983, 1984; Prior and Uglem, 1984), the pneumostome closure rhythm (Prior *et al.*, 1983; Dickinson *et al.*, 1988), water orientation behaviour (Banta *et al.*, 1990), alterations in locomotor activity including circadian locomotor rhythms (Hess and Prior, 1985; Welsford *et al.*, 1990) and modifications of feeding activity (Prior, 1983; Phifer and Prior, 1985; reviews by Riddle, 1983 and Prior, 1985). Because slugs display an array of responses to dehydration stress, they serve as a useful model for the study of multiple regulatory systems.

9.2.1. *Contact-rehydration*

When slugs are dehydrated to the threshold level of 65–70% IBW they display contact-rehydration, during which they move on to a moist pad, assume a characteristic flattened posture and initiate rapid absorption of water (i.e. 7·8 μl/cm^2 per min) through the integument of the foot (Prior, 1984). The rapid absorption of water involves bulk flow of water and solutes through an epithelial paracellular pathway (Prior and Uglem, 1984; Uglem *et al.*, 1985). The rate of water absorption has been found to be quite constant regardless of the size of slug or the level of dehydration. Once a slug has rehydrated to an upper set-point of 93·6 ± 12·2% IBW it moves off the moist surface, thus terminating the behaviour.

Contact-rehydration appears to be controlled by a dual-limit set-point system in which the behaviour is initiated when a slug dehydrates to the lower threshold level and the behaviour is terminated once the slug is rehydrated to the upper set-point. When body hydration is maintained within the tolerable range (approximately 70–100% IBW) contact-rehydration is not initiated.

Contact-rehydration appears to be mediated by the increases in haemolymph osmolality that accompany dehydration. Injection of fully-hydrated slugs with 1·0 M mannitol to increase the haemolymph osmolality to that which occurs during dehydration initiates the behaviour. In addition, dilution of the haemolymph of a dehydrated slug can inhibit expression of the behaviour (Prior, 1984; Prior and Uglem, 1984).

9.2.2. *Neuropeptide control*

Makra and Prior (1985) have shown that mildly dehydrated *Limax* (i.e. 80% IBW), which normally do not exhibit contact-rehydration, exhibit both the behavioural posturing and the rapid absorption of water when injected with 5×10^{-5} M AII (final haemolymph concentration). This effect can be blocked by injection of 5×10^{-5} M saralasin, an inhibitor of AII binding. In addition, injection of fully hydrated slugs with 5×10^{-5} M AII increases integumental water absorption but does not initiate behavioural posturing. Other octapeptides, (AVT, AVP, neurotensin) caused a similar increase in integumental water uptake in fully hydrated slugs, but had no effect on the behavioural posturing of contact-rehydration. Although authentic AII has not been localized in *Limax*, an AVT-like substance has been identified (Sawyer *et al.*, 1984). These results suggest that an AII-like substance may be involved in certain aspects of the regulation of contact-rehydration in *Limax*. The incomplete initiation of contact-rehydration by AII is typical of the effects of injected peptides in gastropods. This may mean that control of these behavioural events involves multiple peptide systems.

9.3. **Locomotor activity**

Dainton (1954) reported that slugs rapidly dehydrated to approximately 80% IBW demonstrated a reduction in overall activity. If, however, slugs are slowly dehydrated on activity wheels, they exhibit marked increases in both the duration and intensity of locomotor activity (Hess and Prior, 1985). As in the case of contact-rehydration, this response is apparently mediated by the increase in haemolymph osmolality that occurs during dehydration. Recently, a similar increase in locomotor activity has been observed in *Limax* during acute dehydration stress (Welsford *et al.*, 1990). Since locomotion results in an increase in water loss due to deposition of a mucus trail during locomotion, the increase in locomotor activity in dehydrating slugs would seem to be somewhat paradoxical. It is likely, however, that the activity may be a trade-off that enhances the probability of dehydrating animals being able to locate moist microhabitats.

When *Limax* are dehydrated to approximately 70% IBW, they demonstrate positive water orientation in a Y maze and a general increase in locomotor activity (Banta *et al.*, 1990). Interestingly, this behaviour is apparently not mediated solely by alterations in haemolymph osmolality or haemolymph volume. Thus, the control of this water regulatory behaviour probably involves a combination of mechanisms. The dehydration-induced increase in locomotor activity, and the observed increase in preference for moist areas, appear to be water regulatory responses which enhance an animal's ability to locate moist microhabitats.

9.3.1. *Neuropeptide control*

Injection of fully hydrated *Limax* with AVT elicits both an increase in locomotor activity and water orientation behaviour (Banta *et al.*, 1990).

This effect is dose-dependent, with a threshold concentration of approximately 10^{-7} M AVT. The effects of dehydration and AVT are additive, in that injection of 10^{-7} M AVT into slugs dehydrated to only 80% IBW elicits water orientation (Banta *et al.*, 1990). The effects of AVT are specific in that other octapeptides (AII and AVP) do not elicit water orientation. However, AII and AVP do cause increases in locomotor activity in animals dehydrated to 80% IBW, without a concomitant increase in water orientation. Thus, while AVP and AII may be involved in the modulation of locomotor activity, the expression of water orientation behaviour appears to require an AVT-like substance.

As described above, injection of mildly dehydrated slugs with 10^{-6} M AII or AVT causes significant increases in locomotor activity, while AVT is capable of increasing locomotor activity even in fully hydrated animals.

9.4. **Feeding behaviour**

Feeding in slugs such as *Limax pseudoflavus* and *L. maximus* is very stereotyped. During the appetitive phase an animal ceases locomotion near a food source, retracts its superior tentacles, everts its lips, and begins rhythmically moving the lips over the surface of the food. In the consummatory phase, rhythmic protraction and retraction of the muscular buccal mass occurs. The cycles of protraction and retraction move the radula over the surface of the food, thereby scraping away food particles which are then ingested (Gelperin *et al.*, 1978). The cyclical radular movements are driven by a discrete pattern of efferent activity from the buccal ganglia known as the feeding motor programme (FMP, see Fig. 9.1; Gelperin *et al.*, 1978).

In *Limax* the feeding motor programme can be initiated by activation of chemosensory pathways by application of food to the lips or electrical stimulation of the lip nerves. In Fig. 9.1 potato extract was applied to the lips of an isolated lip–CNS preparation. Approximately 30 cycles of alternating efferent activity was recorded, which corresponds to the alternate protraction (BR2) and retraction (BR1) of the radula during feeding.

One of the more apparent units in the recording from the salivary nerve (N) in Fig. 9.1 is the fast salivary burster neuron (FSB; Prior and Gelperin, 1977). These bilateral endogenously active motoneurons innervate the paired salivary ducts. Each burst of activity from an FSB causes forceful ejection of saliva into the oesophagus. As seen in Fig. 9.1, during the FMP the fast salivary burster becomes entrained to the late protraction phase. While the FMP is silent the salivary bursters maintain their own endogenous patterns of activity, during which they are seldom synchronous.

When slugs are dehydrated to 65–70% IBW there is a decrease in

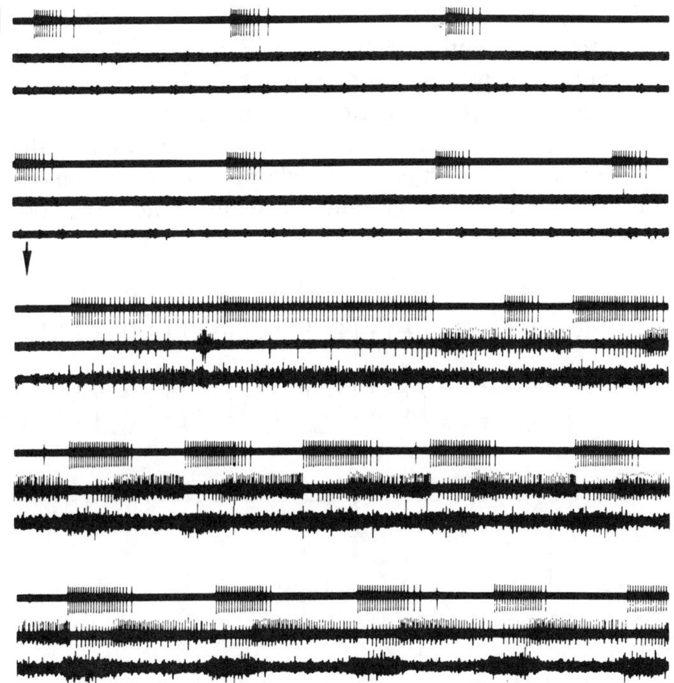

Fig. 9.1. Activation of the feeding motor programme by chemical stimulation (arrow) of the lips in an isolated lip–CNS–buccal ganglion preparation. Recordings from a salivary nerve (SN) show the activity of the fast salivary burster neuron. Recordings from nerves BR2 and BR1 show the efferent activity that underlies the retraction and protraction phases of the FMP respectively (from Prior and Gelperin, 1977).

feeding responsiveness. This decrease is mediated by alterations in haemolymph osmolality (Prior, 1983). There is a similar decrease in the responsiveness of the feeding motor programme recorded from isolated CNS preparations exposed to hyperosmotic saline (Phifer and Prior, 1985). Thus, both intact animals and isolated CNS preparations show a decrease in feeding responsiveness when exposed to an osmotic stress.

9.4.1. *Control by SCPs*

The small cardioactive peptides, SCP_A and SCP_B, increase the responsiveness of the FMP in isolated CNS preparations (Prior and Watson, 1987). It has recently been shown that numerous aspects of the neural networks controlling feeding in gastropods are sensitive to SCP_B and SCP_A (Lloyd, 1978, 1982). For example in *Tritonia* (Willows and Lloyd, 1983), *Aplysia* (see Lloyd, 1982) and *Helisoma* (Murphy *et al.*, 1985) application of 10^{-6} M SCP_B to isolated CNS preparations increases or initiates patterned efferent activity. In *Limax*, SCP_B seldom initiates the FMP but, at concentrations of 10^{-7} or 10^{-6} M, does increase the responsiveness of the central pattern generator (Prior and Watson, 1987). As shown in Fig. 9.2, in the presence of 10^{-6} M SCP_B an otherwise subthreshold stimulation of an isolated CNS preparation is capable of initiating a full feeding motor programme. Thus in *Limax*, low concentrations

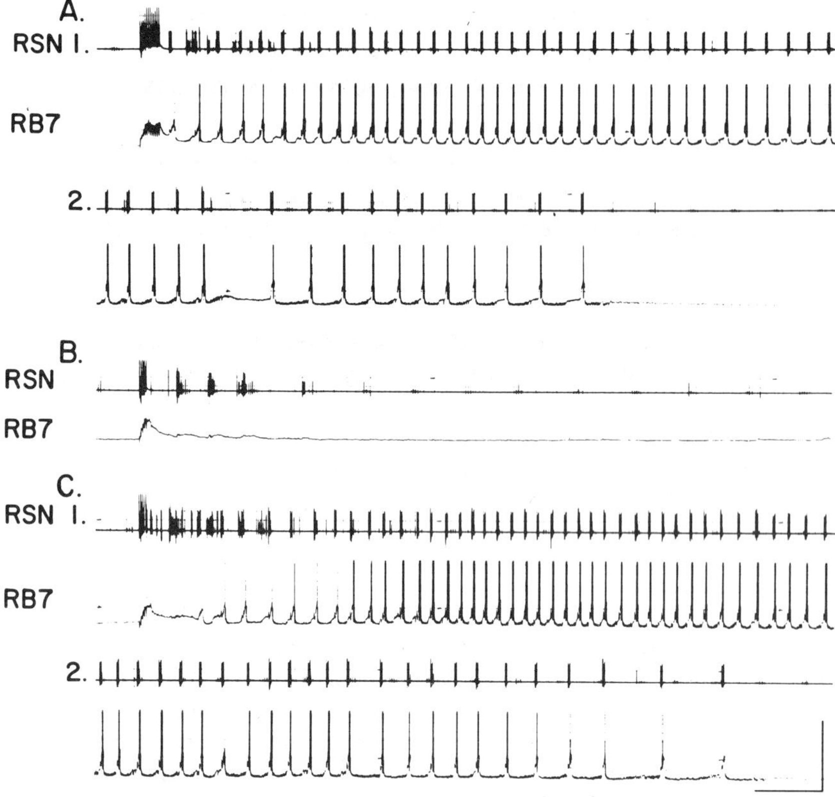

Fig. 9.2. Simultaneous extracellular recordings from the the right salivary nerve (RSN) and intracellular recordings from a protractor motoneuron (RB7) during an FMP. (**A**) 1–2: a train of electrical shocks applied to an external lip nerve initiated an FMP of 37 cycles. The RB7 bursts, each of which was composed of four to seven impulses, were separated by a series of high-frequency IPSPs. (**B**) A subthreshold (fewer stimuli) train of stimuli was applied to the external lip nerve. Although three or four bursts of salivary nerve activity and the corresponding compound EPSPs in RB7 were observed, an FMP was not initiated. After 20 min the preparation was bathed in 2×10^{-6} M SCP_B, and the same subthreshold stimulation used in **B** was repeated. In this case a full FMP was initiated. The calibration bar = 30 (**A**,**B**,**C**) and 60 mV (**C**) (from Prior and Watson, 1987).

of SCP_B can modulate the responsiveness of the central network underlying feeding.

The responsiveness of the fast salivary burster neurons to SCP_B has also been examined. As seen in Fig. 9.3 SCP_B increases the frequency of the FSB in a dose-dependent manner, with an apparent threshold of 10^{-8} M. This increase is not due simply to a depolarization of the neuron, but rather to an increase in the rate of interburst depolarization (Hess and Prior, 1989).

In intact *Limax*, injection of 10^{-5} M SCP_B can initiate the entire sequence of behaviours comprising the initial phase of feeding in the absence of food (Table 9.1). In addition, SCP_B has an excitatory effect on peripheral effectors such as the gut and heart, whose activity in *Limax* has been shown to be affected by feeding (Grega and Prior, 1985; Welsford and Prior, 1987, 1990). It has been shown that feeding behaviour in *Limax* is followed by an increase in heart activity (Grega and Prior, 1985). This co-activation of a central neural network and a peripheral effector organ is representative of integration of multiple physiological systems. As such, this co-regulation of the responsiveness of the FMP and the cardiovascular system has been examined. Using isolated heart or heart–CNS preparations, it was determined that SCP_B increases the force of contraction of the heart in a dose-dependent manner (Fig. 9.4; Welsford and Prior, 1987, 1990). Remarkably the apparent threshold concentration for the increase in heart function (e.g. 10^{-8} M) is very similar to that for increasing the frequency of the FSB (see Fig. 9.3).

9.5. **Buccal neuron B1**

Immunohistochemical studies have demonstrated the existence of buccal neurons which contain SCP_B-like immunoreactive substance (SLIM; Prior and Watson, 1987). A pair of prominent bilaterally positioned buccal neurons (RB1, LB1) were found to be especially reactive to the SCP_B antibody. Thus it was possible that these SLIM-containing neurons could be involved in the SCP_B sensitive coordination of the feeding network and associated modification of heart function.

Intracellular recordings were made from B1 in isolated CNS–heart preparations. The activity of B1 was increased to varying frequencies by intracellular stimulation, while the burst frequency of the ipsilateral FSB was monitored. When B1 was activated at 4–5 impulses/second there was a significant increase in FSB frequency (Fig. 9.5). The excitation of the FSB by B1 stimulation had a delay of several seconds, and continued for several seconds after B1 stimulation had ceased. Similarly there was excitation of the heart following B1 stimulation, again with a threshold frequency of about 4–5 impulses/second (Fig. 9.6). These observations are consistent with the hypothesis that B1 is a multifunctional interneuron

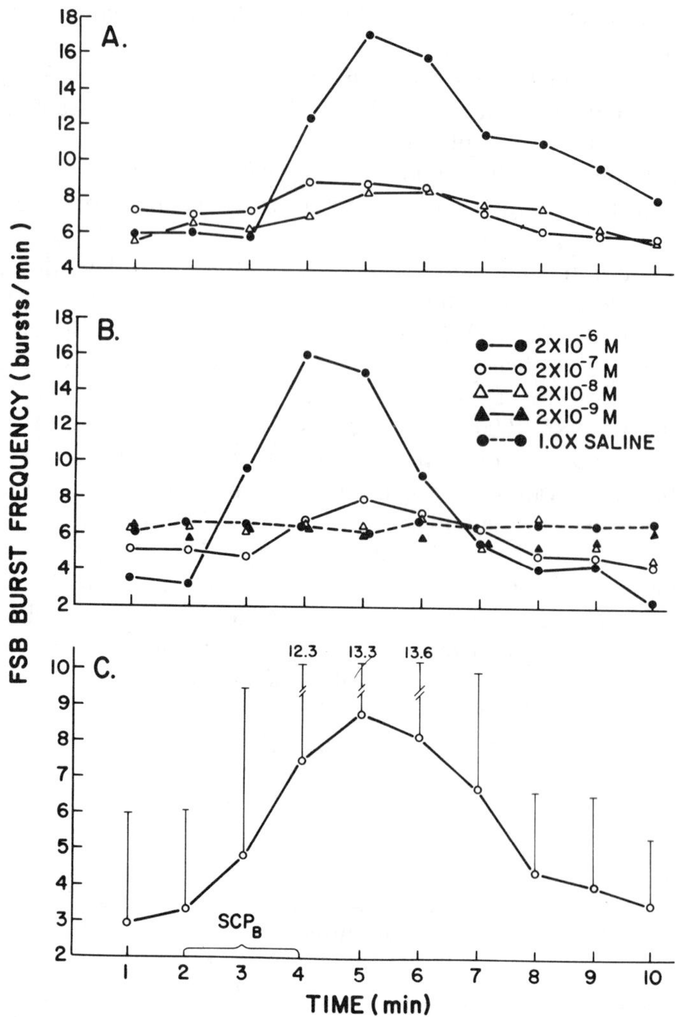

Fig. 9.3. The responses of the fast salivary burster neuron (FSB) to varying concentrations of SCP_B are presented by plotting burst frequency as a function of time during the experiment. In each case SCP_B was superfused over an isolated buccal ganglion–brain preparation between minutes 2 and 4. The preparation was superfused with saline for 20 min between each trial. (**A**) The responses obtained in three trials with the same preparation using various concentrations of SCP_B. Each point represents the burst frequency of the FSB in the preceding 60 s. (**B**) The responses of a second preparation to SCP_B. In this case four different concentrations of SCP_B were used as well as a control saline trial. (**C**) The extent of the variability between preparations is illustrated by plotting the mean ($\pm SD^u$) burst frequency at each time point for 29 trials in 12 preparations during exposure to 2×10^{-6} M SCP_B (from Prior and Watson, 1987).

Table 9.1. Behavioural effects of SCP_B injections in *Limax maximus*

		Behavioural observations		
Treatment	*Locomotion (%)*	*Tentacular retraction (%)*	*Lip eversion (%)*	*Lip movement (%)*
Saline	83	8·0	0·0	0·0
10^{-7} mol l^{-1} SCP_B	83	25	17	8·0
10^{-6} mol l^{-1} SCP_B*	92	33	42	25
10^{-5} mol l^{-1} SCP_B**	17	83	100	58

The percentage of animals that displayed each behaviour is presented in each case.
Each animal received injections of each concentration of SCP_B and the saline control.
The 0·05 probability level was accepted as significant (determined by a Friedman's test and a non-parametric multiple comparisons procedure).
All concentrations of SCP_B are the calculated final haemolymph concentrations.
The results with 10^{-5} and 10^{-6} mol l^{-1} SCP_B injection were significantly different from those with injection of control saline and 10^{-7} mol l^{-1} SCP_B. Furthermore, the results with 10^{-5} mol l^{-1} SCP_B injection were significantly greater than those observed with injection of 10^{-6} mol l^{-1} SCP_B.
* $P < 0{\cdot}05$, ** $P < 0{\cdot}01$. N = 12.
From Shagene *et al.*, 1988.

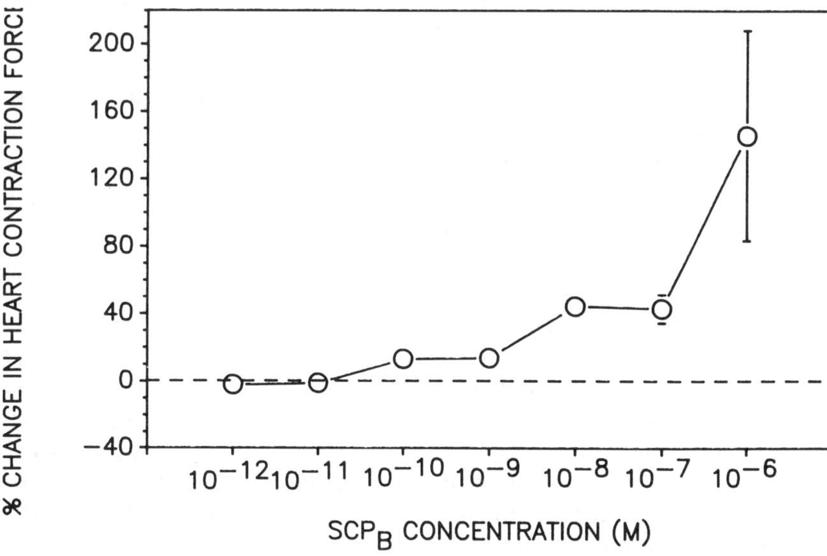

Fig. 9.4. The effect of SCP_B at various doses on the force of heart contractions in isolated heart preparations. Each point is the mean (±SD) of nine preparations. Changes in the force of contraction were normalized to a percentage of the pretreatment (100%) force (from Welsford and Prior, 1990).

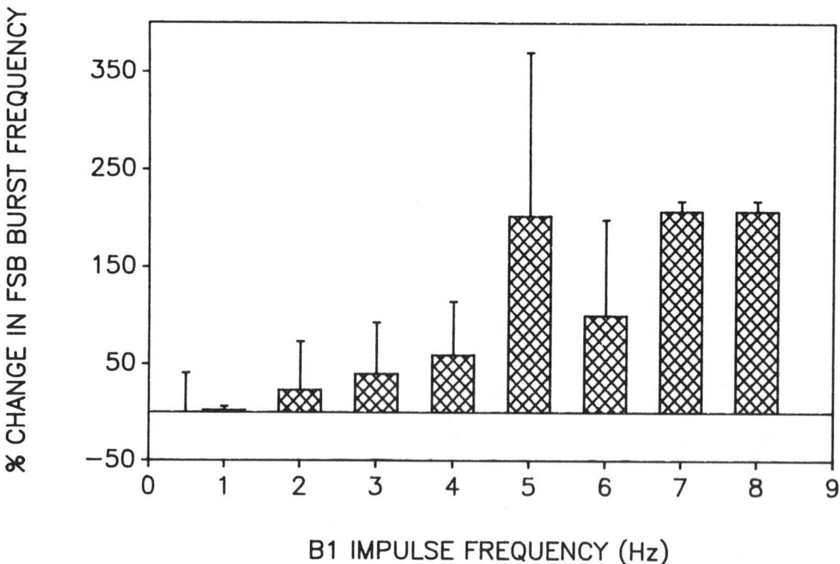

Fig. 9.5. The burst frequency of the fast salivary burster (FSB) in an isolated buccal ganglion–brain preparation of *Limax maximus* is plotted as a function of impulse frequency of a single buccal neuron, B1. B1 was stimulated by pulses of injected current so that 10 impulses at the various frequencies were elicited (from Prior, 1989).

which is involved in integration of both peripheral and central targets. In this case the activity of the central pattern generator for feeding and the cardiovascular system are simultaneously modified. Such a multifunctional peptidergic neuron could serve to simultaneously coordinate the responsiveness of physiological and behavioural functions.

9.6. **Conclusions**

In certain cases the role of particular neuropeptides in the control of water regulatory behaviours is quite specific. For example, integumental water uptake is increased by AII, AVT, AVP and neurotensin, although AII is the only one of these peptides which can initiate the complete array of responses underlying contact-rehydration. In addition, AVT, AII and AVP each cause increases in locomotor activity, although AVT is the only peptide that elicits water orientation.

Thus although certain behavioural responses can be elicited by specific peptides, it appears that the control of most of the behaviours studied involves multiple peptide systems. This complexity is furthered by the finding that, in *Limax*, buccal neuron B1 contains a second peptide (e.g. FMRFamide) that has a strong inhibitory effect on both the feeding

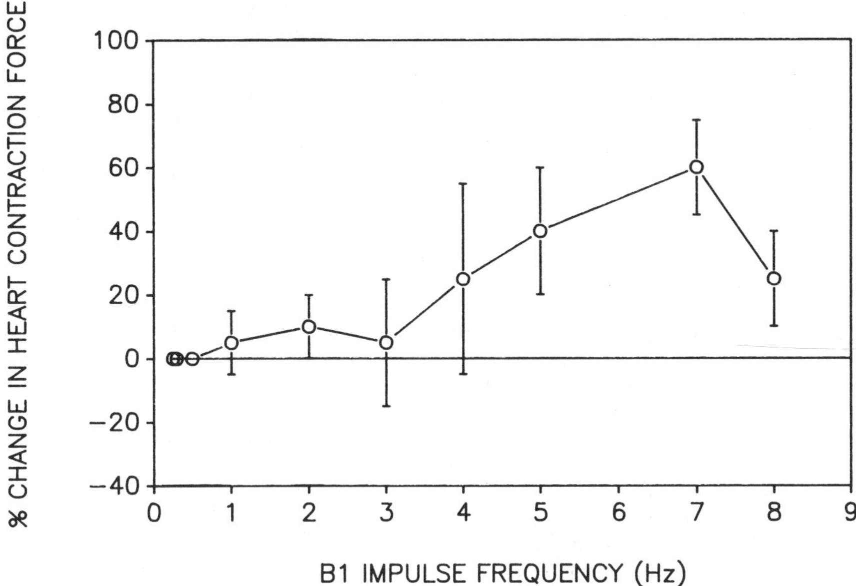

Fig. 9.6. Cumulative frequency response curve for the effect of B1 stimulation on heart contractile force in *Limax maximus*. Data are from nine heart/CNS preparations. Single B1 neurons were driven at each instantaneous spike frequency in each preparation. The number of action potentials was held constant at 10 for each frequency. Changes in heart contractile force are expressed as a percentage of the prestimulation value within each preparation (from Welsford and Prior, 1990).

network and the heart (Welsford and Prior, 1990). Thus this single multifunctional neuron has both central and peripheral targets and contains both excitatory and inhibitory neuropeptides. Clarification of the role of multiple peptide containing neurons in the simultaneous regulation of central and peripheral targets will be an important issue in future work.

Acknowledgements

The work reviewed in this paper was supported in part by the National Science Foundation, National Institutes of Health and the Arizona Disease Control Research Commission. This is contribution number 222 from the Tallahassee, Sopchoppy Gulf Coast Marine Biological Association.

References

Banta, P. A., Welsford, I. G. and Prior, D. J. (1990). Water orientation behavior in the terrestrial gastropod *Limax maximus*: the effects of dehydration and arginine-vasotocin (AVT). *Physiol. Zool.*, **63(4)**, 683–96.

Dainton, B. H. (1954). The activity of slugs. I. The induction of activity by changing temperatures. *J. Exp. Biol.*, **31**, 165–87.

Dickinson, P. S., Prior, D. J. and Avery, C. (1988). The pneumostome closure rhythm in slugs: a response to dehydration controlled by hemolymph osmolality and peptide hormones. *Comp. Biochem. Physiol.*, **89A**(4), 579–86.

Gelperin, A., Chang, J. J. and Reingold, S. C. (1978). Feeding motor program in *Limax*. I. Neuromuscular correlates and control by chemosensory input. *J. Neurobiol.*, **9**, 285–300.

Grega, D. S. and Prior, D. J. (1985). The effects of feeding on heart activity in the terrestrial slug, *Limax maximus*: central and peripheral control. *J. Comp. Physiol.*, **156A**, 539–45.

Hess, S. D. (1982). Dehydration-induced changes in body volume, hemolymph osmolality, and activity of identified pedal ganglion neurons in the Terrestrial slug, *Limax maximus*. M.S. thesis, University of Kentucky.

Hess, S. D. and Prior, D. J. (1985). Locomotor activity of the terrestrial slug, *Limax maximus*: response to progressive dehydration. *J. Exp. Biol.*, **116**, 323–30.

Hess, S. D. and Prior, D. J. (1989). Small cardioactive peptide B modulates feeding motoneurons in *Limax maximus. J. Exp. Biol.*, **142**, 473–8.

Lloyd, P. E. (1978). Distribution and molecular characteristics of cardioactive peptides in the snail, *Helix aspersa. J. Comp. Physiol.*, **128A**, 269–76.

Lloyd, P. E. (1982). Cardioactive neuropeptides in gastropods. *Fed. Proc.*, **41**, 2948–52.

Machin, J. (1964). The evaporation of water from *Helix aspersa*. I. The nature of the evaporating surface. *J. Exp. Biol.*, **41**, 759–69.

Makra, M. E. and Prior, D. J. (1985). Angiotensin II can initiate contact-rehydration in terrestrial slugs. *J. Exp. Biol.*, **119**, 385–8.

Murphy, A. D., Lukowiak, K. and Stell, W. K. (1985). Peptidergic modulation of patterned motor activity in identified neurons of *Helisoma. Proc. Natl. Acad. Sci. USA*, **82**, 140–4.

Phifer, C. B. and Prior, D. J. (1985). Body hydration and haemolymph osmolality affect feeding and its neural correlate in the terrestrial gastropod, *Limax maximus. J. Exp. Biol.*, **118**, 405–21.

Prior, D. J. (1981). Hydration-related behaviour and the effects of osmotic stress of motor function in the slugs. *Limax maximus* and *Limax pseudoflavus*. In: Salanki, J. (ed.), *Advances in Physiological Sciences*, Vol. 23: *Neurobiology of Invertebrates*. Pergamon Press, Oxford, pp. 131–45.

Prior, D. J. (1983). Hydration-induced modulation of feeding responsiveness in terrestrial slugs. *J. Exp. Zool.*, **227**, 15–22.

Prior, D. J. (1984). Analysis of contact-rehydration in terrestrial gastropods: osmotic control of drinking behaviour. *J. Exp. Biol.*, **111**, 63–73.

Prior, D. J. (1985). Water-regulatory behaviour in terrestrial gastropods. *Biol. Rev.*, **60**, 403–24.

Prior, D. J. (1989). Neuronal control of osmoregulatory responses in gastropods. In: Gilles, R. (ed.), *Advances in Comparative and Environmental Physiology*. Springer Verlag, New York.

Prior, D. J. and Gelperin, A. (1977). Autoactive molluscan neuron: reflex function and synaptic modulation during feeding in the terrestrial slug, *Limax maximus. J. Comp. Physiol.*, **114**, 217–32.

Prior, D. J. and Uglem, G. L. (1984). Analysis of contact-rehydration on terrestrial gastropods. Absorption of ^{14}C-inulin through the epithelium of the foot. *J. Exp. Biol.*, **111**, 75–80.

Prior, D. J. and Watson, W. H. (1987). The molluscan neuropeptide, SCP_B, increases the responsiveness of the feeding motor program of *Limax maximus*. *J. Neurobiol.* **19**(1), 87–105.

Prior, D. J., Hume, M., Varga, D. and Hess, S. D. (1983). Physiological and behavioural aspects of water balance and respiratory function in the terrestrial slug, *Limax maximus*. *J. Exp. Biol.*, **104**, 111–27.

Riddle, W. A. (1983). Physiological ecology of land snails and slugs. In: Russell-Hunter, W.D. (ed.), *The Mollusca*, Vol. 6: *Ecology*. Academic Press, New York, pp. 431–61.

Sawyer, W., Deyrup-Olsen, I. and Martin, A. W. (1984). Immunological and biological characteristics of the vasotocin-like activity in the head ganglia of gastropod molluscs. *Gen. Comp. Endocrinol.*, **54**, 97–108.

Shagene, K. A., Welsford, I. G., Prior, D. J. and Banta, P. A. (1988). Injection of small cardioactive peptide (SCP_B) can initiate the appetitive phase of feeding in the slug, *Limax maximus*. *J. Exp. Biol.*, **143**, 553–7.

Treherne, J. E. (1980). Neuronal adaptations to ionic and osmotic stress. *Comp. Biochem. Physiol.*, **67B**, 455–63.

Uglem, G. L., Prior, D. J. and Hess, S. D. (1985). Analysis of contact-rehydration in terrestrial gastropods: estimation of the pore size and molecular sieving of the integumental paracellular pathway. *J. Comp. Physiol.*, **156**, 285–9.

Welsford, I. G. and Prior, D. J. (1987). The effect of SCP_B application and buccal neuron, B1, stimulation on heart activity in the slug, *Limax maximus*. *Am. Zool.*, **27**(4), 138A.

Welsford, I. G. and Prior, D. J. (1990). Modulation of heart activity in the terrestrial slug, *Limax maximus* by the feeding motor program, the small cardioactive peptides and buccal neuron B1 stimulation. (In press).

Welsford, I. G., Banta, P. A. and Prior, D. J. (1990). Size-dependent responses to dehydration in the terrestrial slug, *Limax maximus*: locomotor activity and huddling behavior. *J. Exp. Zool.*, **253(2)**, 229–34.

Willows, A. O. D. and Lloyd, P. E. (1983). Synthetic SCP_B elicits patterned neural feeding activity from buccal ganglia of *Tritonia*. *Soc. Neurosci. Abst.*, **9**, 386.

10 *Cornelis Janse, W. C. Wildering and Marcel van der Roest*

Peptidergic neurons in the ageing molluscan brain

10.1. Introduction

Neuropeptides occur in a wide diversity and have a great variety of actions on the nervous system (cf. other chapters in this volume; de Kloet *et al.*, 1987). In many cases these substances exert wide-acting effects because they are released either at many sites in the nervous system, or into the interneuronal space or the blood system. Consequently, changes in peptidergic neurons may have profound effects on the functioning of the nervous system. Such changes may thus contribute significantly to ageing processes in the nervous system. At present there is an increasing interest in the role of neurohormonal factors, including neuropeptides, in ageing processes in the nervous system (Landfield, 1987). This chapter deals with peptidergic neurons in the ageing pond snail *Lymnaea stagnalis*.

10.2. Ageing in molluscs

The occurrence of ageing in an animal population is characterized by an increase of the age-specific death rate with age (Comfort, 1979). In natural populations the death rate is in general independent of age. In such populations the number of survivors decreases exponentially with age (Fig. 10.1A). Many animal populations maintained in laboratory conditions show an age-related increase in the age-specific death rate. In these populations the survival curve becomes more rectangular as ageing gains importance as a death factor (Fig. 10.1B). This demonstrates that ageing is a generally occurring phenomenon in the animal kingdom.

Pulmonates are molluscs which can be cultured in the laboratory relatively easily (cf. Janse and Joosse, 1989) and a number of survival studies on these animals are known (Comfort, 1979). Recently survival studies in *Lymnaea* kept under standard and well-controlled laboratory conditions demonstrated that the age-specific death rate increases with age (Slob and Janse, 1988) indicating that ageing occurs in this species. In culture conditions (Janse *et al.*, 1988; Slob and Janse, 1988) the survival curve in this species resembles the rectangular survival curve of ageing animals (Fig. 10.2).

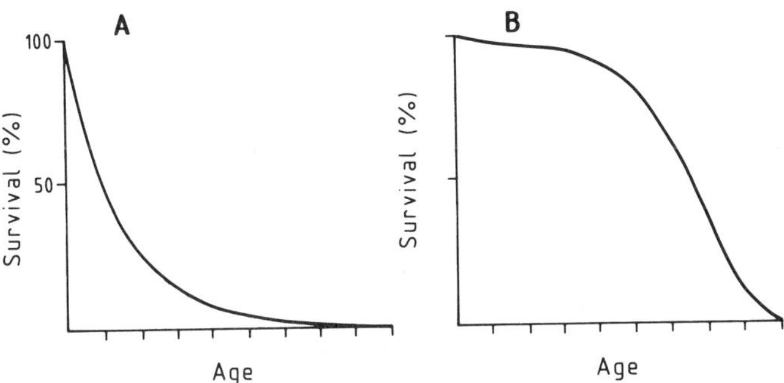

Fig. 10.1. Survival in natural (**A**) and laboratory (**B**) conditions (after Comfort, 1979).

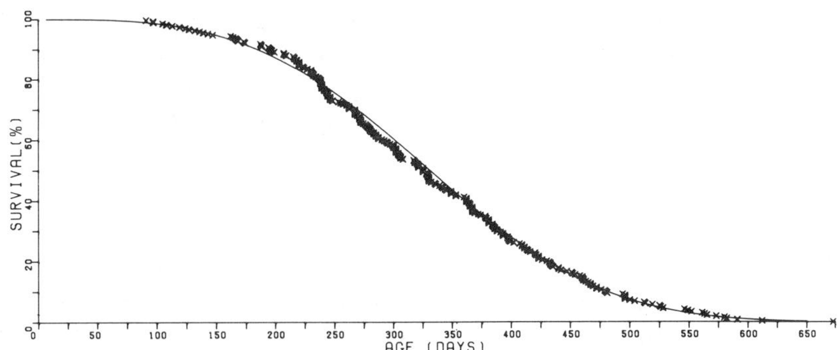

Fig. 10.2. Survival of *Lymnaea stagnalis* in laboratory conditions. The continuous line is drawn according to a mathematical model of survival (after Slob and Janse, 1988). Mortality increases at an age of about 8 months, the median life span is reached at about 12 months and maximum life span at about 22 months.

10.3. **Peptidergic neurons in molluscs**

In molluscs, as in vertebrates, neuropeptides have a wide variety of functions. Neuropeptides in molluscs play a role in neuroendocrine control systems (Joosse, 1986), in modulation of neuronal networks underlying behaviour (Murphy *et al.*, 1985; van der Wilt *et al.*, 1988), in the induction of complex behaviours such as egg-laying in *Lymnaea* and *Aplysia* (Geraerts *et al.*, 1988), in synaptic control of muscles (Cottrell *et al.*, 1983, Schot *et al.*, 1983) or neurons (Boer *et al.*, 1984) and in neurite outgrowth and synapse formation (Bulloch, 1987). The molluscan central nervous system (CNS) contains a great number of identifiable peptidergic neurons. This makes molluscs suitable for the study of age-related changes in peptidergic systems, and for studying the consequences of these changes for the functioning of the nervous system.

10.4. The fate of peptidergic neurons in the *Lymnaea* brain during ageing

In the CNS of the freshwater snail *Lymnaea stagnalis* a great number of different peptidergic systems have been described (Boer and van Minnen, 1988; Joosse, 1986). Many of the peptides contained in the *Lymnaea* nervous system are, immunocytochemically, vertebrate-like (Boer and van Minnen, 1988). In *Lymnaea* several peptidergic systems are at present the subject of morphological, physiological, behavioural and molecular studies (Joosse, 1986; Boer and van Minnen, 1988; Geraerts *et al.*, 1988; van der Wilt *et al.*, 1988). Ageing studies in *Lymnaea* are at present focused on two different peptidergic systems. One system, the caudo dorsal cell (CDC) system consists of many neurons and is involved in control of female reproductive activity (Geraerts *et al.*, 1988). The second system consists of two neurons (VD1 and RPD2*) and is involved in modulation of respiratory functions (van der Wilt *et al.*, 1988). The CDCs constitute a peptidergic system which is periodically active, whereas activity of VD1 and RPD2 is modulated by input related to the external oxygen level around the animal.

10.4.1. *The CDC system*

The CDCs are located in two bilateral clusters of 50 electrotonically coupled cells in the cerebral ganglia. Their neurohaemal area is located in the cerebral commissure (Fig. 10.3A). CDCs are normally silent in mature animals; however, periodically, just prior to ovulation and egg-laying, the CDCs become electrically active (ter Maat *et al.*, 1986). During electrical activity they secrete a hormone (CDCH) and multiple other peptides into the haemolymph (Geraerts *et al.*, 1985). This induces a stereotyped pattern of behaviours which terminates in the deposition of an egg mass (Fig. 10.3B) (see also Geraerts *et al.*, 1988).

In *Lymnaea* longitudinal studies (Janse *et al.*, 1989) showed that after an initial increase, egg-laying activity starts to decrease at an age of about 250 days. Figure 10.4 shows that in a laboratory culture of about 500 days most animals have ceased egg-laying activity. Cross-sectional studies in which groups of animals were studied simultaneously confirmed this. Measurements on isolated animals showed that the decline in egg-laying activity is composed of a gradual decrease of the egg-laying frequency of individual animals and a gradual decrease of the percentage of animals that lay eggs. Similarly, the initial increase in egg-laying activity in young animals consisted of a gradual increase in frequency of egg-laying activity

* VD1 and RPD2 refer to identified neurons which are named according to their position in the CNS of *Lymnaea* (Benjamin and Winlow, 1981).

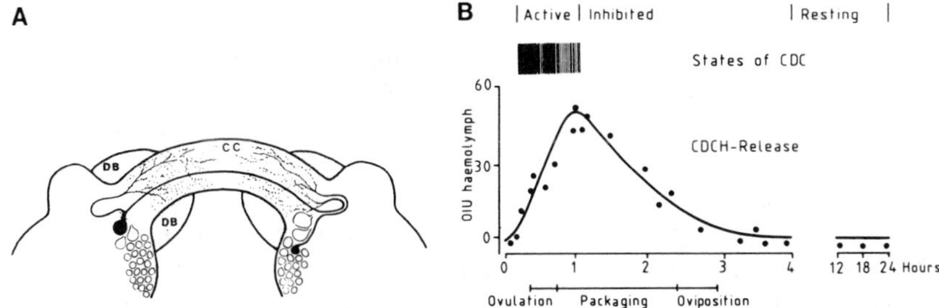

Fig. 10.3. **A**: Ventral view of the cerebral ganglia of *Lymnaea* with details of the morphology of two different types of CDC's. **B**: Interrelations of electrical activity of CDCs (states), concentration of CDCH (indicated as number of ovulation-inducing units, OIU) in the blood and egg-laying (after Janse and Joosse, 1989). DB: dorsal body; CC: cerebral commissure.

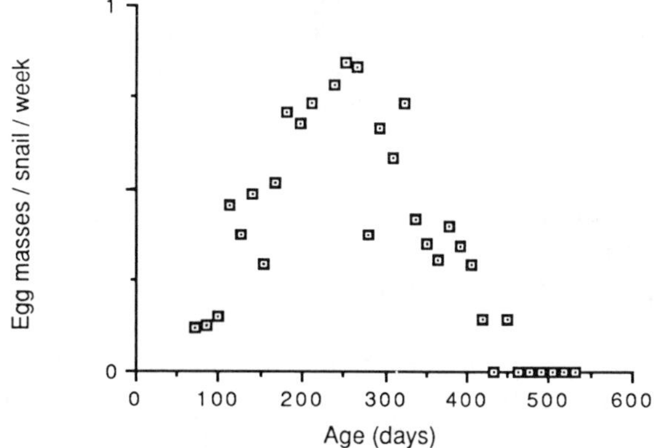

Fig. 10.4. Egg-laying activity of a *Lymnaea* culture as plotted against age.

of individuals and an increase of the number of animals that start female reproduction. These gradual changes indicate that transition from the reproductive state into the postreproductive state, as well as maturation in *Lymnaea*, occur gradually and with great individual variability. This resembles the situation in vertebrates (Finch *et al.*, 1984; Steger *et al.*, 1981).

Recent experiments (unpublished) showed that implantation of cerebral ganglia of young animals including the dorsal bodies* (DB) enhanced oviposition frequency in old animals. This indicates that the decrease in oviposition frequency originates in higher control centres. In some mam-

* In the pond snail *Lymnaea stagnalis* only one endocrine organ is known, the paired dorsal bodies (DB). The DB are located on the cerebral ganglia, and control oocyte growth and maturation and secretory activity of the female accessory sex organs (cf. Joosse, 1988).

mals reproductive cessation has been related to changes in neuroendocrine control systems which in turn probably also originate from changes in higher brain centres (Finch, 1976; Finch *et al.*, 1984; Meites, 1988).

Recent physiological experiments confirmed the idea that cessation of egg-laying is due to changes in higher control centres. Neurophysiological experiments showed that CDCs of senescent postreproductive snails can still be electrically activated. Moreover, injections of homogenates of CDCs of old postreproductive animals showed that these CDCs still contain a substance which can induce ovulation in young reproductive snails. Finally, senescent postreproductive snails still lay eggs upon injection with synthetic CDCH (Janse *et al.*, 1990). Obviously, immediately after the start of female reproductive cessation, old non-laying snails still contain CDCs which are in principle functionally intact.

It is known that chemical and electrical communication between the CDCs is of importance for synchronous activation of the CDC system (Ter Maat *et al.*, 1988). Preliminary experiments in our laboratory showed that electrical and chemical communication between CDCs of postreproductive animals still occurs. This means that inactivation of the CDC system is probably not due to interruption of communication between the CDCs. Most likely activation of the CDCs is impaired in old animals.

Morphological studies showed that in old postreproductive *Lymnaea* degeneration of CDCs occurs (Joosse, 1964; Janse *et al.*, 1987b). It is known that, in the developing vertebrate nervous system, integrity and growth of neurons depend on active use (Greenough, 1988). This may also apply to the ageing nervous system (Smith, 1984). Impairment of functioning of nerve cells due to reduced use has been demonstrated in *Aplysia* (Peretz and Zolman, 1987). It is conceivable that in *Lymnaea* degeneration of CDCs occurs after these cells have been inactive for a long period of time. It is indeed known that *Lymnaea* can live several months after reproductive cessation (unpublished observations).

In addition to the changes reported above immunocytochemical studies (Gesser and Larsson, 1985) suggested that peptide contents of CDCs change with age. CDCs of young *Lymnaea* showed enkephalin-like immunoreactivity. In *Lymnaea* of intermediate age the CDCs showed enkephalin-like and gastrin/cholecystokinin-like immunoreactivity. Old *Lymnaea*, however, showed only gastrin/cholecystokinin-like immunoreactivity in their CDCs. How age-related changes in peptide contents in CDCs relate to changes in reproductive activity of *Lymnaea* is unknown.

10.4.2. *The VD1/RPD2 system*

VD1 and RPD2 are two giant electrotonically coupled peptidergic neurons located in the visceral and right parietal ganglion, respectively.

Immunocytochemical studies suggested that the cells contain a peptide related to ACTH (Boer *et al.*, 1979). The cells branch extensively in the CNS and the peripheral nerves. Moreover, they send many fine fibres in the connective tissue sheath of the CNS. Figure 10.5A shows the main branching pattern of the cells as revealed with immunocytochemical methods and with Lucifer Yellow fillings.

These neurons play a modulatory role in the control of respiratory functions (van der Wilt *et al.*, 1987, 1988). Neurophysiological studies showed that the neurons receive input from oxygen-sensitive receptors in the mantle and lung area (Janse *et al.*, 1985). This and the extensive branching pattern of the cells led us to the idea that the neurons probably play an important, more general, role in the regulation of metabolic processes which are related to the respiratory condition of the animal.

Recent observations revealed that, in addition to common synaptic input, VD1 and RPD2 each receive separate synaptic input (Wildering *et al.*, unpublished). Integration of the different inputs of the two neurons is probably mediated by the tight electrical junction between the neurons. The junction is located in the left parieto-visceral connective (Benjamin and Pilkington, 1986). Normally the neurons indeed fire their action potentials in close synchrony (Janse *et al.*, 1986b).

With age VD1 and RPD2 increase their diameter (Janse *et al.*, 1986b). Such neuronal growth has also been observed in other molluscs (Rattan and Peretz, 1981). Neuronal growth increases the membrane surface area and thereby decreases the input resistance of the neurons (Janse *et al.*, 1986b). One of the direct consequences of input resistance decrease for a neuronal system will be a decreased sensitivity to synaptic input. In the VD1/RPD2 system this was studied by testing the sensitivity of the cells

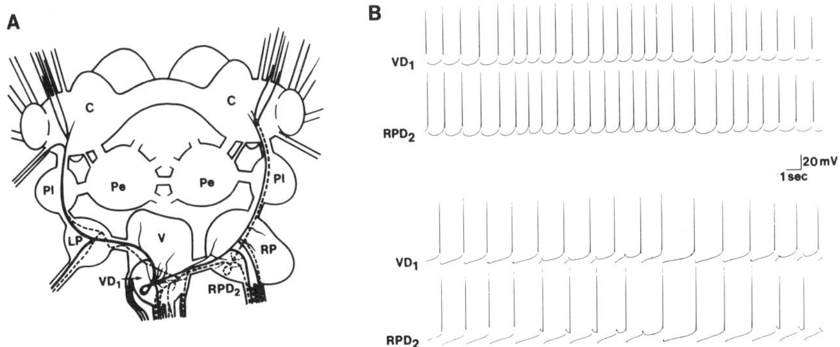

Fig. 10.5. **A**: CNS of *Lymnaea* with branching pattern of VD1 and RPD2 (after Boer *et al.*, 1979). **B**: Simultaneous recordings of the two cells in young (3 months) (upper recordings) and old (12 months) (lower recordings). Note the disturbances in synchrony in lower recordings. C, Pe, Pl, LP, RP, and V: cerebral, pedal, pleural, left and right parietal, and visceral ganglion, respectively.

to the putative transmitter histamine. Despite the decrease in input resistance, sensitivity to histamine in these cells increased with age (Wildering *et al.*, unpublished). A similar increase of sensitivity of *Lymnaea* nerve cells to putative transmitters with age have been reported by Frolkis *et al.* (1984).

In addition to the changes described above synchrony of firing of the neurons changes (Janse *et al.*, 1986b, 1987a). Figure 10.5B shows the firing pattern of the neurons in young and old animals. With age the coupling efficiency between the cells also decreases. It has been suggested that this is mainly due to an increase of junctional resistance between the neurons (Janse *et al.*, 1986b). Recent studies confirmed that the junctional resistance increases with age, but also showed that in animals of 16–20 months of age the junction between the two cells is as effective as in young animals. In these old animals the coupling of the cells is comparable to those of young animals.

Figure 10.6 shows that in old animals individual variance of the junctional resistance is very low. This may be due to an increase of mortality in animals with a high junctional resistance in conjunction with individual variability in the rate of ageing. Still another possibility is that recovery occurs in old animals. The idea that communication between VD1 and RPD2 is related to survival is in line with the idea that the cells are involved in the regulation of metabolic processes related to the respiratory condition of the animal. The second possibility, that recovery of the electrical junction between VD1 and RPD2 occurs in very old animals, might be related to hormonal changes, e.g. accompanying reproductive cessation.

Recent experiments (Wildering *et al.*, unpublished) indicate that active membrane properties of VD1 and RPD2 change with age. Taken with the changes reported above this may indicate that within one neuron several

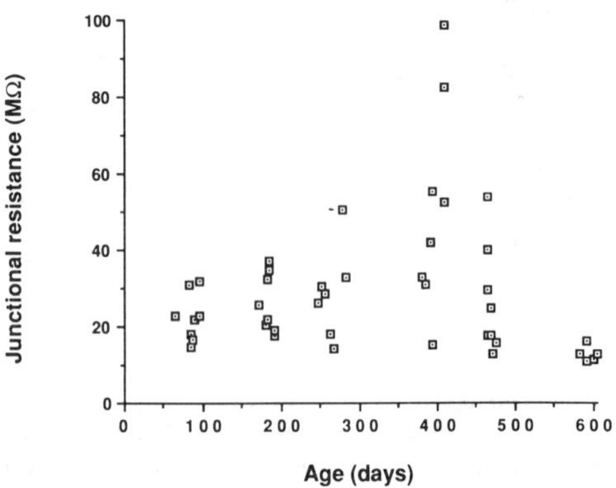

Fig. 10.6. Junctional resistance between VD1 and RPD2 as plotted against age.

properties change with age simultaneously. It is as yet unknown whether these changes have a common origin.

10.5. **Possible roles of peptides in ageing of the *Lymnaea* nervous system**

One of the theories on ageing in the nervous system is that the concentrations of humoral factors to which nerve cells are exposed change (Cuttler, 1982). Such changes may lead to a cascade of changes in the nervous system, eventually resulting in severe damage (Finch *et al.*, 1984; Sapolsky *et al.*, 1986; Landfield, 1987). These changes may consist of an increase in the concentration of a substance which damages neurons (e.g. steroids in mammals, cf. Sapolsky *et al.*, 1986). On the other hand changes may also consist of a decrease in the concentration of substances of importance to outgrowth and/or maintenance of neurons (e.g. trophic factors, cf. Landfield, 1987).

In *Lymnaea* several indications exist that changes occur with age in humoral factors, including peptides. Firstly, as reported above, female reproduction starts to cease in *Lymnaea* at an age of about 250 days. *Lymnaea* can live several months after cessation of reproduction (unpublished observations). This implies that in old *Lymnaea* CDCs are inactive and do not secrete their contents for a considerable period of time. This in turn probably means that in old postreproductive *Lymnaea* the periodic massive release of CDC peptides has ceased.

Secondly, during reproductive decline in old snails differences in sensitivity of female sex organs to CDCH also occurred (Janse *et al.*, 1990). Injections of CDCH showed that old *Lymnaea* that had ceased female reproductive activity responded less to CDCH injection than old, still-reproducing, animals of the same age. No differences were found between old and young reproductive animals. This indicates that differences in sensitivity of the female sex organs to CDCH are not related to differences in age, but to differences in sexual activity. Previously, it has been suggested that in *Lymnaea* sensitivity of female reproductive organs to CDCH is controlled by hormone secreted by the endocrine dorsal bodies (DBH) (Joosse, 1988). It is conceivable that in old animals DBH levels in the blood decrease due to deterioration of the DBs. Restoration of oviposition frequency after implantation of cerebral ganglia (including DBs) of young animals into old animals indeed occurred (Janse *et al.*, unpublished). Moreover, degeneration of DBs in old animals has been reported (Joosse, 1964). The nature of DBH is at present the subject of study in our laboratory.

One of the conspicuous features of the ageing brain both in higher animals and in man is that there is a reduced capacity to compensate for brain damage (Cotman and Schiff, 1979; Scheff *et al.*, 1984; McWilliams, 1988). Invertebrates growth capacity is influenced by hormone levels

(Goudsmit *et al.*, 1988) and age-related changes might result from changes in such levels. In pulmonate snails it has been shown that somatostatin influences axon outgrowth and synapse formation (Bulloch, 1987) and that there are neurons which are immunoreactive to somatostatin (Boer and van Minnen, 1988). Recently, in *Lymnaea* brain an insulin-like neuropeptide has been identified which has a stimulatory effect on neuronal outgrowth in culture conditions (Smit *et al.*, 1988; Kits *et al.*, 1990).

Capacity of the molluscan nervous system to repair damage has also been shown to be dependent of the age of the animal. In *Lymnaea* restoration of the tentacle-withdrawal reflex after damage of sensory input is impaired in old animals. Evidence exists that this is due to impairment of restoration of synaptic connections in the CNS (Janse *et al.*, 1986a). It is conceivable that in old *Lymnaea* levels of peptides which influence formation of nervous connections decrease. In *Aplysia* (Schacher and Flaster, 1987) indications exist that there is neuronal release of growth-promoting factors. Synapse formation between neurons isolated from old animals is less effective than those from young animals, and is stimulated by the presence of young neurons.

10.6. **Concluding remarks**

In both the vertebrate and molluscan nervous system peptides serve many functions and may have wide-acting effects. It can thus be expected that changes in peptidergic neurons during ageing may have important implications for the functioning of the brain. The molluscan CNS contains giant neurons, among which are many peptidergic neurons. This permits the combination of physiological, biochemical, morphological and molecular techniques in ageing studies on individual peptidergic neurons. This chapter showed that in *Lymnaea* properties of peptidergic neurons change with age. This may even result in cessation of peptide release. Concentration changes may also occur in humoral factors, resulting in plasticity changes of nerve cells. In view of the basic similarities between molluscan and vertebrate nervous systems (see also Janse and Joosse, 1989) ageing studies in the *Lymnaea* CNS may thus contribute to a general understanding of the role of peptides in cellular and subcellular ageing mechanisms in the nervous system.

References

Benjamin, P. R. and Pilkington, J. B. (1986). The electrotonic location of low-resistance intercellular junctions between a pair of giant neurons in the snail *Lymnaea*. *J. Physiol.*, **370**, 111–26.

Benjamin, P. R. and Winlow, W. (1981). The distribution of three wide-acting synaptic inputs to identified neurons in the isolated brain of *Lymnaea stagnalis* (L.). *Comp. Biochem. Physiol.*, **70A**, 293–307.

Boer, H. H., Schot, L. P. C., Roubos, E. W., ter Maat, A., Lodder, J. C., Reichelt, D. and Swaab, D. F. (1979). ACTH-like immunoreactivity in two electrotonically coupled giant neurons in the pond snail *Lymnaea stagnalis. Cell Tiss. Res.*, **202**, 231–40.

Boer, H. H. and van Minnen, J. (1988). Immunocytochemistry of hormonal peptides in molluscs: optical and electron microscopy and the use of monoclonal antibodies. In: Thorndyke, M. C. and Goldsworthy, G. J. (eds), *Society for Experimental Biology Seminar Series*, Vol. 33: *Neurohormones in Invertebrates*, Cambridge University Press, Cambridge, pp. 20–41.

Boer, H. H., Schot, L. P. C., Reichelt, D., Brand, H. and ter Maat, A. (1984). Ultrastructural immunocytochemical evidence for peptidergic neurotransmission in the pond snail *Lymnaea stagnalis. Cell Tiss. Res.*, **238**, 197–201.

Bulloch, A. G. M. (1987). Somatostatin enhances neurite outgrowth and electrical coupling of regenerating neurons in *Helisoma. Brain Res.*, **412**, 6–17.

Comfort, A. (1979). *The Biology of Senescence*, 3rd edn. Churchill Livingstone, Edinburgh and London.

Cotman, C. W. and Schiff, S. W. (1979). Synaptic growth in aged animals. In: Cherkin, A. *et al.* (eds), *Physiology and Cell Biology of Aging*, Vol. 8. Raven Press, New York, pp. 109–20.

Cottrell, G. A., Schot, L. P. C. and Dockray, G. J. (1983). Identification and probable role of a single neurone containing the neuropeptide *Helix*-FMRF-amide. *Nature*, **304**, 6348.

Cuttler, R. G. (1982). The dysdifferentiative hypothesis of mammalian aging and longevity. In: Giacobini, E. *et al.* (eds), *Aging*, Vol. 20: *The Aging Brain: cellular and molecular mechanisms of aging in the nervous system*. Raven Press, New York, pp. 1–19.

Finch, C. E. (1976). The regulation of physiological changes during mammalian aging. *Quart. Rev. Biol.*, **51**, 49–83.

Finch, C. E., Felicio, L. S., Mobbs, C. V. and Nelson, J. F. (1984). Ovarian and steroidal influence on neuroendocrine aging processes in female rodents. *Endocrine Rev.* **5**, 467–97.

Frolkis, V. V., Stupina, A. S., Martinenko, O. A., Tóth, S. and Timchenko, A. I. (1984). Aging of neurons in the mollusc *Lymnaea stagnalis* structure, function and sensitivity to transmitters. *Mech. Ageing Dev.*, **25**, 91–102.

Geraerts, W. P. M., Vreugdenhil, E., Ebberink, R. H. M. and Hogenes, Th. M. (1985). Synthesis of multiple peptides from a larger precursor in the neuroendocrine caudo-dorsal cells of *Lymnaea stagnalis. Neurosci. Lett.*, **56**, 241–6.

Geraerts, W. P. M., ter Maat, A. and Vreugdenhil, E. (1988). The peptidergic neuroendocrine control of egg laying behavior in *Aplysia* and *Lymnaea*. In: Laufer, H. and Downer, G. H. (eds), *Endocrinology of Selected Invertebrate Types*, Vol. II, Alan R. Liss, New York, pp. 141–231.

Gesser, B. P. and Larsson, L. (1985). Changes from enkephalin-like to gastrin/cholecystokin-like immunoreactivity in snail neurons. *J. Neurosci.*, **5**, 1412–17.

Goudsmit, E., Fliers, E. and Swaab, D. F. (1988). Testosteron supplementation restores vasopressin innervation in the senescent rat brain. *Brain Res.*, **473**, 306–13.

Greenough, W. T. (1988). The turned-on brain: developmental and adult responses to the demands of information storage. In Easter, S. S. *et al.* (eds), *From Message to Brain. Directions in developmental neurobiology*. Sinauer Assoc., Massachusetts, pp. 288–302.

Janse, C., van der Wilt, G. J., van der Plas, J. and van der Roest, M. (1985). Central and peripheral neurones involved in oxygen perception in the pulmonate snail *Lymnaea stagnalis* (Mollusca, Gastropoda). *Comp. Biochem. Physiol.*, **82A**, 459–69.

Janse, C., van Beek, A., van Oorschot, I. and van der Roest, M. (1986a). Recovery of damage in a molluscan nervous system is impaired with age. *Mech. Ageing Dev.*, **35**, 179–83.

Janse, C., van der Roest, M. and Slob, W. (1986b). Age-related decrease in electrical coupling of two identified neurones in the mollusc *Lymnaea stagnalis. Brain Res.*, **376**, 208–12.

Janse, C., van der Roest, M. Bedaux, J. J. M. and Slob, W. (1987a). The pond snail *Lymnaea stagnalis*: an animal model for aging studies at the neuronal level. In: Boer, H. H. *et al.* (eds), *Neurobiology. Molluscan Models*. North Holland, Amsterdam, pp. 335–40.

Janse, C., Wildering, W. C., van Minnen, J., van der Roest, M. and Roubos, E. W. (1987b). Ageing in the pond snail *Lymnaea stagnalis. Proc. 28th Dutch Fed. Meeting*, p. 239.

Janse, C., Slob, W., Popelier, C. M. and Vogelaar, J. W. (1988). Survival characteristics of the mollusc *Lymnaea stagnalis* under constant culture conditions: effects of aging and disease. *Mech. Ageing Dev.*, **42**, 263–74.

Janse, C. and Joosse, J. (1989). Ageing in molluscan nervous and neuroendocrine systems. In: Schreibman, M. P. and Scanes, C. G. (eds), *Development, Maturation and Senescence of Neuroendocrine Systems: A Comparative Approach*, Academic Press, New York, pp. 43–61.

Janse, C., Wildering, W. C. and Popelier, C. M. (1989). Age-related changes in female reproductive activity and growth in the mollusc *Lymnaea stagnalis. J. Gerontol.*, **44**, B148–55.

Janse, C., Maat, A. ter and Pieneman, A. W. (1990). Molluscan ovulation hormone containing neurons and age-related reproductive decline. *Neurobiol. Ageing*, **11**, 457–63.

Joosse, J. (1964). Dorsal bodies and dorsal neurosecretory cells of the cerebral ganglia of *Lymnaea stagnalis*. L. *Arch. Néerl. Zool.*, **15**, 1–103.

Joosse, J. (1986). Neuropeptides: peripheral and central messengers of the brain. In: Ralph, Ch. (ed.), *Comparative Endocrinology: Developments and Directions*. Alan R. Liss, New York, pp. 13–32.

Joosse, J. (1988). The hormones of molluscs. In: Laufer, H. and Downer, G. H. (eds), *Endocrinology of selected invertebrate types*, vol. II. Alan R. Liss, New York, pp. 89–140.

Kits, K. S., de Vries, N. J. and Ebberink, R. H. M. (1990). Molluscan insulin-related neuropeptide promotes neurite outgrowth in dissociated neuronal cell cultures. *Neurosci. Lett.*, **109**, 253–8.

Kloet, E. R. de, Wiegant, M. and de Wied, D. (1987). Neuropeptides and brain function. *Progress in Brain Research*, vol. 72. Elsevier, Amsterdam.

Landfield, P. W. (1987). Modulation of brain aging correlates by long-term alterations of adrenal steroids and neurally-active peptides. In: de Kloet, E. R. *et al.* (eds), *Progress in Brain Research*, Vol. 72: *Neuropeptides and Brain Function*, Elsevier, Amsterdam, pp. 279–300.

Maat, A. ter, Dijks, F. A. and Bos, N. P. A. (1986). *In vivo* recordings of neuroendocrine (caudo-dorsal cells) in the pond snail. *J. Comp. Physiol.*, **158**, 853–9.

Maat, A. ter, Geraerts, W. P. M., Jansen, R. F. and Bos, N. P. (1988). Chemically mediated positive feed back generates long-lasting afterdischarge in a molluscan neuroendocrine system. *Brain Res.*, **438**, 853–9.

McWilliams, J. R. (1988). Age-related decline in anatomical plasticity and axonal sprouting. In: Petit, T. L. and Ivy, G. O. (eds), *Neurology and Neurobiology*, Vol. 36: *Neural Plasticity: a life span approach*. Alan R. Liss, New York, pp. 329–49.

Meites, J. (1988). Neuroendocrine basis of aging in the rat. In: Everitt, A. V. and Walton, J. R. (eds), *Interdisciplinary Topics in Gerontology*, Vol. 24: *Regulation of Neuroendocrine Aging*. Karger, Basel, pp. 37–50.

Murphy, A. D., Lukowiak, K. L. and Stell, W. (1985). Peptidergic modulation of patterned motor activity in identified neurons of *Helisoma*. *Proc. Natl. Acad. Sci. USA*, **82**, 7140–4.

Peretz, B. and Zolman, J. F. (1987). Regular activation contributes to differential aging of motor neurons in *Aplysia*. In: Boer, H. H. *et al.* (eds), *Neurobiology. Molluscan Models*. North Holland, Amsterdam and New York, pp. 330–4.

Rattan, K. S. and Peretz, B. (1981). Age-dependent behavioral changes and physiological changes in identified neurons in *Aplysia californica*. *J. Neurobiol.*, **12**, 469–78.

Sapolsky, R. M., Krey, L. C. and McEwen, B. S. (1986). The neuroendocrinology of stress and aging: the glucocorticoid cascade hypothesis. *Endocrine Rev.*, **7**, 284–301.

Schacher, S. and Flaster, M. S. (1987) Formation of chemical synapses by adult *Aplysia* neurons *in vitro* is facilitated by the presence of juvenile neurons. In: Seil, F. J. *et al.* (eds), *Progress in Brain Research*, Vol. 71: *Neuronal Regeneration*. Elsevier, Amsterdam, pp. 281–9.

Scheff, S. W., Anderson, K. and DeKosky, S. T. (1984). Morphological aspects of brain damage in aging. In: Scheff, S. W. (ed.), *Aging and Recovery of Function in the Central Nervous System*. Plenum Press, New York, pp. 57–85.

Schot, L. P. C., Boer, H. H. and Wijdenes, J. (1983). Localization of neurons innervating the heart of *Lymnaea stagnalis* studied immunocytochemically with anti-FMRFamide and anti-vasotocin. In: Lever, J. and Boer, H. H. (eds), *Molluscan Neuro-endocrinology*. North Holland, Amsterdam, pp. 203–8.

Slob, W. and Janse, C. (1988). A quantitative method to evaluate the quality of interrupted animal cultures in aging studies. *Mech. Ageing Dev.*, **42**, 275–90.

Smit, A. B., Vreugdenhil, E., Ebberink, E. H. M., Geraerts, W. P. M., Klootwijk, J. and Joosse, J. (1988). Growth controling neurons produce the precursor of an insulin related peptide. *Nature*, **331**, 535–8.

Smith, C. B. (1984). Aging and changes in cerebral metabolism. *TINS*, **7**, 203–8.

Steger, R. W., Huang, H. and Meites, J. (1981). Reproduction. In: Massoro, E. J. (ed.), *Handbook of Physiology of Aging*. CRC Press, Boca Raton, FA, pp. 33–382.

Wilt, G. J. van der, van der Roest, M. and Janse, C. (1987). Neuronal substrates of respiratory behaviour and related functions in *Lymnaea stagnalis*. In: Boer, H. H. *et al.* (eds), *Neurobiology. Molluscan Models*. North Holland, Amsterdam, pp. 292–6.

Wilt, G. J. van der, van der Roest, M. and Janse, C. (1988). The role of two peptidergic giant neurons in modulation of respiratory behaviour in the pond snail *Lymnaea stagnalis*. In: Salanki, J. and S.-Rózsa, K. (eds), *Symposia Biologica Hungarica*, Vol. 36: *Neurobiology of Invertebrates. Transmitters, Modulators and Receptors*. Akadémiai Kiadó, Budapest, pp. 377–86.

11 *Stacia B. Moffett**

Neural control of male reproductive function in pulmonate and opisthobranch gastropods

11.1. Introduction

Reproduction is, evolutionarily speaking, the most crucial behaviour to perfect, so it is no surprise that its hormonal and neural control is complex. Female reproductive function in gastropods is one of the best-characterized examples of peptide neuroendocrine control (Rothman *et al.*, 1985; Geraerts *et al.*, 1988; Blankenship *et al.*, 1989). Control of male reproductive behaviour is less well understood. This review focuses on the use of neurotransmitter immunocytochemistry to identify relevant central neurons and target organ innervation. Background information on the opisthobranch *Aplysia*, the basommatophoran pulmonates *Lymnaea*, *Melampus* and *Helisoma* and the stylommatophoran pulmonate *Helix* is also presented.

11.2. Background

11.2.1. *Anatomy and muscular responses of the male system*

Pulmonates and opisthobranchs are hermaphroditic, with sperm and ova arising in adjacent cell lines in the ovotestis and travelling through a common duct (spermoviduct) before diverging either into separate ducts (vas deferens and oviduct) or separate paths within the same duct (Tompa, 1984; Geraerts and Joosse, 1984; Hadfield and Switzer-Dunlap, 1984). During copulation in *Aplysia*, sperm is moved from storage in the vesicula seminalis by peristalsis and then by cilia along the internal duct and external genital groove (Thompson and Bebbington, 1969). Excision of a portion of the vas deferens that runs through the body wall in *Lymnaea* blocks expression of male reproductive behaviour (van Duivenboden and ter Maat, 1988), an intriguing observation that cannot currently be explained. Fertilization in gastropods is internal and, with the exception of many land slugs, the male system includes a penis, which is

* Current address: Department of Zoology, Washington State University, Pullman, WA 99164, USA.

normally retained within the body in a sheath (preputium) by a retractor muscle. The male reproductive tract is quite complicated, with some tubular segments that are lengthened by circular muscle contraction and shortened by longitudinal muscle contraction. There are also whole-body responses during mating that remain to be elucidated. Increases in body wall pressure may be important in penile eversion (Audesirk and Audesirk, 1985; *Aplysia*: Koester, personal communication). Such changes could be controlled by hormonal or direct neural activation of body wall musculature.

The penis retractor muscle has been employed in tests of neurotransmitter function because it is a relatively discrete muscle with parallel spindle-shaped unstriated fibers (Wabnitz, 1976; Blankenship *et al.*, 1977). The retractor muscle and other parts of the penial complex (preputium, vas deferens and penis) contract both spontaneously and in response to nerve stimulation (Goddard, 1962; Jaegar, 1962, 1963; Duncan, 1964; Blankenship *et al.*, 1977). The retractor muscle of *Helix* exhibits brief contractions that are unaffected by isolation from the CNS and longer contractions that require intact central neural connections and can be mimicked by penis nerve stimulation (Wabnitz, 1976). The spontaneous activity remaining after isolation of the muscle is not attributed to myogenicity, but rather to clusters of neurons associated with the muscle (Wabnitz, 1976; see also Fig. 11.4H).

11.3. **Neurons important for male reproductive behaviour**

Asymmetric clusters of neurons associated with innervation of the penial complex are found ipsilateral to the penis in the otherwise bilaterally symmetrical cerebral and pedal ganglia. In *Aplysia* the penis develops on the right side. Cobalt chloride backfills of the two penis nerves that project from the right pedal ganglion fill pedal neurons that innervate the penis and which, upon intracellular activation, produce excitatory junction potentials in the retractor muscle (Rock *et al.*, 1977). However, both excitatory and inhibitory postsynaptic potentials were recorded from the muscle (Blankenship *et al.*, 1977).

In pulmonates the penis nerve is a cerebral nerve that contains axons of neurons in the ipsilateral cerebral and pedal ganglia. In *Helix*, cobalt chloride backfills of the penis nerve revealed cells in cerebral ganglia, pedal ganglia, and also the right cerebropedal connective (Eberhardt and Wabnitz, 1979). Activation of one identified right pedal cluster neuron was correlated with retraction; another identified cell appeared to be sensory. Neurons in the right pedal cluster of *Lymnaea*, the I cluster (Slade *et al.*, 1981) are homologous to the right pedal clusters of *Melampus*, and the left pedal cluster of *Helisoma*, a planorbid snail (see Figs 11.2, 11.3). The cells are electrically coupled in *Lymnaea*, and stimulation of individual neurons causes contraction of the penis retractor (Fig. 11.1).

The cerebral asymmetric cluster is located ventrally in the posterior-lateral margin of the right ganglion of *Melampus* (Ridgway *et al.*, 1988) and *Lymnaea* (E cluster: Khennak and McCrohan, 1988) and in the corresponding region in the left ganglion of *Helisoma* (Fig. 11.2). These clusters are probably not homologous with the *Helix* cerebral cluster that is backfilled from the penis nerve.

In addition to cerebral neurons that send axons out of the penis nerve, there are asymmetrical clusters of cerebral neurons that project to the pedal ganglion ipsilateral to the penis. In *Helix*, stimulation of meso-cerebral neurons elicits movement of the dart sac or penis, presumably by activating pedal neurons (Chase, 1986). Homology of these cells with *Aplysia* H cluster neurons (nomenclature of Fredman and Jahan-Parwar, 1975) or the asymmetric clusters identified by antibody immunoreactivity in *Melampus* and *Lymnaea* remains to be determined (see sections 11.4.4 and 11.4.5, below).

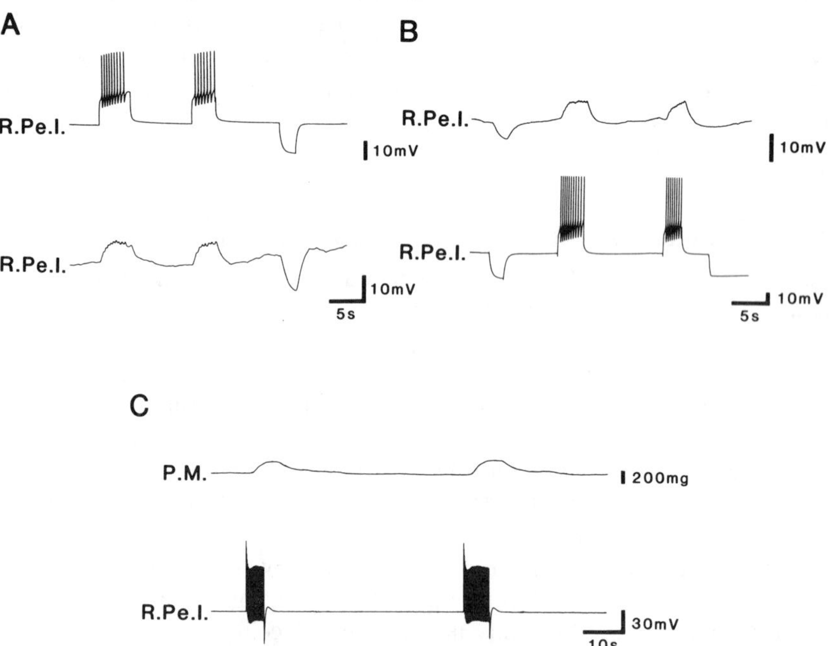

Fig. 11.1. **A**,**B**: Electrical coupling between right pedal ganglion I (R.Pe.I.) cluster neurons of *Lymnaea stagnalis*. Intracellular injection of depolarizing or hyperpolarizing current into the normally silent cells spreads through non-rectifying synapses. **C**: Simultaneous intracellular recording from R.Pe.I. neuron and penis retractor muscle tension: injection of the neuron with depolarizing current results in muscle contraction. (N. I. Syed and R. L. Ridgway, unpublished).

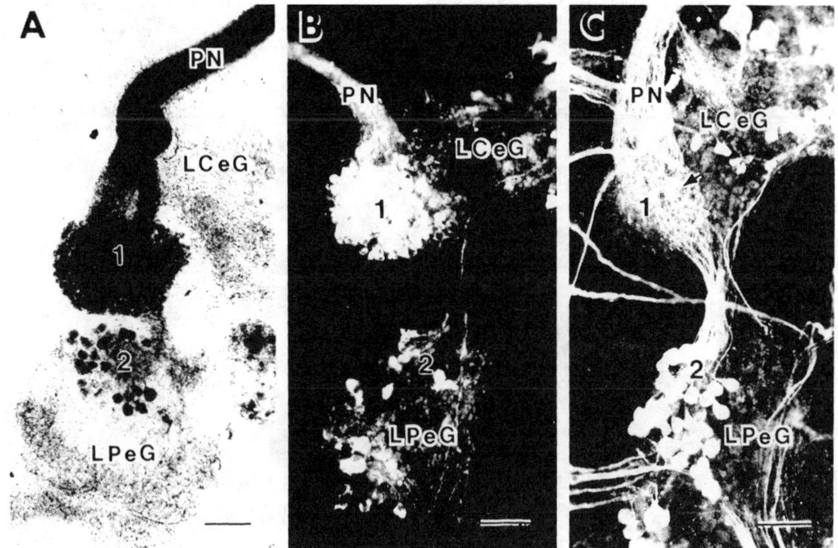

Fig. 11.2. Neuron clusters that project into the penis nerve in *Helisoma trivolvis*. **A**: Cobalt chloride backfill of left cerebral pars ventralis cluster near cerebropedal connective and left pedal cluster in dorsolateral region near the connective. **B**: FMRFamide-like immunoreactivity of the cerebral cluster and a few neurons in the area of the pedal cluster (antibody courtesy of J. Biship, NIH, Bethesda, MD). **C**: Serotonin-like immunoreactivity in some cells of the pedal cluster (anti-serotonin: INC Star Corp., Stillwater, MN); note neuropilar projections and varicosities (arrowhead) in region occupied by cerebral cluster. Calibration = 100 μm. (R. L. Ridgway, T. Culver and N. I. Syed, unpublished data.)

11.4. **Neurons implicated in male reproductive function: immunocytochemical mapping**

11.4.1. *FMRFamide family*

FMRFamide was the first peptide transmitter localized in both the male reproductive system and CNS of snails (Greenberg, 1983; Lehman and Greenberg, 1987, Lehman and Price, 1987). Among gastropods there is a phylogenetic split in the distribution of peptides in the FMRFamide family: FMRFamide is apparently ubiquitous, and opisthobranchs may possess FLRFamide but not heptapeptides, whereas in pulmonates several heptapeptides are found, with pQDPFLRFamide typically present in stylommatophorans and the heptapeptide containing the Gly analogue (pGDPFLRFamide) in the basommatophorans (Price *et al*., 1987).

In *Lymnaea*, antibodies generated against FMRFamide revealed an asymmetric cluster (E cluster) in the right cerebral ganglion (Schot and Boer, 1982; Boer *et al*., 1984). A different antibody to FMRFamide

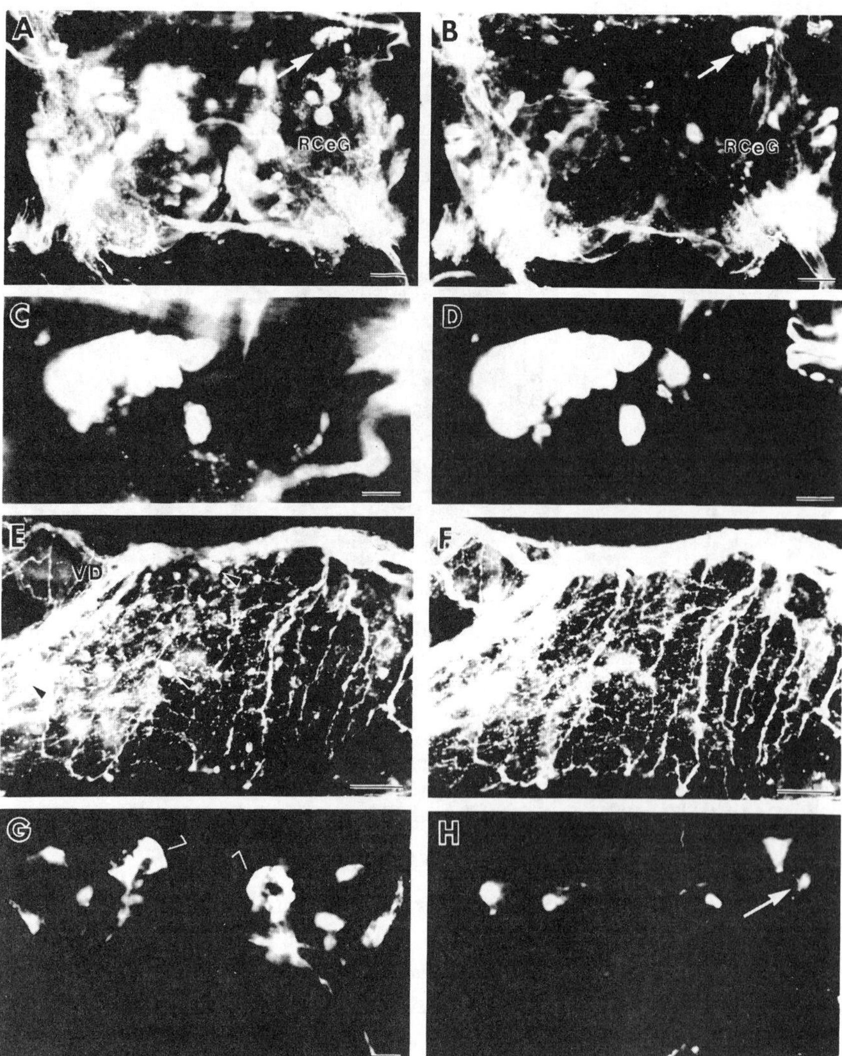

Fig. 11.3. **A**,**B**: Double immunofluorescent labelling of wholemount of *Melampus bidentatus* pedal ganglia, ventral focus on asymmetric right pedal cluster (arrows). **A**: Serotonin-like immunofluorescence (anti-serotonin in goat, INC Star Corp, Stillwater, MN; rhodamine donkey anti-goat, Chemicon, El Segundo, CA). **B**: CARP-like immunofluorescence in the same wholemount (anti-CARP in rabbit, courtesy of Y. Muneoka and Y. Terano; fluorescein donkey anti-rabbit, Chemicon). **C**: Higher magnification of the right pedal cluster in **A** (serotonin), and **D**, the same cluster, showing CARP immunofluorescence. **E**,**F**: Penis sheath and vas deferens (VD) double fluorescent labelling showing partial colocalization of serotonin-like immunofluorescence (**E**) and CARP-like fluorescence (**F**). Arrowheads indicate cells in **C** not present in **D**. **G**: Buccalin-like immunofluorescence in wholemount of *Melampus bidentatus* cerebral ganglia. Arrowheads indi-

identified the corresponding cluster of cerebral neurons in *Helisoma*, where it corresponds to the cerebral cluster identified in penis nerve backfills (Ridgway, Culver, and Syed, unpublished; Fig. 11.2A,B).

11.4.2. *Serotonin (5-HT)*

In *Lymnaea*, *Melampus*, *Helisoma* and *Helix*, neurons exhibiting serotonin immunoreactivity have been detected in an asymmetric pattern in the pedal ganglion ipsilateral to the penis (Croll and Chiasson, 1989; Kemenes *et al.*, 1989; Ridgway *et al.*, 1988; Ridgway, unpublished; Hernádi *et al.*, 1989; Fig. 11.2A; 11.3A,C). In *Lymnaea* the I cluster, located posterolateral to the superior pedal nerve in each pedal ganglion, forms a distinct lobe on the right side. The number of neurons increases as the snail grows, and two subpopulations make up the cluster. Some of the neurons send axons to the cerebral ganglion and out to the penis sheath. In *Melampus*, specific reinnervation of the sheath by these pedal neurons occurs after removal of the cerebral ganglion (Ridgway *et al.*, 1988). Antibody immunoreactivity also suggests serotonergic innervation of the penis sheath (Fig. 11.3E), the penis retractor muscle in *Aplysia* and *Helix* (not shown), and the preputium of *Limax* (Fig. 11.4A).

11.4.3. *CARP*

The catch-relaxing peptide (CARP) was isolated from pedal ganglia of the mussel *Mytilus* (Hirata *et al.*, 1989a,b). It is a heptapeptide related to myomodulin isolated from *Aplysia* (Cropper *et al.*, 1987). CARP-like immunoreactivity colocalized with serotonin-like immunoreactivity in both the pedal neuron cluster of *Melampus* and in the peripheral ramifications of those neurons (Fig. 11.3A–F). In contrast, individual neuron somata within the penial complex exhibit either serotonin or CARP-like immunofluorescence but not colocalization. CARP-like immunofluorescence is also found in the penis retractor muscle and penis sheath of *Aplysia*, *Helix* and *Limax* (Fig. 11.4G–I).

11.4.4. *Buccalin*

Buccalin is one member of a peptide family that includes parabuccalin and other peptides (Kupfermann *et al.*, 1988). Buccalin-like immuno-

cate clusters that project to the visceral and right parietal ganglia and from there to male and female portions of the reproductive tracts (anti-buccalin in rabbit, courtesy of Mark Miller, Columbia University). **H**: Wholemount of *Melampus* cerebral ganglia showing immunoreactivity to alpha bag cell antibodies; arrow indicates right cerebral asymmetric cluster that projects in the cerebropedal connective. Calibration = 200 μm in **A**, **B**, **E**, **F**; 50 μm in **C** and **D**.

fluorescence has been found in *Aplysia* H cluster neurons in the right cerebral ganglion and also in the retractor muscle and other parts of the reproductive tract (Kupfermann *et al.*, 1988). In *Melampus*, buccalin-like immunofluorescence revealed no asymmetry in the cerebral ganglia. However, symmetrical clusters in approximately the same region as the H cluster project to structures associated with the gonad via the right parietal and visceral ganglia (Fig. 11.3G). In *Melampus* the penis sheath and a portion of the efferent duct are innervated by axons which exhibit buccalin-like immunofluorescence and which could be traced from the visceral ring and pedal ganglia (Fig. 11.4D,E). The reproductive tract of *Limax* (Fig. 11.4B) shows more extensive buccalin-like innervation than that of *Melampus*.

11.4.5. *ELH and CDCH*

Egg-laying in gastropods is controlled by peptidergic systems (Geraerts *et al.*, 1988). The egg-laying hormone, ELH, present in the bag cells of *Aplysia*, is now recognized as one of several products of a gene family that is active in both neural and non-neural tissues (Blankenship *et al.*, 1989). Antibodies to alpha bag cell extract labelled not only the bag cell clusters but also ectopic bag cells in the abdominal and right pleural ganglia, and a bilateral cluster of smaller immunoreactive cells in the cerebral ganglia (Painter *et al.*, 1990). In *Melampus* the same antibodies reacted with a symmetrically distributed cluster and pair of cells in the cerebral and other ganglia, and also an asymmetric cluster in the right cerebral ganglion (Fig. 11.3H). These neurons are located ventrally close to, but more medial than, the cerebral cluster backfilled from the penis nerve, and they project into the cerebropedal connective. A role in control of male reproductive function is suggested by their possible homology with the mesocerebral neurons of *Helix*, which are more numerous and larger on the right side (Chase, 1986).

In *Lymnaea* the cerebral caudodorsal cells, CDCs, are homologous to the bag cells (Geraerts *et al.*, 1988). They are somewhat asymmetrically distributed, with the right ganglion possessing more neurons (Joosse, 1964). Caudodorsal cell hormone (CDCH)-like immunoreactivity is exhibited in three groups of cerebral cells in addition to the caudodorsal cells (Minnen and Vreugdenhil, 1987). One of these groups is decidedly asymmetrical, with the larger right cluster of approximately 60 neurons found in the same location as the unpaired cluster that exhibits alpha bag cell extract immunoreactivity in *Melampus*. An additional asymmetry was found in the pedal ganglia of *Lymnaea*, with the extra cluster located contralateral to the penis, on the left side.

Neuron clusters which may function in male reproduction are summarized in Table 11.1. The roles of the transmitters identified with these

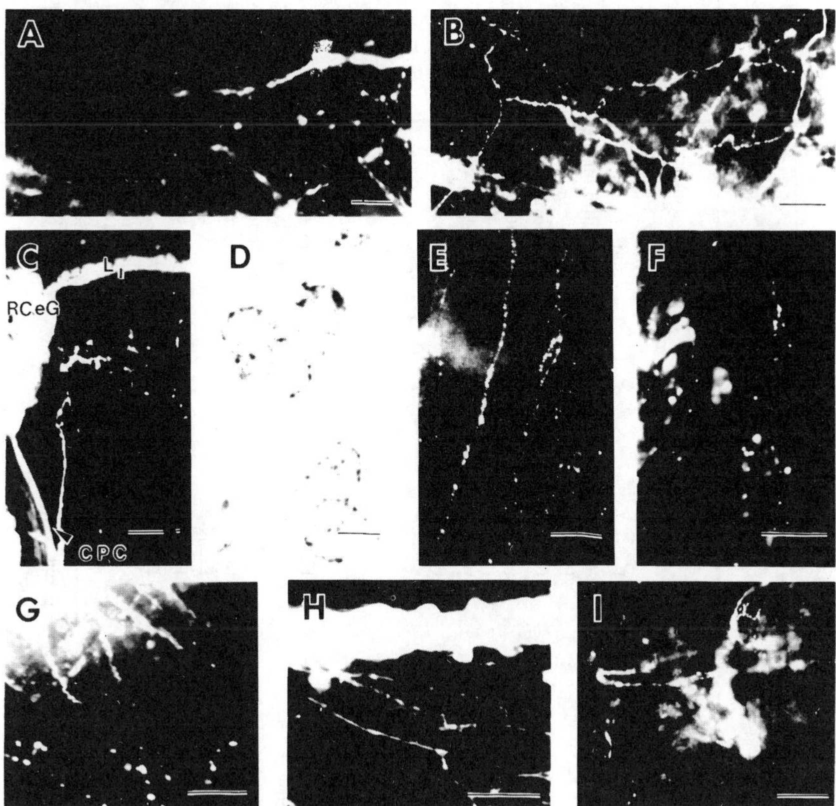

Fig. 11.4. Double immunofluorescent labelling of *Limax maximus* preputium showing serotonin-like reactivity (**A**) and buccalin-like reactivity (**B**). **C**–**F**: Buccal-in-like immunoreactivity. **C**: Axons projecting along cerebropedal connective (CPC) from pedal and visceral ring neurons to penis sheath in *Melampus*. **D**: Processes, shown in negative contrast, encircle cell-like shapes in region where peripheral neurons are localized in *Melampus* penis sheath. **E**: Common duct in *Melampus*. **F**: Penis retractor muscle of *Aplysia californica*. **G**–**I**: CARP-like immunofluorescence; **G**: *Helix aspersa* penis sheath; **H**: Peripheral neurons form a strand from which axons radiate onto the preputium of *Limax maximus*; **I**: Varicosities on surface of penis retractor muscle fibres of *Aplysia californica*. Calibration = 100 μm in **A** and **B**; 200 μm in **C** and **E**–**I**; 25 μm in **D**.

neurons in the coordination of sperm development and transport and in copulation behaviour must now be examined.

11.5. **Responses to peptides and other transmitter substances**

In *Melampus* the penis retractor muscle and the penis and penis sheath (preputium) have been tested for responses to some of the neurotransmit-

Table 11.1 Distribution and transmitter immunoreactivity of asymmetric neuron clusters in selected gastropods

Genus	*Location*	*Transmitter family*	*Axon projections*	*Reference*
Cerebral neuron clusters relevant to male reproduction				
Lymnaea	r vent-lat	FMRF	?	Schot and Boer, 1982
	r vent-lat	ELH-CDCH	?	van Minnen and Vreugdenhil, 1987
	r ant lobe	?	Penis N.	Khennak and McCrohan, 1988
Helisoma	l ventral	FMRF	Penis N.	Ridgway, unpublished
Melampus	r ventral	?	Penis N.	Moffett, 1989
	r ventral	ELH-CDCH	C-Pd Conn.	Moffett, unpublished
	r&l mid-dorsal	Buccalin	C-Pl Conn.	Moffett, unpublished
Helix	cer.-pd. conn	?	Penis N.	Eberhardt and Wabnitz, 1979
	r dorsal	?	Penis N.	Eberhardt and Wabnitz, 1979
	r mesocerebrum	?	C-Pd Conn.	Chase, 1986
Aplysia	r ventral	ELH-CDCH	?	Painter *et al.*, 1990
	r dorsal (H)	Buccalin	?	Kupfermann *et al.*, 1988
Pedal neuron clusters relevant to male reproduction				
Lymnaea	r ant vent	Serotonin	Penis N.	Croll and Chiasson, 1989
	r ant vent	Serotonin	?	Kemenes *et al.*, 1989
	l gang (?)	ELH-CDCH	?	van Minnen and Vreugdenhil, 1987
Helisoma	l ant-vent	Serotonin	Penis N.	Ridgway *et al.*, unpublished
Melampus	r ant-vent	Serotonin	Penis N.	Moffett, 1989
	r ant-vent	CARP	Penis N.	Moffett, 1989
Helix	r vent-lat	?	Penis N.	Eberhardt and Wabnitz, 1979
	r&l ant-vent	Serotonin	?	Hernádi *et al.*, 1989
Aplysia	r mid-vent	?	Penis N.	Rock *et al.*, 1977

Abbreviations: ant lobe, anterior lobe; ant-vent, anterioventral; C-Pd Conn, cerebropedal connective; C-Pl Conn, cerebropleural connective; cer. pd., cerebropedal; mid-vent, midventral; vent-lat, ventrolateral. For transmitter abbreviations, see text.

ters identified by immunocytochemical mapping. Agents were added to the bath in concentrations indicated on the figures. Acetylcholine (ACh), which is colocalized with peptide transmitters in many molluscan muscles, was also tested.

Both FMRFamide and ACh have an excitatory effect on the penial complex of *Melampus* (Moffett, 1989; see also Fig. 11.5). FMRFamide acts at a lower concentration. Tension recordings show a prolonged contraction with gradually declining tension in response to both transmitters. The effect of the two transmitters is additive. Similar responses were recorded from the penis retractor muscle of *Aplysia*. FMRFamide or ACh also cause contraction of the male reproductive tract of *Helix* (Lehman and Greenberg, 1987); the penis retractor muscle of *Lymnaea* (Duncan, 1964; Geraerts and Joosse, 1984) and the pulmonate *Strophocheilos* (Jaeger, 1962).

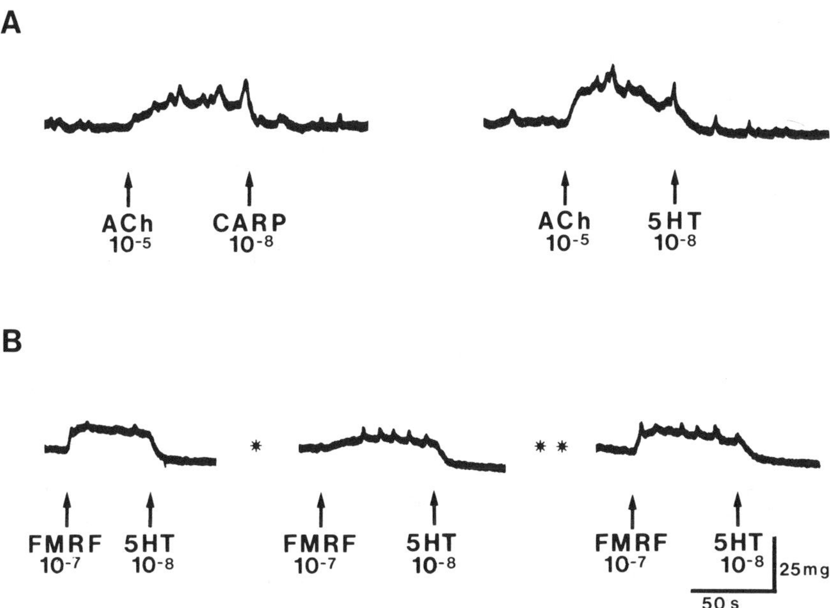

Fig. 11.5. **A,B**: Tension recordings from penis retractor muscle and associated penis sheath and efferent duct (vas deferens) of *Melampus bidentatus*. At the arrows, neurotransmitter was applied; the bath concentration is given. **A**: Both CARP (peptide courtesy of I. Kubota) and serotonin (5-HT) relax contractions produced by application of ACh. **B**: FMRFamide application causes contraction at a lower concentration than ACh. Both serotonin and CARP relax FMRFamide contractions (CARP response not shown). Exposure of the muscle preparation to 10^{-8} CARP for 20 min (*) reduces response to FMRFamide but not relaxing effect of serotonin. The response to FMRFamide is restored by a washing and 20 min rest (**).

In *Melampus*, two transmitters have been tested which relax the contractions caused by FMRFamide and ACh: serotonin and CARP (Moffett, 1989). Serotonin is more effective than CARP at a given concentration, and in some muscle preparations is able to drop the baseline tension below the resting level (Fig. 11.5A,B). CARP has an additional effect which was not seen with serotonin: chronic application of low levels of CARP reduce the response of the muscle preparation to FMRFamide (Fig. 11.5B). Serotonin is also able to relax the penis retractor muscle of *Strophocheilos* (Jaeger, 1963). Like FMRFamide and serotonin (Painter and Greenberg, 1982), CARP is likely to exert multiple actions on target cells. In *Helix* neurons it affects Ca^{2+} permeability (Kiss, 1988). In the snail *Achatina*, twitch contractions of the penis retractor muscle were not inhibited by CARP, but these contractions were inhibited by the hexapeptide *Mytilus* inhibitory peptides (Hirata *et al.*, 1988). Thus male reproductive tract may be regulated by this additional family of peptides.

11.6. **Future directions**

Our current knowledge concerning the innervation of the male reproductive tract of gastropods has many gaps, and one of the easiest ways to remedy the situation is to apply the technique of immunocytochemistry to more gastropods. The asymmetrical distribution of neurons involved in control of the penial complex facilitates their identification. However, it is not necessary to perform mapping studies on all species once a pattern is established, and species that are ideal for mapping may not be the best ones in which to pursue biochemical analysis of transmitter structure or behaviour, and its underlying neural circuitry.

The demonstration that a particular peptide is present in neurons that innervate a part of the reproductive tract is only a starting point for a variety of approaches that can lead to an understanding of function. We should also keep in mind that a particular neurotransmitter family may not always turn up in the homologous neurons of different species. Peptides may mediate the fine-tuning of basic machinery, and be utilized in different combinations in the different contexts in which the neurons have evolved. However, the findings summarized in Table 11.1 indicate that immunocytochemical mapping is a useful tool in the analysis of innervation of the male reproductive system of pulmonates and opisthobranchs.

Acknowledgements

I am grateful to S. D. Painter, J. E. Blankenship, R. L. Ridgway and D. F. Moffett, Jr for suggestions and criticisms on early stages of this manuscript, to N. I. Syed, T. Culver, and R. L. Ridgway for sharing work in progress, and to all who shared their peptides and antibodies with me. Most of this work was accomplished while I was on sabbatical in the laboratory of J. E. Blankenship, and was supported in part by NS27314 to J.E.B. and NIH NS22896 to S.B.M.

References

Audesirk, T. and Audesirk, G. (1985). Behavior of gastropod molluscs. In Wilbur, K. M. (ed.-in-chief), *The Mollusca: Neurobiology and Behavior*, Part 1, Vol. 8. Academic Press, Orlando, pp. 1–94.

Blankenship, J. E., Rock, M. K. and Hill, J. (1977). Physiological properties of the penis retractor muscle of *Aplysia*. *J. Neurobiol.*, **8**, 549–68.

Blankenship, J. E., Nagle, G. T., Painter, S. D. and Jong-Brink, M. de (1989). Multiple roles of a genetically related family of neuroendocrine and exocrine peptides that regulate reproductive behavior in *Aplysia*. In: Lakoski, J. M. *et al.* (eds), *Neural Control of Reproductive Function*, Alan R. Liss, New York, pp. 273–84.

Boer, H. H., Schot, L. P. C., Reichelt, D., Brand, H. and Maat, A. ter (1984). Ultrastructural immunocytochemical evidence for peptidergic neurotransmission in the pond snail *Lymnaea stagnalis*. *Cell Tiss. Res.*, **238**, 197–201.

Chase, R. (1986). Brain cells that command sexual behavior in the snail *Helix aspersa*. *J. Neurobiol.*, **17**, 669–79.

Croll, R. P. and Chiasson, B. J. (1989). Postembryonic development of serotonin-like immunoreactivity in the central nervous system of the snail, *Lymnaea stagnalis*. *J. Comp. Neurol.*, **280**, 122–42.

Cropper, E. C., Tenenbaum, R., Kolks, M. A. G., Kupfermann, I. and Weiss, K. R. (1987). Myomodulin, a bioactive neuropeptide present in an identified cholinergic buccal motor neuron of *Aplysia*. *Proc. Natl. Acad. Sci. USA*, **84**, 5483–6.

Duivenboden, Y. A. van and Maat, A. ter (1988). Mating behavior of *Lymnaea stagnalis*. *Malacologia*, **28**, 53–64.

Duncan, C. J. (1964). Rhythmic activity in an isolated penis preparation from the freshwater snail *Limnaea stagnalis*. *Zeit. Vergl. Physiol.*, **48**, 295–301.

Eberhardt, B. and Wabnitz, R. W. (1979). Morphological identification and functional analysis of central neurons innervating the penis retractor muscle of *Helix pomatia*. *Comp. Physiol. Biochem.*, **63A**, 599–613.

Fredman, S. M. and Jahan-Parwar, B. (1975). Synaptic connections in the cerebral ganglion of *Aplysia*. *Brain Res.* **100**, 209–14.

Geraerts, W. P. M. and Joosse, J. (1984). Freshwater snails (Basommatophora). In: Wilbur, K. M., (ed.-in-chief), *The Mollusca: Reproduction*, Vol. 7. Academic Press, Orlando, pp. 142–207.

Geraerts, W. P. M., Maat, A. ter and Vreugdenhil, E. (1988). The peptidergic neuroendocrine control of egglaying behavior in *Aplysia* and *Lymnaea*. In: Laufer, H. and Downer, R. (eds), *Endocrinology of Molluscs: Invertebrate Endocrinology*, Vol. 2. Alan R. Liss, New York, pp. 144–231.

Goddard, C. K. (1962). Function of the penial apparatus of *Helix aspersa* Muller. *Austral. J. Biol. Sci.*, **15**, 218–32.

Greenberg, M. J. (1983). The responsiveness of molluscan muscles to FMRF-amide, its analogs and other neuropeptides. In: Lever, J. and Boer, H. H. (eds), *Molluscan Neuroendocrinology*. North Holland, Amsterdam, pp. 190–5.

Hadfield, M. G. and Switzer-Dunlap, M. (1984). Opisthobranchs. In Wilbur, K.

M. (ed.-in-chief), *The Mollusca: Reproduction*, Vol. 7. Academic Press, Orlando, pp. 209–350.

Hernádi, L., Elekes, K. and S.-Rózsa, K. (1989). Distribution of serotonin-containing neurons in the central nervous system of the snail *Helix pomatia*. Comparison of immunocytochemical and 5,6-dihydroxytryptamine labelling. *Cell Tissue Res.*, **257**, 313–23.

Hirata, T., Kubota, I, Iwasawa, N., Takabatake, I., Ikeda, T. and Muneoka, Y. (1988). Structures and actions of *Mytilus* inhibitory peptides. *Biochem. Biophys. Res. Commun.*, **152**, 1376–82.

Hirata, T., Kubota, I., Imada, M. and Muneoka, Y. (1989a). Pharmacology of relaxing response of *Mytilus* smooth muscle to the catch-relaxing peptide. *Comp. Physiol. Biochem.*, **92C**, 289–95.

Hirata, T., Kubota, I., Imada, M., Muneoka, Y. and Kobayashi, M. (1989b). Effects of catch-relaxing peptide on molluscan muscles. *Comp. Biochem. Physiol.*, **92C**, 283–8.

Jaegar, C. P. (1962). Physiology of mollusca. III – Action of acetylcholine on the penis retractor muscle of *Strophocheilos oblongus. Comp. Biochem. Physiol.*, **7**, 63–69.

Jaegar, C. P. (1963). Physiology of mollusca. IV – Action of serotonin on the penis retractor muscle of *Strophocheilos oblongus. Comp. Biochem. Physiol.*, **8**, 131–136.

Joosse, J. (1964). Dorsal bodies and dorsal neurosecretory cells in the cerebral ganglia of *Lymnaea stagnalis*. L. *Arch. Neerl. Zool.*, **15**, 1–103.

Kemenes, G. Y., Elekes, K., Hiripi, L. and Benjamin, P. R. (1989). A comparison of four techniques for mapping the distribution of serotonin and serotonin-containing neurons in fixed and living ganglia of the snail *Lymnaea*. *J. Neurocytol.*, **18**, 193–208.

Khennak, M. and McCrohan, C. R. (1988). Cellular organisation of the cerebral anterior lobes in the central nervous system of *Lymnaea stagnalis*. *Comp. Biochem. Physiol.*, **91A**, 387–94.

Kiss, T. (1988). Catch-relaxing peptide (CARP) decreases the Ca-permeability of snail neuronal membrane. *Experientia*, **44**, 998–1000.

Kupfermann, I., Cropper, E. C., Miller, M. W., Alezivos, A., Tenenbaum, R. and Weiss, K. R. (1988). Buccalin: distribution of immunoreactivity in the *Aplysia* nervous system and biochemical localization to identified motor neuron B16. *Soc. Neurosci. Abs.*, **14**, 177.

Lehman, H. K. and Greenberg, M. J. (1987). The actions of FMRFamide-like peptides on visceral and somatic muscles of the snail *Helix aspersa*. *J. Exp. Biol.*, **131**, 55–68.

Lehman, H. K. and Price, D. A. (1987). Localization of FMRFamide-like peptides in the snail *Helix*. *J. Exp. Biol.*, **131**, 37–53.

Minnen, J. van and Vreugdenhil, E. (1987). The occurrence of gonadotrophic hormones in the central nervous system and the reproductive tract of *Lymnaea stagnalis*. An immunocytochemical and *in situ* hybridization study. In: Boer, H. H. *et al.* (eds), *Neurobiology. Molluscan Models*. North Holland, Amsterdam, pp. 194–9.

Moffett, S. B. (1989). A physiological and immunocytochemical study of innervation of the penial complex in the snail *Melampus bidentatus*. *Soc. Neurosci. Abs.*, **15**, 737.

Painter, S. D. and Greenberg, M. J. (1982). A survey of responses of bivalve hearts to the molluscan neuropeptide FMRFamide and to 5-hydroxytryptamine. *Biol. Bull.*, **162**, 311–332.

Painter, S. D., Kalman, V. K., Nagle, G. T. and Blankenship, J. E. (1990).

Localization of immunoreactive alpha-bag-cell peptide in the central nervous system of *Aplysia. J. Comp. Neurol.* (In press).

Price, D. A., Davies, N. W., Doble, K. E. and Greenberg, M. J. (1987). The variety and distribution of the FMRFamide-related peptides in molluscs. *Zool. Sci.*, **4**, 395–410.

Ridgway, R. A., Bailey, R. M. and Moffett, S. B. (1988). Regeneration of serotonergic axons projecting to the penial complex in the salt marsh pulmonate snail *Melampus bidentatus. Soc. Neurosci. Abs.*, **14**, 1055.

Rock, M. K., Blankenship, J. E. and Lebeda, F. J. (1977). Penis-retractor muscle of *Aplysia*: excitatory motor neurons. *J. Neurobiol.*, **8**, 569–79.

Rothman, B. S., Mayeri, E. and Scheller, R. H. (1985). The bag cell neurons of *Aplysia* as a possible peptidergic multitransmitter: from genes to behavior. In: Zomzely-Neurath, C. and Walker, W. A. (eds), *Gene Expression in Brain*. John Wiley & Sons, New York, pp. 235–74.

Schot, L. P. C. and Boer, H. H. (1982). Immunocytochemical demonstration of peptidergic cells in the pond snail *Lymnaea stagnalis* with an antiserum to the molluscan cardioactive tetrapeptide FMRFamide. *Cell Tiss. Res.*, **225**, 347–54.

Slade, T. C., Mills, J. and Winlow, W. (1981). The neuronal organisation of the paired pedal ganglia of *Lymnaea stagnalis* (L.). *Comp. Biochem. Physiol.*, **69A**, 789–803.

Tompa, A. S. (1984). Land snails (Stylommatophora). In: Wilbur, K. M. (ed.-in-chief), *The Mollusca: Reproduction*, Vol. 7. Academic Press, Orlando, pp. 48–140.

Thompson, T. E. and Bebbington, A. (1969). Structure and function of the reproductive organs of three species of *Aplysia* (Gastropoda: Opisthobranchia). *Malacology*, **7**, 347–80.

Wabnitz, R. W. (1976). Mechanical and electromyographic study of the penis retractor muscle (PRM) of *Helix pomatia. Comp. Biochem. Physiol.*, **55A**, 253–9.

12 *Harry H. Boer, Jan van Minnen and Ron M. Kerkhoven*

Evolutionary aspects of morphology and function of peptidergic systems, with particular reference to the pond snail *Lymnaea stagnalis*

Neuropeptides are widely distributed in the animal kingdom (Scharrer, 1987). Based on their primary amino acid structure they can be arranged into families, which seem to be represented in all regions of the evolutionary scale (Lynch and Snyder, 1986; Thorndyke, 1986). This phenomenon can be explained by considering the possible mutagenesis (gene duplication, gene conversion, point mutations) of genes coding for neuropeptide precursors (Wilson, 1985; Hadley, 1988). During this process the neuropeptides will have diverged structurally, but it may be expected that they will still show similarities. With regard to the diversity, tissue-specific expression of related neuropeptide genes and differences in neuropeptide precursor processing can also be considered. An important question is whether the available evidence has any bearing on theories to be proposed on the functions that structurally related peptides serve in a particular species or animal group (e.g. Nieuwenhuis, 1985).

12.1. Structural relations of neuropeptides across the species border

The proposed mechanism for the history of neuropeptides is sustained by reports indicating the relatedness of these molecules in evolutionarily distant animal groups. In some cases the molecules even appear to be identical in structure: e.g. the head activator of *Hydra* occurs in identical form in mammals (Bodenmüller and Schaller, 1981), and the enkephalins, originally extracted from mammals, were shown to be present in the nervous tissues of the molluscs *Mytilus edulis* and *Lymnaea stagnalis* (e.g. Leung *et al.*, 1990). As an example of neuropeptides that are not fully identical, but structurally closely related, the 'oxytocins–vasopressins' of the vertebrate classes (cf. Hadley, 1988) of insects (Proux *et al.*, 1987) and of gastropods (Sawyer *et al.*, 1984; Ebberink and Joosse, 1985; Cruz *et al.*, 1987) can be mentioned. Furthermore, in numerous investigations where antibodies to vertebrate or invertebrate neuropeptides were used, immunocytochemical relations were reported in animals unrelated to those whose neuropeptides served as immunogens (e.g. Miller and Benzer, 1983; Schot *et al.*, 1984; Veenstra *et al.*, 1985; Conway and Gainer, 1987; Bjenning and Holmgren, 1988). These observations also suggest structural relatedness, although it must be admitted that immunocytochemical methods are not sufficient to establish the structural identity of molecules, due to the possibility of cross-reactions (Landis, 1985).

Several neuropeptidergic systems of the snail *L. stagnalis* have received attention. The focus has been on the Caudo-Dorsal Cell system (CDC), on the Light-Green Cell system (LGC) including the Canopy Cells (CC), and to a lesser extent on the 'ACTH-cells' (Figs 1, 3, 5, 9). With antibodies to neuropeptides of these systems and with *in situ* hybridization the question was addressed whether related peptides (peptidergic systems) occur in other species.

12.1.1. *The CDC*

The neuroendocrine CDC are located in two paired clusters of about 50 neurons each in the cerebral ganglia (Fig. 12.3). They play a central role in the control of egg-laying and egg-laying behaviour (Geraerts *et al.*, 1988). They produce the 36 amino acid caudo-dorsal cell hormone (CDCH; ovulation hormone). In addition, other bioactive peptides derive from the CDC (Vreugdenhil *et al.*, 1988). The CDC peptides are synthesized as components of high molecular weight precursor molecules. These are encoded by a small multigene family consisting of two to four members. This family can be subdivided into two classes, the CDCH-I and CDCH-II genes (Fig. 12.2) (Vreugdenhil *et al.*, 1988). The genes code for different CDCH molecules and additional CDC peptides. For some of the additional CDC peptides a biological function has been established, e.g., calfluxin, a 14 amino acid peptide induces Ca^{2+} fluxes in the albumen gland cells (Dictus *et al.*, 1987). Furthermore, the ovulation hormone in cooperative action with α-CDCPs (full length α-CDCP: 11 amino acids and two fragments, 10 and 9 amino acids) cause auto-excitation of the CDC (Brussaard *et al.*, 1990). The CDC release their products into the haemolymph at the periphery of the cerebral commissure (Roubos, 1984) and non-synaptically from the CDC collateral system (Fig. 12.1) in the neuropile of the commissure (Schmidt and Roubos, 1987).

With antibodies specific for the expression of the CDC-I (aCDCH-I) or the CDC-II gene (aCDCH-II) and with labelled oligonucleotides (Fig. 12.2) it was demonstrated that all CDC and 1–4 cells in the pleural ganglia (ectopic CDC) express both CDCH genes (see Fig. 12.1; van Minnen *et al.*, 1988). Furthermore, in each cerebral ganglion a group of 5–10 small neurons express the CDCH-I gene, but not the CDCH-II gene. These neurons have no neurohaemal area. Their axons seem to contact the CDC synaptically (Fig. 12.1).

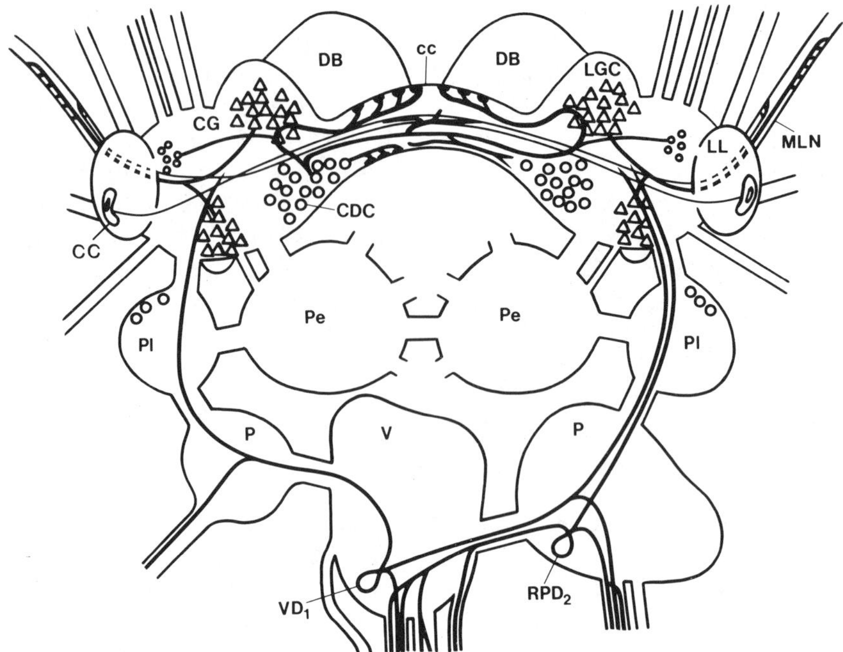

Fig. 12.1. Schematic representation of the localization of three neuropeptidergic systems in the CNS of *L. stagnalis*. The CDC axon endings are located in the periphery of the cerebral commissure (cc); furthermore the CDC collateral system is located in the cc. Axons of a lateral group of small CDC-like cells contact the CDC. The LGC axons run to the ipsilateral median lip-nerve (MLN); the CC axons run to the contralateral median lip-nerve. The giant VD_1 and RPD_2 branch extensively in the CNS; furthermore axon branches leave the CNS through several visceral nerves. CG, cerebral ganglion; DB, dorsal body; LL, lateral lobe; P, parietal; Pe, pedal; Pl, pleural; V, visceral ganglion.

In the opisthobranch *Aplysia californica* the egg-laying hormone is produced by the neuroendocrine bag cells (cf. Geraerts *et al.*, 1988). When the egg-laying hormone precursor molecules of *L. stagnalis* and of *Aplysia* are compared, certain areas show a high degree of sequence homology at the amino acid level. This homology is particularly high in the β-bag cell peptide (*Aplysia*)/β-CDC-peptide (*L. stagnalis*) region. One peptide (amino acids RLRFH) is present on the precursors of both species (Fig. 12.2). It was considered that antibodies to this peptide might be used to identify egg-laying hormone systems in other molluscs (the basommatophoran snails *L. ovata*, *Bulinus truncatus*, *Biomphalaria glabrata* and *Planorbarius corneus* and the stylommatophoran slugs *Agriolimax reticulatus*, *Arion ater*, *Arion hortensis* and *Limax maximus* were studied).

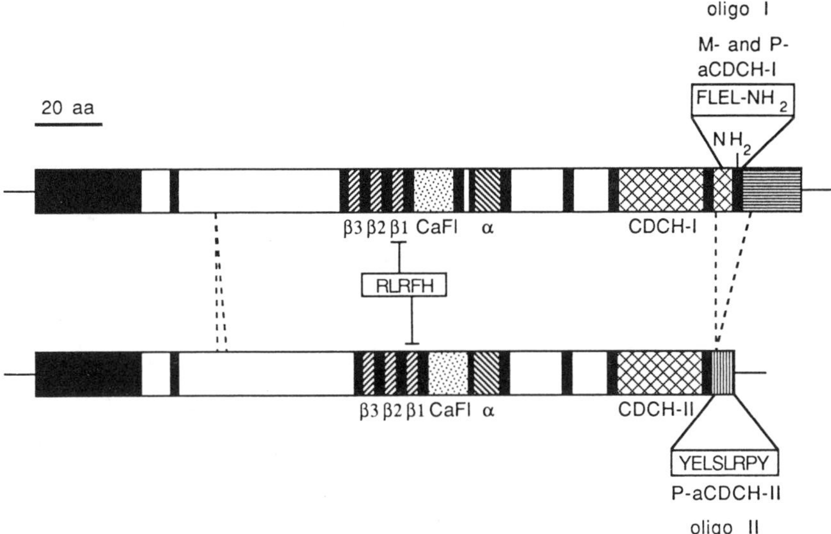

Fig. 12.2. Schematic representation of the CDCH-I and CDCH-II preprohormones. Each of the proteins is initiated by a methionine followed by a hydrophobic signal sequence (black). The neuropeptides CDCH-I, CDCH-II, α-CDCP, β_1, β_2 and β_3 CDCP, calfluxin (CaFl) and the completely diverged carboxy-terminal peptide of the CDCH-II preprohomone are shown. Deletions are indicated by broken lines. Specific oligonucleotides and antibodies for the respective CDCH preprohormones are indicated. aa, Amino acids.

In the basommatophoran snails the antibody-marked neurons appeared to have the same morphology and localization as the CDC of *L. stagnalis*, i.e. they are located in the cerebral ganglia and have their neurohaemal area in the periphery of the cerebral commissure (cf. Roubos and van de Ven, 1987). Furthermore a group of small neurons (10–20) was observed in the lateral part of each cerebral ganglion. These seem to be homologous to the small cerebral neurons of *L. stagnalis* that express the CDCH-I gene. In *A. ater*, *A. hortensis* and *L. maximus* in the cerebral ganglia only neurons homologous to the small CDCH-I gene expressing neurons are present (a group of 20–40 neurons at the lateral side of each ganglion). The neurons homologous to the CDC of *L. stagnalis* are most likely located in the visceral ganglion of the slugs. In this ganglion the antibody appeared to mark a group of 20–30 neurons with diameters ranging from 60 to 80 μm (Fig. 12.4). We were unable to trace axon tracts of these neurons to a neurohaemal area.

Biochemical and molecular biological techniques will be required to characterize the peptides produced by the immunoreactive neurons and to establish their relationship with peptides from the egg-laying controlling systems of *A. californica* and *L. stagnalis*.

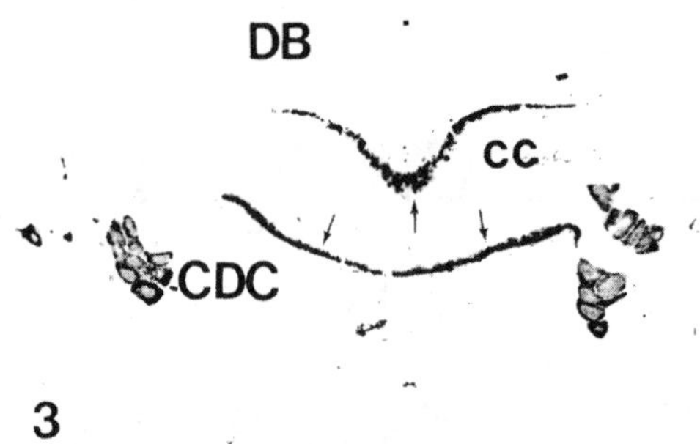

Fig. 12.3. *L. stagnalis*. The CDC cell bodies and the CDC axon endings in the periphery of the cerebral commissure (cc) are stained with anti-CDCH. DB, dorsal body. ×60.

Fig. 12.4. *Arion ater*, visceral ganglion. Cells (arrows) stained with anti-β-CDCP. ×150.

12.1.2. *The LGC and the CC*

LGC are located in a medio- and laterodorsal cluster (about 50 neurons per cluster) in each cerebral ganglion (Fig. 12.5). From each cluster an axon bundle runs to the median lip-nerve, where they form a neurohaemal area (e.g. Roubos and van de Ven, 1987). There is evidence that the cells are involved in the regulation of various processes related to growth (cf. Joosse, 1988). Each CC is located in one of the lateral lobes, small ganglia attached to the cerebral ganglia (Fig. 12.5). The cells send their axon via the cerebral commissure to the contralateral lip-nerve (van

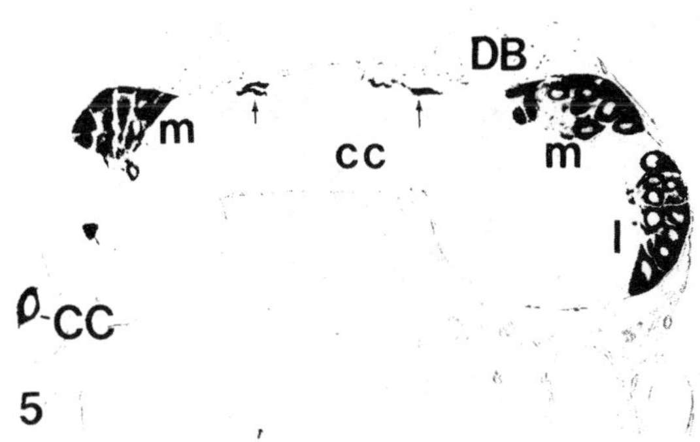

Fig. 12.5. The medial (m) and lateral (l) LGC and the CC in the cerebral ganglia are strongly immunoreactive to the MIP-C antibody. Arrows indicate axon processes of the CC in the cerebral commissure (cc). DB, dorsal body. ×60.

Minnen *et al.*, 1979). The CC have the same ultrastructural and histochemical characteristics as the LGC.

In other pulmonates cells homologous to the LGC have been demonstrated (Wijdenes *et al.*, 1980; van Minnen and Sokolove, 1981; Roubos and van de Ven, 1987). In the slug *A. reticulatus* their involvement has been shown in the regulation of body growth (Wijdenes and Runham, 1977). In *Helix aspersa* a second function has been attributed to the cells, *viz.* inhibition, in a synaptic fashion, of the activity of the dorsal bodies (DB), endocrine organs involved in the control of reproduction (Wijdenes *et al.*, 1987).

Recently, in *L. stagnalis* the primary structure of a pre-proinsulin-related protein was identified (Smit *et al.*, 1988). Apparently there are at least five closely related insulin-like genes (cf. van Minnen *et al.*, 1989). By means of *in situ* hybridization it was demonstrated that transcription of the genes takes place in the LGC and the CC (Fig. 12.6).

With affinity-purified antibodies raised to fragments of one of the C-peptides of the molluscan insulin-related peptides (MIPs), the occurrence of identical or related peptides was studied in other gastropods (Basommatophora: *L. ovata*, *B. truncatus*, *B. glabrata*, *P. corneus*; Stylommatophora: *H. aspersa*, *L. maximus*; Opisthobranchia: *A. californica*); furthermore the *Lymnaea* cDNA probe was used for *in situ* hybridization studies (van Minnen and Schallig, 1990).

With the latter technique negative results were obtained in all species studied, except in *L. stagnalis*. With anti-MIP-C immunoreactive neurons were observed in the cerebral ganglia of all basommatophorans (Fig. 12.8). The cells are localized in the same area in the cerebral ganglia as

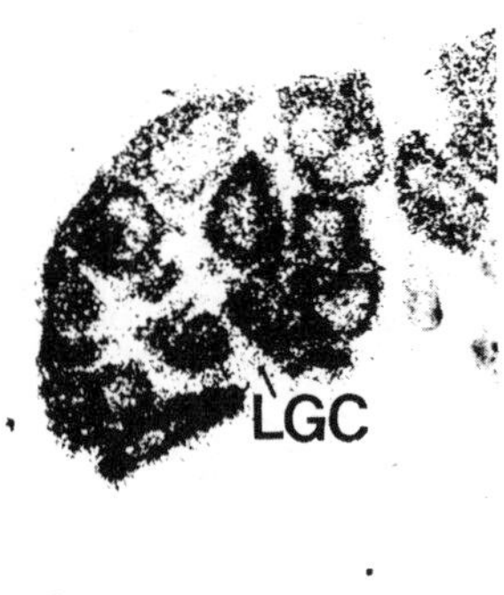

Fig. 12.6. *L. stagnalis. In situ* hybridization using a cDNA probe marking the LGC. ×150.

the LGC of *L. stagnalis* (Fig. 12.1). The axons run towards the median lip-nerves, where they form a neurohaemal area. Outside the cells homologous to the LGC, the CC are also immunoreactive. Furthermore, only in *B. glabrata* (in the pedal and left parietal ganglia) about five immunoreactive neurons were found.

In the stylommatophorans many strongly immunoreactive neurons were found at the mediodorsal side of the cerebral ganglia. It is highly probable that they are homologous to the LGC of *L. stagnalis*. In addition, in *H. aspersa* about 50 large (diameter 60 μm), faintly marked neurons are present in this area (Fig. 12.7). In *L. maximus* a total of about 10 immunoreactive cells are present in the pleural and pedal ganglia. In both species studied immunoreactive fibre tracts were observed in the pleuro–parietal–visceral ganglion complex. Furthermore, from a fibre tract traversing the cerebral commissure immunoreactive fibres branch off and penetrate the perineurium, giving rise to an extensive network of immunoreactive varicose fibres. Part of the varicosities were found on, or in the immediate vicinity of, the DB cells (Fig. 12.7), others on muscle fibres.

Anti-MIP-C immunoreactive neurons were observed in all the ganglia of the CNS of *A. californica*, the highest number in the abdominal ganglion (about 50, diameter 30–150 μm). Furthermore, in the pedal ganglia a cluster of approximately 40 neurons (diameter 20–40 μm) is located. In the other ganglia 10–20 immunoreactive neurons were observed. Fibre tracts were found in all ganglia and in many nerves.

12.1.3. *The 'ACTH'-cells (VD_1 and RPD_2)*

With respect to the occurrence of '*Lymnaea* peptides' across the species border another approach was chosen. Monoclonal antibodies were raised

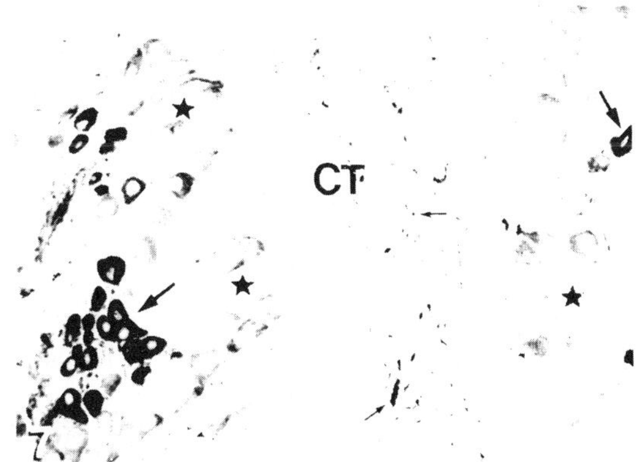

Fig. 12.7. *H. aspersa*, anti-MIP-C staining. In the cerebral ganglia both strongly (arrow) and moderately marked (asterisks) cells are present. In the connective tissue surrounding the cerebral ganglia (CT) many immunoreactive fibres are present (thin arrows). ×60.

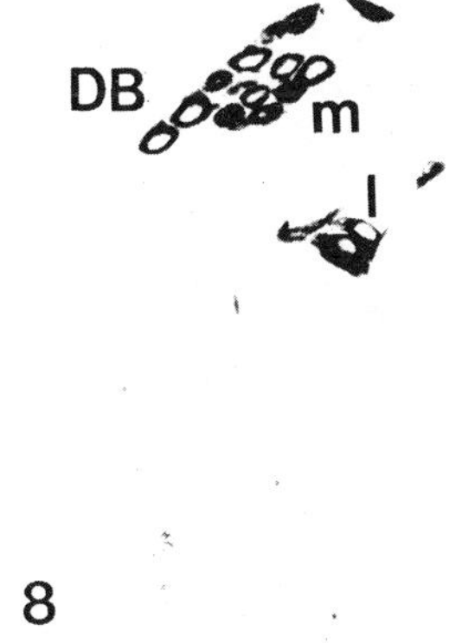

Fig. 12.8. *B. glabrata*, anti-MIP-C staining of the mediodorsal (m) and laterodorsal (l) LGC in the cerebral ganglia are strongly marked. DB, dorsal body. ×75.

to homogenates of whole brains of *L. stagnalis* (van Minnen and Boer, 1987; Boer and van Minnen, 1988) or to a fraction containing only the molecules with a molecular weight in the neuropeptide range (< 30 kD; Kerkhoven *et al.*, 1990). The emerging hybridoma cell lines of the latter fusion experiment were selected in three immunocytochemical screening steps. In the first step the culture supernatant of each hybridoma was applied to tissue sections (7 μm) of one whole CNS of *L. stagnalis*. Those hybridomas that showed a neuron-specific staining were, in the second step, tested on sections of one whole CNS of the guppy. The third step was performed with culture media that were immunopositive with neurons from *L. stagnalis* as well as from the guppy. These were screened on sections of the CNS of the Wistar rat, the wall lizard (*Gekko gecko*), the cockroach (*Periplaneta americana*) and on human brain sections of the cortical areas 4, 10, 17/18, the hypothalamus and of the cerebellum.

The fusion yielded 297 hybridoma cell cultures of which 66 produced antibodies directed to single neurons or to groups of neurons (e.g. the CDC) of *L. stagnalis*. Three of the culture media appeared to react with neurons of the guppy. Of these three one monoclonal antibody (Mab4H5) gave positive results with neurons of all other species tested (Kerkhoven *et al.*, 1990).

In the CNS of *L. stagnalis* Mab4H5 stains two giant neurons (100 μm), one in the visceral and one in the right parietal ganglion (Fig. 12.9). The cells have axons extending into the other ganglia (Fig. 12.1). Ultrastructurally (immunolabelling) the reaction product appeared to be present in small secretory vesicles. The giant neurons were identified as VD_1 and RPD_2 (cf. Soffe and Benjamin, 1980). It has previously been shown that these electrotonically coupled cells are immunoreactive to anti-ACTH (Boer *et al.*, 1979). Physiological studies indicate that they form part of the circuitry involved in the regulation of the respiratory system (van der Wilt, 1990). In the guppy Mab4H5 reacts with neurons in the reticular formation and with fibres in the optical tectum. In the cockroach immunostaining (cell bodies and fibres) was observed in the abdominal ganglia and the suboesophageal ganglia. In the wall lizard neurons in the cerebellum, tectum and telencephalon and many fibres showed a positive reaction. In the rat immunostaining was observed in Purkinje neurons in the cerebellum (Fig. 12.11), in neocortical pyramidal neurons (Fig. 12.10), in mitral cell dendrites and in many nerve fibres. In the human brain immunopositive Purkinje neurons were observed in the cerebellum (Fig. 12.12) and in the cortical areas many pyramidal neurons appeared to be stained.

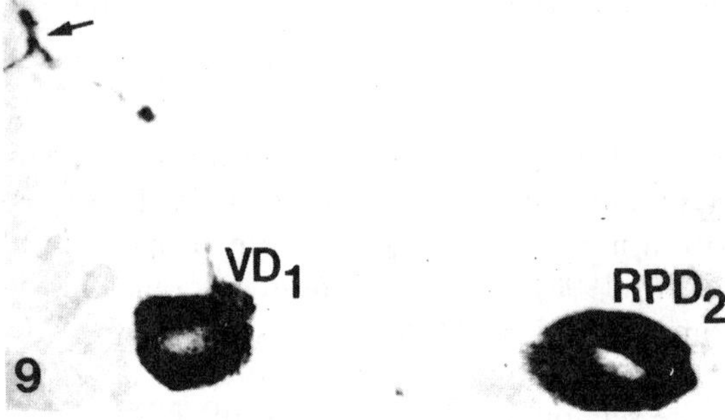

Fig. 12.9. *L. stagnalis*. VD_1 and RPD_2 are stained with Mab4H5. Arrow indicates stained axon. ×170.

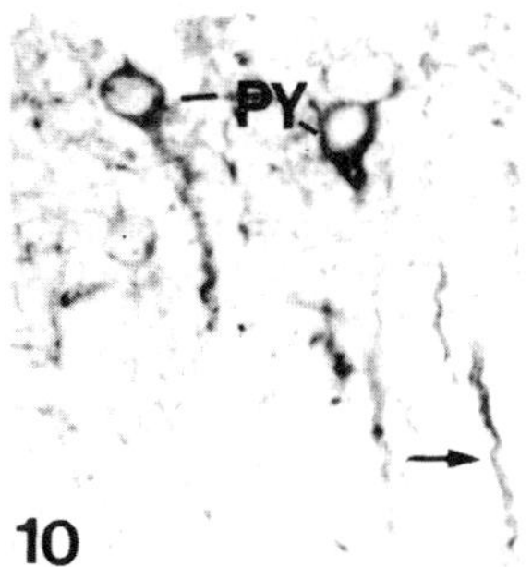

Fig. 12.10. Rat cortex (lamina V) pyramidal neurons (PY) and their axons (arrows) stained with Mab4H5. ×150.

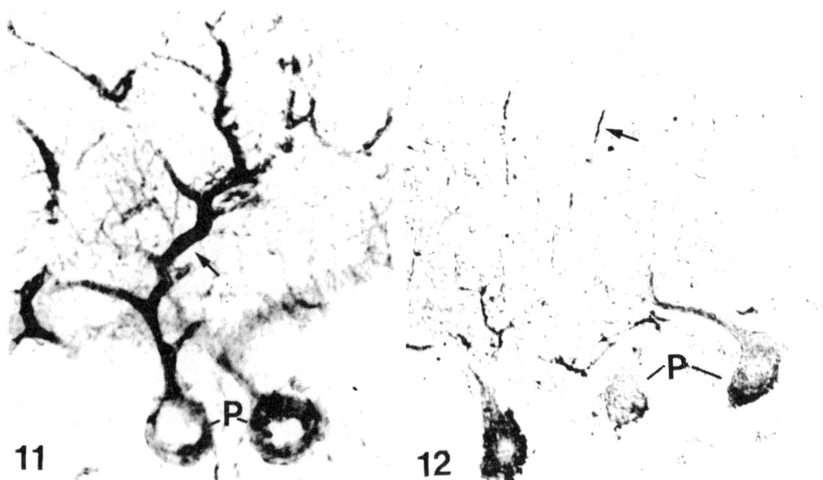

Fig. 12.11 and 12.12. Purkinje cells (P) in the cerebellum of the rat (Fig. 12.11) and man (Fig. 12.12) are stained with Mab4H5. ×190.

The results strongly suggest that Mab4H5 reacts with neuropeptides (e.g. immunostaining only of particular, well-localized cells; presence of antigen in secretory vesicles in *L. stagnalis*). This hypothesis is supported by preliminary observations showing that a peptide isolated with Mab4H5 (affinity chromatography) from rat brain has an exciting effect on the 'ACTH cells' of *L. stagnalis* (electrophysiological recordings). It is not yet clear to what extent the immunoreactive molecules of the different species are structurally related. Nonetheless, the results show that they share the antigenic determinant to Mab4H5.

12.2. **Tissue-specific expression of neuropeptide genes**

Tissue-specific expression of related neuropeptide genes and differential neuropeptide precursor processing leads to the production and release in various parts of the body of structurally different, but related (sets of) neuropeptides. In subsequent sections attention will be paid to peripheral neuropeptide gene expression in *L. stagnalis*.

12.2.1. *CDC-CDCH-I, CDCH-II*

By means of *in situ* hybridization and immunocytochemistry two different peripherally located cell populations were found to express the CDCH genes, namely peripheral neurons and epithelial (exocrine) cells.

(*a*) *Peripheral neurons.* Neurons exclusively reacted with the aCDCH-I antiserum, demonstrating that only the CDCH-I gene is expressed. The neurons and their fibre tracts in peripheral tissues were almost exclusively found throughout the female reproductive tract, with the exception of the albumen gland. Their numbers in the female reproductive organs varied considerably. High numbers were found in the area where the spermoviduct divides into the male and female duct (carrefour area), in the efferent duct of the muciparous gland, in the oothecal gland (Fig. 12.13) and the vagina. The neurons are located in the connective tissue surrounding the female tract, as well as in between the glandular epithelial cells of the tract. With *in situ* hybridization all these neurons showed a hybridization signal. Many immunoreactive varicosities were observed on muscle fibres surrounding the female tract, on glandular cells and on ciliated epithelial cells. This suggests that the aCDCH-I immunoreactive substances in the neurons function as neurotransmitters. Furthermore,

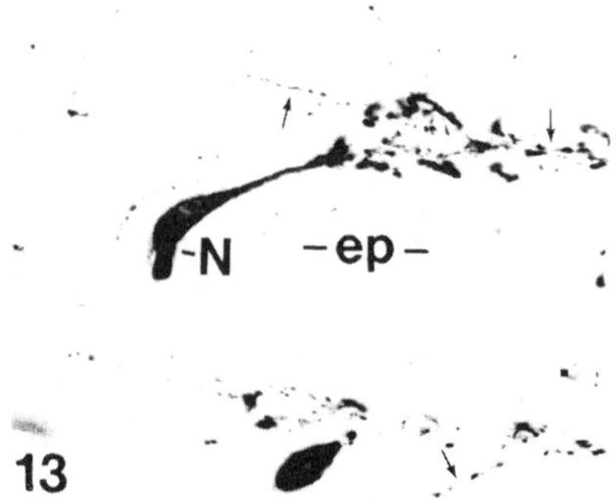

Fig. 12.13. Neuron (N) stained in the oothecal gland by aCDCH-I. Many varicosities (arrows) are present at the basal and lateral sides of the secretory cells of the epithelium (ep). ×340.

aCDCH-I immunoreactive giant neurons (cell size up to 100 μm) were found in low numbers in the gonadal branch of the intestinal nerve. In addition, outside the reproductive tract, low numbers of neurons were found in the connective tissue underneath the skin of the mantle edge and the head. These neurons have the appearance of primary sensory neurons (cf. Zijlstra, 1972).

(*b*) *Epithelial cells*. Epithelial secretory exocrine cells of the male part of the reproductive tract were found to react either with aCDCH-I or aCDCH-II, but in no case with both antisera. The antisera react with the secretory granules. In the sperm duct almost exclusively aCDCH-I reactive cells were found, while in the prostate gland, besides non-reactive cells, both aCDCH-I and aCDCH-II immunoreactive cells (Fig. 12.14) and in the penis only aCDCH-II cells were observed. Also with *in situ* hybridization the epithelial cells showed a hybridization signal (only the sperm duct was studied). In the lumina of the sperm duct, prostate gland and vas deferens many aCDCH-I and aCDCH-II immunoreactive granules were observed. After copulation, immunoreactive material was present in the vagina of the female copulant. This observation demonstrates that CDCH genes derived material from the male accessory sex glands is transferred during copulation.

12.2.2. *LGC-MIPs*

Sections of whole *L. stagnalis* were screened with anti-MIP-C for the possible expression of MIP genes in peripheral tissues. In none of the tissues were immunoreactive neurons or epithelial cells observed. Yet it seems highly likely that in the alimentary tract an insulin-like substance is produced. First, previous observations (unpublished) using an anti-bovine insulin antibody have shown epithelial cells to react in the pyloric area of

Fig. 12.14. *L. stagnalis*, prostate gland showing aCDCH-I (cells with grey granules) and aCDCH-II (cell with black granules) staining. Asterisk: unstained cells. ×190.

the stomach, and partially purified extracts of the digestive tract appeared to stimulate the uptake of 2-deoxyglucose by rat fat cells and to compete with pork insulin (Hemminga, 1984). Furthermore, HPLC studies have shown structural similarity of *L. stagnalis* gut insulin and pork insulin (Ebberink and Joosse, 1985).

12.2.3 *'ACTH cells' – Mab4H5*

Fibres of VD_1 and RPD_2 leave the CNS through the pallial nerve complex (Fig. 12.1). They spread out in the connective tissue sheath surrounding the CNS, where release of the product is thought to take place. One of the targets of the pallial nerves is the osphradium, which would measure the acidity and the oxygen content of the environment. In the osphradium ganglion two neurons appear to be immunoreactive to Mab4H5. Their fibres leave the ganglion and innervate the surface of the osphradium epithelium, where they seem to have sensory endings extending to the lumen of the osphradium tubule.

The salivary glands contain seven cell types of which one, the pseudochromosome cell (Boer *et al.*, 1967) is positive to Mab4H5. The product of the salivary glands is released into the alimentary tract. The epithelium of the tract receives innervation with Mab4H5 positive fibres. In some preparations neurons in the mantle and tentacle react positively.

12.3. **Functional significance of central and peripheral expression of neuropeptide genes**

There are examples indicating functional similarity of structurally related peptides in essentially unrelated species. Thus, as in the vertebrates the enkephalins would also in molluscs be involved in suppressing the effects of noxious stimuli (Stefano *et al.*, 1987) and the vasopressins of insects in the regulation of the water and ion balance (Proux *et al.*, 1987). On the other hand, in many instances evidence is lacking, or opposes the idea of some kind of unity in function. It is not clear what might be the function of the head activator of *Hydra* in the human brain, whereas the adipokinetic hormones of insects and the red pigment-concentrating hormone of crustaceans, although structurally closely related, obviously serve different functions.

As no hypoglycaemic factor was found in *L. stagnalis* (Hemminga, 1984), the insulin-related peptides of this snail would also seem to fall into the latter category. However, similarities in function become apparent if we look at the MIPs in a broader context, i.e. if we consider that the MIPs, just like the vertebrate insulins and insulin-related peptides, are involved in the regulation of various aspects of body growth (cf. Hadley, 1988; Joosse, 1988). The fact that in other pulmonates neurons that have been implicated in the regulation of growth (Wijdenes and Runham, 1977; Roubos and van de Ven, 1987) react with anti-MIP-C (van Minnen and Schallig, 1990) supports the idea of functional similarities of peptides across the species border. Whether in *Aplysia* MIP-like peptides also play a role in growth regulation is not clear, as the anti-MIP-C positive neurons of this species have not yet been studied physiologically.

The Mab4H5 antibody stains a variety of cells in different species. On the basis of comparing the functions of immunoreactive cell types (pyramidal neurons, Purkinje cells, VD_1 and RPD_2 of *L. stagnalis*), little can be said at present about possible unity in function of the immunopositive peptide(s).

With respect to the CDC-like peptides, on the other hand – so far studied only in molluscs – unity in function seems clear, *viz*. involvement in the regulation of egg-mass formation and egg-laying behaviour

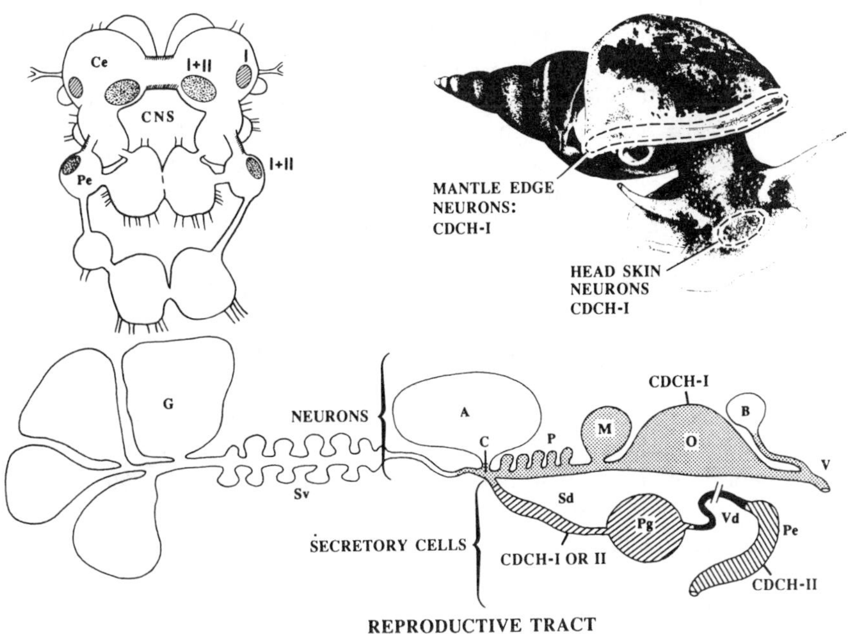

Fig. 12.15. Expression of the CDCH-genes in *L. stagnalis*. The CDCH genes are expressed in neurons in the central nervous system (CNS), in the skin (head region) and mantle edge, and in the female part of the reproductive tract (dotted areas), while expression in epithelial (exocrine) cells is found in the male part of the reproductive tract (hatched areas). A, Albumen gland; B, bursa copulatrix; C, carrefour area; Ce, cerebral ganglion; G, gonad; M, muciparous gland; O, oothecal gland; P, pars contorta; Pe, penis; Pg, prostate gland; Pl, pleural ganglion; Sd, sperm duct; Sv, seminal vesicles; V, vagina; Vd, vas deferens.

(Geraerts *et al.*, 1988). The morphological distribution of the cells expressing the CDCH genes elucidates and supports this hypothesis (Fig. 12.15): expression of the genes is virtually restricted to cells known to regulate reproduction (CDC) and to cells that can be expected to be involved in the regulation of this process, i.e. cells in the reproductive organs. Furthermore, the small cerebral neurons expressing CDCH-I can also be implicated in the regulation of the reproduction process: their axons contact the CDC synaptically. The CDCH-I positive sensory neurons in the head and mantle epidermis may well play a role in conveying from the environment signals related to egg-laying behaviour.

The results (especially those on the CDCH system) support the theory that members of neuropeptide families in concerted action regulate various aspects of a particular general physiological process (CDCH system: reproduction; MIP-LGC system: growth). In the vertebrates several other examples of this principle can be mentioned, e.g. the central and peripheral expression of neuropeptide genes whose products are involved in the regulation of feeding and digestion (CCK/gastrin, secretin/glucagon) of reproduction and reproductive behaviour (e.g. LHRH-like peptides) of water and ion homeostasis (vasopressins, neurotensins) and of adaptation and stress (POMC, proenkephalin derived peptides) (Hadley, 1988; van Minnen, 1988). Apparently the neuropeptides can act as neurohormones (those released into the circulation) or as neurotransmitters (those released at synapses or non-synaptically in nervous or other tissue). Some may even have a pheromone-like function, such as the LHRH-like peptides released into the milk and semen of the rat (Baram *et al.*, 1977; van Minnen, 1988) and the CDCH-I and II derived peptides of the semen of *L. stagnalis*.

The theory appears to have important implications for further research into functional aspects of neuropeptidergic systems. Firstly, it indicates that investigations should not be restricted to the central nervous system and to the classical endocrine organs, as obviously expression of neuropeptide genes in peripheral organs can give important clues about the functions of particular gene products. Secondly, it should be considered that general physiological processes such as feeding and digestion, reproduction, adaptation and stress, regulation of water balance and ion balance, growth and development, are intimately linked. The fact that anti-MIP-C positive cerebral growth-regulating neurons of *H. aspersa* inhibit DB activity (i.e. inhibit reproduction) may serve as an example to show linkage (antagonism) of growth and reproduction in this snail. It would seem difficult to attribute functions within the framework of a particular general physiological process to peptides that serve primarily to establish the links between the processes or to peptides that regulate phenomena related to all main physiological processes (e.g. changes in heart beat frequency, blood pressure, respiration, etc.).

References

Baram, T., Koch, Y., Hazum, E. and Fridkin, M. (1977). Gonadotrophin releasing hormone in milk. *Science*, **198**, 300–2.

Bjenning, C. and Holmgren, S. (1988). Neuropeptides in the fish gut. *Histochemistry*, **88**, 155–63.

Bodenmüller, H. and Schaller, H. C. (1981). Conserved amino acid sequence of a neuropeptide, the head activator, from the coelenterates to humans. *Nature*, **293**, 579–80.

Boer, H. H. and Minnen, J. van (1988). Immunocytochemistry of hormonal peptides in molluscs: optical and electron miscroscopy and the use of monoclonal antibodies. In: Thorndyke, M. C. and Goldsworthy, G. J. (eds.), *Neurohormones in Invertebrates*. Cambridge University Press, Cambridge, pp. 19–41.

Boer, H. H., Schot, L. P. C., Roubos, E. W., Maat, A. ter, Lodder, J. C., Reichelt, D. and Swaab, D. F. (1979). ACTH-like immunoreactivity in two electrotonically coupled giant neurons in the pond snail *Lymnaea stagnalis*. *Cell Tiss. Res.*, **202**, 231–40.

Boer, H. H., Wendelaar Bonga, S. E. and Rooyen, N. van (1967). Light and electron microscopical investigations on the salivary glands of *Lymnaea stagnalis* L. *Z. Zellforsch.*, **76**, 228–47.

Brussaard, A. B., Ebberink, R. H. M., Schluter, N. C. M., Kits, K. S. and Maat, A. ter (1990). Discharge induction in molluscan peptidergic cells requires a specific set of four autoexcitatory neuropeptides. *Neuroscience*. (In press).

Conway, K. M. and Gainer, H. (1987). Immunocytochemical studies of vasotocin, mesotocin, and the neurophysins in the *Xenopus* hypothalamo-neurohypophysial system. *J. Comp. Neurol.*, **264**, 494–508.

Cruz, L. J., Santos, P. de, Zafaralla, G. C., Ramillo, C. A., Zeikus, R., Gray, W. R. and Oliveira, B. M. (1987). Neuropeptides in the fish gut. *J. Biol. Chem.*, **262**, 15821–4.

Dictus, W. J. A. G., Jong-Brink, M. de and Boer, H. H. (1987). A neuropeptide (calfluxin) is involved in the influx of calcium into mitochondria of the albumen gland of the freshwater snail *Lymnaea stagnalis*. *Gen. Comp. Endocrinol.*, **65**, 439–50.

Ebberink, R. H. M. and Joosse, J. (1985). Molecular properties of various snail peptides from brain and gut. *Peptides*, **6** (Suppl. 3), 451–7.

Geraerts, W. P. M., Maat, A. ter and Vreugdenhil, E. (1988). The peptidergic neuroendocrine control of egg-laying behaviour in *Aplysia* and *Lymnaea*. In: Laufer, H. and Downer, R. G. H. (eds). *Invertebrate Endocrinology*, Vol 2. Alan R. Liss, New York, pp. 141–231.

Hadley, M. E. (1988). *Endocrinology*. Prentice-Hall, Inglewood Cliffs, NJ.

Hemminga, M. A. (1984). Regulation of glycogen metabolism in the freshwater snail *Lymnaea stagnalis*. Ph.D. thesis, Free University, Amsterdam.

Joosse, J. (1988). The hormones of molluscs. In: Laufer, H. and Downer, R. G. H. (eds), *Invertebrate Endocrinology*, Vol. 3, *Endocrinology of Selected Invertebrate Types*. Alan R. Liss, New York, pp. 89–140.

Kerkhoven, R. M., Minnen, J. van and Boer, H. H. (1990). Neuron specific monoclonal antibodies raised against homogenate of the pond snail *Lymnaea stagnalis* immunoreact with neurones in the central nervous system of the cockroach, the guppy, the wall lizard, the rat and man. *J. Chem. Neuroanat.* (In press).

Landis, M. D. (1985). Promise and pitfalls in immunocytochemistry. *TINS*, July, pp. 312–17.

Leung, M. K., Boer, H. H., Minnen, J. van, Lundy, J. and Stefano, G. B.

(1990). Evidence for an enkephalinergic system in the nervous system of the pond snail *Lymnaea stagnalis*. *J. Brain Res.* (In press).

Lynch, D. R. and Snyder, S. H. (1986). Neuropeptides: multiple molecular forms, metabolic pathways, and receptors. *Ann. Rev. Biochem.*, **55**, 773–99.

Miller, G. A. and Benzer, S. (1983). Monoclonal antibodies cross-reactions between *Drosophila* and human brain. *Proc. Natl. Acad. Sci. USA*, **80**, 7641–5.

Minnen, J. van (1988). Production and exocrine secretion of LHRH-like material by the male rat reproductive tract. *Peptides*, **9**, 515–18.

Minnen, J. van and Boer, H. H. (1987). Generation and application of monoclonal antibodies raised against homogenates of whole central nervous systems of the pond snail *Lymnaea stagnalis*. *Proc. Kon. Ned. Akad. Wetensch.*, Ser. C, **90**, 193–201.

Minnen, J. van and Schallig, H. (1990). Demonstration of insulin-related substances in the central nervous systems of pulmonates and *Aplysia californica*. *Cell Tiss. Res.*, **260**, 38–86.

Minnen, J. van and Sokolove, P. G. (1981). Neurosecretory cells in the central nervous system of the giant garden slug *Limax maximus*. *J. Neurobiol.*, **12**, 297–301.

Minnen, J. van, Haar, Ch. van de, Raap, A. K. and Vreugdenhil, E. (1988). Localization of ovulation hormone-like neuropeptide in the central nervous system of the snail *Lymnaea stagnalis* by means of immunocytochemistry and *in situ* hybridization. *Cell Tiss. Res.*, **251**, 477–84.

Minnen, J. van, Reichelt, D. and Lodder, J. C. (1979). An ultrastructural study of the neurosecretory canopy cell of the pond snail *Lymnaea stagnalis*, with the use of the horseradish peroxidase tracer technique. *Cell Tiss. Res.*, **204**, 453–62.

Minnen, J. van, Smit, A. B. and Joosse, J. (1989). Central and peripheral expression of genes coding for egg-laying and insulin-related peptides in a snail. *Arch. Histol. Cytol.*, **52** (Suppl.), 241–52.

Nieuwenhuis, R. (1985) *Chemoarchitecture of the Brain*. Springer-Verlag, Berlin.

Proux, J. P., Miller, C. A., Li, J. P., Carney, R. L., Girardie, A., Delaage, M. and Schooley, D. A. (1987). Identification of an arginine vasopressin-like diuretic hormone from *Locusta migratoria*. *Biochem. Biophys. Res. Commun.*, **149**, 180–6.

Roubos, E. W. (1984). Cytobiology of the ovulation hormone producing neuroendocrine caudodorsal cells of *Lymnaea stagnalis*. *Int. Rev. Cytol.*, **89**, 295–347.

Roubos, E. W. and Ven, A. M. H., van de (1987). Morphology of neurosecretory cells in basommatophoran snails homologous with egg-laying and growth hormone-producing cells of *Lymnaea stagnalis*. *Gen. Comp. Endocrinol.*, **67**, 7–23.

Sawyer, W. H., Deyrup-Olsen, I. and Martin, A. W. (1984). Immunological and biological characteristics of the vasotocin-like activity in the head ganglia of gastropod molluscs. *Gen. Comp. Endocrinol.*, **54**, 97–108.

Scharrer, B. (1987). Neurosecretion. Beginnings and new directions in neuropeptide research. *Ann. Rev. Neurosci.*, **10**, 1–17.

Schmidt, E. D. and Roubos, E. W. (1987). Morphological basis for non-synaptic communication within the central nervous system by exocytotic release of secretory material from the egg-laying stimulating neuro-endocrine caudodorsal cells of *Lymnaea stagnalis*. *Neuroscience*, **20**, 247–57.

Schot, L. P. C., Boer, H. H. and Montagne-Wajer, C. (1984). Characterization of multiple immunoreactive neurons in the central nervous system of the pond snail *Lymnaea stagnalis*. *Histochemistry*, **81**, 373–8.

Smit, A. B., Vreugdenhil, E., Ebberink R. H. M., Geraerts, W. P. M., Klootwijk, J. and Joosse, J. (1988). Growth-controlling molluscan neurons produce the precursor of an insulin-related peptide. *Nature*, **331**, 535–8.

Soffe, S. R. and Benjamin, P. R. (1980). Morphology of two electrotonically-coupled giant neurosecretory neurons in the snail, *Lymnaea stagnalis*. *Comp. Biochem. Physiol.*, **67A**, 35–46.

Stefano, G. B., Sardesai, R., Ndubuka, C., Brown, D., Pratt, S., Braham, E., Leung, M. K. and Hiripi, L. (1987). Opiates influence food consumption and thermal responsiveness in the land snails *Helix aspersa* and *Helix pomatia*. In: Boer, H. H. *et al.* (eds), *Neurobiology. Molluscan Models*. North Holland, Amsterdam, pp. 261–4.

Thorndyke, M. C. (1986). Immunocytochemistry and evolutionary studies with particular reference to peptides. In: Polak, J. M. and Noorden, S. van (eds), *Immunocytochemistry*. J. Wright and Sons, Bristol. pp. 300–27.

Veenstra, J. A., Romberg-Privee, H. M., Schooneveld, H. and Polak, J. M. (1985). Immunocytochemical localization of peptidergic neurons and neurosecretory cells in the neuroendocrine system of the Colorado potato beetle with antisera to vertebrate regulatory peptides. *Histochemistry*, **82**, 9–18.

Vreugdenhil, E., Jackson, J. F., Bouwmeester, T., Smit, A. B., Minnen, J. van, Heerikhuizen, H. van, Klootwijk, J. and Joosse, J. (1988). Isolation, characterization and evolutionary aspects of a cDNA clone encoding multiple neuropeptides involved in a stereotyped egg-laying behaviour of the freshwater snail *Lymnaea stagnalis*. *J. Neurosci.*, **81**, 4184–91.

Wijdenes, J. and Runham, N. W. (1977). Studies on the control of growth in *Agriolimax reticulatus* (Mollusca, Pulmonata). *Gen. Comp. Endocrinol.*, **31**, 154–6.

Wijdenes, J., Minnen, J. van and Boer, H. H. (1980). A comparative study on neurosecretion demonstrated by the alcian blue-alcian yellow technique in three terrestrial pulmonates (Stylommatophora). *Cell Tiss. Res.*, **210**, 47–56.

Wijdenes, J., Schluter, N. C. M., Gomot, L. and Boer, H. H. (1987). In the snail *Helix aspersa* the gonadotropic hormone-producing dorsal bodies are under inhibitory nervous control of putative growth hormone-producing neuroendocrine cells. *Gen. Comp. Endocrinol.*, **68**, 224–9.

Wilson, A. C. (1985). The molecular basis of evolution. *Sci. Am.*, **253**(4), 164.

Wilt, G. J. van der (1990). The neuronal basis of respiratory behaviour of the freshwater pulmonate snail *Lymnaea stagnalis*. Ph.D. thesis, Free University, Amsterdam.

Zijlstra, U. (1972). Distribution and ultrastructure of epidermal sensory cells in the freshwater snails *Lymnaea stagnalis* and *Biomphalaria pfeifferi*. *Neth. J. Zool.*, **22**, 283–98.

Part III

Arthropod studies

13 *Christian Walther, Klaus E. Zittlau, Harald Murck and Ronald J. Nachman*

Peptidergic modulation of synaptic transmission in locust skeletal muscle

13.1. **Introduction**

Since the discovery of aminergic modulation of neuromuscular transmission in the jumping muscle of the locust (Hoyle, 1975; Evans and O'Shea, 1977) an unexpected multiplicity of peptidergic modulation in this preparation has been found. The scope of this article is to outline briefly our work on pre- and postsynaptic neuropeptide effects in a restricted fibre population of this muscle (for a previous review on modulation in the jumping muscle see Evans and Myers, 1986a). Of the two excitatory motoneurons innervating this muscle one serves for slow movements and postural control ('slow excitor'; SETi), the other for strong, fast contractions (e.g. during jumping; 'fast extensor'; FETi; terminology after Hoyle, 1978). Peptidergic fine control is quite effective on contractions evoked by SETi and apparently much less so on those evoked by FETi.

Three levels at which modulatory influences may occur can be distinguished: transmitter release, postsynaptic membrane potential changes and excitation contraction coupling. In the case of certain FMRFamide-like peptides actions at all three levels have been demonstrated (e.g. Walther *et al.*, 1984; Schiebe and Walther, 1988, 1989). With regard to the action of the excitatory transmitter glutamate (cf. Usherwood, 1978) to our knowledge no involvement of neuropeptide-mediated modulation has been described yet. The same applies to the depolarization-activated Ca^{2+} current commonly found in insect skeletal muscles (e.g. Ashcroft and Stanfield, 1982) though quite recently, in lobster skeletal muscle, an effect of proctolin on single Ca^{2+} channels has been reported (Krouse *et al.*, 1988). Our knowledge of the modulation of excitation contraction coupling is still rather limited. We therefore compare here the peptide effects studied at only two levels, i.e. transmitter release, evoked by SETi-stimulation, and resting conductance of the muscle membrane in the locust (*Schistocerca gregaria*) with an emphasis on the putative peptide receptors involved.

13.2. **Overview of the peptides investigated**

Before presenting our findings some background information about the investigated peptides is given in brief (see Table 13.1 for the amino acid sequences and the relative effectiveness of the various peptides). Procto-

Table 13.1 Amino acid sequences and relative physiological potency of peptides*

Proctolin	Arg-Tyr-Leu-Pro-Thr	−/+
FMRFamide	Phe-Met-Arg-Phe-NH_2	+/+
FLRFamide	Phe-Leu-Arg-Phe-NH_2	+/+
YGGFMRFamide	Tyr-Gly-Gly-Phe-Met-Arg-Phe-NH_2	+/+
YGGFLRFamide	Tyr-Gly-Gly-Phe-Leu-Arg-Phe-NH_2	+/+
Met^5-Enk-Arg^6-Phe^7 (YGGFMRF)	Tyr-Gly-Gly-Phe-Met-Arg-Phe	+/−
Metenkephalin	Tyr-Gly-Gly-Phe-Met	−/−
Leucomyosuppressin (LMS)	pGlu-Asp-Val-Asp-His-Val-Phe-Leu-Arg-Phe-NH_2	+/+
SchistoFLRFamide	Pro-Asp-Val-Asp-His-Val-Phe-Leu-Arg-Phe-NH_2	n.t.
Leucosulfakinin (LSK)	Glu-Gln-Phe-Glu-Asp-Tyr-Gly-His-Met-Arg-Phe-NH_2	−/−
Leucosulfakinin II (LSKII)	pGlu-Ser-Asp-Asp-Tyr-Gly-His-Met-Arg-Phe-NH_2	+/+
CCK-Octapeptide	-Met-Gly-Trp-Met-Asp-Phe-NH_2	−/−
Adipokinetic hormone (AKH)	pGlu-Leu-Asn-Phe-Thr-Pro-Asn-Trp-Gly-Thr-NH_2	−/−
Catch-relaxing peptide (CARP)	Ala-Met-Pro-Met-Leu-Arg-Leu-NH_2	+/+
Crustacean cardioactive peptide (CCAP)	Pro-Phe-Cys-Asn-Ala-Phe-Thr-Gly-Cys-NH_2	+/−

* + (−) in front and behind the slash in column 3, indicate whether (or not) the peptide at a concentration of 10^{-6} M is active at the pre- and post-synaptic side, respectively.

n.t.: not tested

Relative potencies of peptides effective at 10^{-6} M:

(a) Potentiation of transmitter release:
YGGFM(L)RFamide $\gg$ FM(L)RFamide $\simeq$ LSKII > CARP > CCAP $\simeq$ YGGFMRF

(b) Reduction of muscle nonsynaptic membrane resistances:
Proctolin $\gg$ YGGFM(L)RFamide > LMS > FM(L)RFamide > CARP > LSKII

lin has been demonstrated in certain motoneurons of insects (e.g. Bishop *et al.*, 1981) and seems to occur in SETi (Worden *et al.*, 1985). In cockroach muscle co-release with glutamate from excitatory nerve endings has been shown (Adams and O'Shea, 1983). It potentiates neurally evoked contractions and causes contractures at concentrations $\geq 10^{-9}$ M (e.g. May *et al.*, 1979; O'Shea and Adams, 1986).

FMRF- and FLRFamide-like peptides, in particular the synthetic analogue YGGFM(L)RFamide, potentiates SETi-evoked contractions (Walther *et al.*, 1984; Evans and Myers, 1986b). The presence of several FM(L)RFamide-related peptides in locust CNS seems likely from immunohistological investigations (e.g. Myers and Evans, 1985a,b; Walther and Schäfer, 1988) and work with extracts from neurohaemal organs on the extensor tibiae preparation (Walther and Schiebe, 1987; Schiebe *et al.*, 1989). Quite recently a first, N-terminally extended, FLRFamide-peptide, termed SchistoFLRFamide, has been isolated from locust CNS (Robb *et al.*, 1989). No FM(L)RFamide-like immunoreactivity could be demonstrated within motor axons of the locust, and it is quite likely that such peptides circulate within the haemolymph from neurohaemal organs to the skeletal musculature.

Leucomyosuppressin (LMS) has been isolated from cockroach brain (Holman *et al.*, 1986) and is unique among a dozen other cockroach neuropeptides in that it depresses spontaneous contractions of the hindgut while all others are powerful stimulants. Likewise it depresses neurally evoked contractions (at 10^{-7}–10^{-6} M) in locust skeletal muscle (Cuthbert and Evans, 1989). The amino acid sequence of LMS is homologous with that of SchistoFLRFamide except for the N-terminal amino acid.

The leucosulphakinins (LSK, LSKII) from cockroach brain (Nachman *et al.*, (1986a,b) contain the C-terminal sequence-HMRFamide. They are closely related to the FMRFamide family (Nachman *et al.*, 1988). In the change from phenylalanine to histidine the aromatic character is maintained at that position.

Adipokinetic hormone (AKH) from locust corpus cardiacum (Stone *et al.*, 1976) or related peptides seem to occur in locust CNS (Schooneveld *et al.*, 1983). In the most proximal part of the jumping muscle the rhythm of myogenic slow contractions is accelerated by AKH (O'Shea, 1985).

Crustacean cardioactive peptide (CCAP; kindly supplied by R. Keller), first isolated from pericardial organs of a crab (Stangier *et al.*, 1987) and recently demonstrated in the locust (Stangier *et al.*, 1989), enhances hindgut contractions and seems to circulate as a hormone in the locust.

Catch-relaxing peptide (CARP; kindly supplied by Y. Muneoka) was isolated from the bivalve mollusc *Mytilus* (Hirata *et al.*, 1987). It is biologically active on various molluscan muscles and neurons (e.g. Kiss, 1988; Hirata *et al.*, 1989).

13.3. **Effects on transmitter release**

These were investigated by voltage-clamping synaptic currents evoked by SETi stimulation and applying peptides at a concentration of 10^{-6} M (unless indicated otherwise). Amplitude changes of synaptic current ('ejc', for 'excitatory junctional current' in arthropod muscle; cf. Fig. 13.1) are indicative of changes in presynaptic transmitter output. If the average quantal content of an ejc is increased or decreased the coefficient of variation (CV; i.e. variance/mean of the ejc amplitude) changes inversely (Del Castillo and Katz, 1954). With the peptides tested we gener-

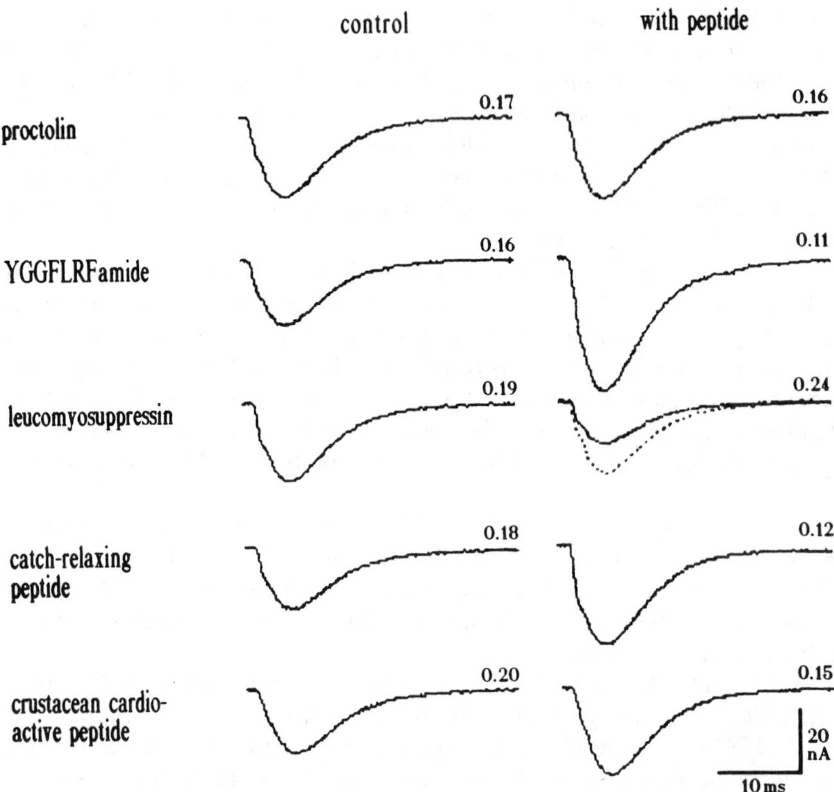

Fig. 13.1. Effects of various peptides on excitatory synaptic current. Currents were measured immediately before and ≥2 min after application of peptides. For leucomyosuppressin only, the measurement was started ~ 1 min after application; a second one was made 6 min later (interrupted trace). Averages of 20 currents measured in voltage clamp are shown. The numbers above each trace indicate the coefficient of variation for the mean amplitude from 50 signals (cf. text). Holding potential −70 mV. $(Ca^{2+})_0$ = 1·25 mM; $(Mg^{2+})_0$ = 12 mM. Stimulation rate 1 Hz. Concentration of proctolin 10^{-10} M, of YGGFLRFamide 10^{-8} M and of all others 10^{-6} M. Measurements from one fibre.

ally observed concomitant changes of CV (cf. Fig. 13.1) together with the changes in mean ejc-amplitude ('Δejc') though with small amplitude-differences considerably larger sample sizes might have revealed the change in CV more clearly.

With proctolin, tested at 10^{-10} M, a concentration which already leads to a significant change in muscle membrane resistance (cf. below) and potentiation of SETi-evoked contractions, no change in ejc amplitude and CV occurs (Fig. 13.1). Similarily, in a crayfish neuromuscular preparation, where proctolin strongly enhances the tension generated by a given level of depolarization, nanomolar concentrations of this peptide had no effect on the excitatory junction potentials (ejps; Bishop *et al.*, 1987). It cannot be ruled out that transmitter release is affected by higher concentrations, but a direct test of this was prohibited by the rather strong SETi-evoked twitches or spontaneous contractures which occurred.

With YGGFMRFamide it has been shown, by direct application of glutamate, that it does not affect the postsynaptic transmitter sensitivity (Walther *et al.*, 1984). It enhances quantal content of focally evoked and extracellularly recorded ejcs (Schiebe and Walther, 1988). Consistent with this, the CV of ejps was found to be reduced when these were potentiated by YGGFMRFamide (Walther and Schiebe, 1987). YGGFMRFamide and YGGFLRFamide were equally effective, raising the ejc amplitude by ~ +100% at 10^{-8} M (Fig. 13.1). They were the most potent of all peptides tested. FM(L)RFamide was clearly less potent (Δejc ~ +120%) and Met^5-Enk-Arg^6-Phe^7 (YGGFMRF) was hardly active (Δejc ~ +25%). Both findings are in line with previous investigations on SETi-evoked contractions (e.g. Walther *et al.*, 1984).

Leucomyosuppressin causes a transient depression of release (by ~ −50%; Fig. 13.1). Depression by LMS has been demonstrated before in a mealworm neuromuscular preparation (Yamamoto *et al.*, 1988). That the effect in our preparation begins to fade after a few minutes corresponds to the finding that high concentrations (10^{-7}–10^{-6} M) of SchistoFLRFamide (cf. above) cause a transient reduction of SETi-evoked contractions (Robb *et al.*, 1989). The effects of the other peptides tested here did not, however, show signs of tachyphylaxis within 5–10 min, and in the case of YGGFMRFamide and proctolin were demonstrated to persist practically undiminished for at least 30 min in the continuous presence of peptide. That transmitter release can be modified in opposite directions by different members of the FM(L)RFamide peptide family is analogous to the finding in locust heart of both accelerating and inhibitory FMRFamide effects which may be due to the presence of two types of receptors (Cuthbert and Evans, 1989).

While unsulphated leucosulphakinin (LSK) was ineffective, unsulphated LSKII (Δejc ~ +100%) was about as effective as FM(L)RFamide. Cholecystokinin (CCK), which bears some structural homology to the leukosulphakinins, was ineffective (the C-terminal octapeptide was tested). Met-enkephalin and adipokinetic hormone were ineffective, whereas catch-relaxing peptide (CARP) and crustacean cardioactive peptide (CCAP) gave moderate potentiations (Δejc ~ +50% and ~ +25%, respectively; cf. Fig. 13.1). As for the latter, this accords with the observation that CCAP, at least at 10^{-5} M, causes potentiation of SETi-evoked contractions (M. Schiebe, personal communication).

From a comparison of the various peptide actions several considerations with regard to the presynaptic peptide receptors spring to mind. (1) Only C-terminally amidated peptides are potent modulators of transmitter release. (2) There are probably at least two different populations of receptors for FM(L)RFamide-like peptides, since both potentiating and depressant effects are produced by different members of this peptide family. The specific nature of the action of LMS cannot be due to its possessing the sequence FLRF instead of FMRF since (YGG)FMRFamide and (YGG)FLRFamide are equally effective potentiators. It seems, therefore, that the extended N-terminal sequence is relevant. Likewise the high potency of YGGFM(L)RFamide (compared to FM(L)RFamide) and the difference between the two leucosulphakinins emphasize the discriminative relevance of the amino acid sequence near the N-terminal end.

The fact that CARP and CCAP are effective leads to the question of whether their effects involve a receptor population distinct from those postulated for the FM(L)RFamide-like peptides. One could, for example, assume that there is a common receptor for FM(L)RFamide-like peptides, distinct from that for LMS, which recognizes the sequence Met-Arg-X-NH_2 where X could be phenylalanine, leucine (both hydrophobic) and perhaps certain other similar amino acids but not tyrosine (slightly negatively charged because of –OH). This would be consistent with the results with CCK8, LSK/LSKII and CARP. The effectiveness of CCAP – though clearly lower than that of CARP – is not so readily explained by this scheme. Perhaps there is a third type of presynaptic peptide receptor, or alternatively a weak cross-reactivity by CCAP with a –FM(L)RFamide receptor must be considered a possibility. This would be biologically irrelevant if physiological CCAP concentrations are considerably lower than 10^{-6} M.

13.4. **Effects on the muscle membrane conductance**

YGGFMRFamide causes a decrease in membrane resting conductance and thus prolongs the SETi-evoked excitatory junction potential (Walther and Schiebe, 1987). In parallel with this, a depolarization is observed (Walther *et al.*, 1984; Schiebe and Walther, 1988) which, however, is partially artificial since depolarizing current enters the fibre through the leaks around the microelectrodes. The physiological significance of this primarily non-synaptic membrane effect may be seen in the longer and probably slightly higher depolarization *beyond the electromechanical threshold* attained by the excitatory junction potential in the presence of peptide. The resting conductance consists mainly of a K^+ and a Cl^- conductance. The conductance decrease is not caused by an effect on the latter (Schiebe and Walther, 1988) but due to inactivation of a K^+-leak conductance. This is characterized by its susceptibility to block by divalent cations such as Co^{2+} or Cd^{2+} (Walther and Zittlau, 1988). There is indirect evidence that the peptide effect is mediated via G-proteins (Murck et al., 1989) but it is not yet clear whether or not a second messenger is also involved.

The tests with the various peptides were again performed with a standard concentration of 10^{-6} M except for those substances which are particularly potent. The effects on input conductance ('ΔG_{in}') were determined from voltage current curves obtained both in the presence and absence of peptide by voltage-clamp measurements. The reversal potential of the peptide effect and the magnitude of the conductance change were determined. It was found that with all peptides which were active the conductance was always decreased, and the reversal potential always close to or identical with the K^+ equilibrium potential. Hence it may well be that all these peptides act on the same, not voltage-gated K^+ conductance (Walther and Zittlau, 1988). This decrease leads to an increased inward current if the membrane potential is more positive than E_{K^+} (Fig. 13.2). This illustrates that any depolarizing influence will act more

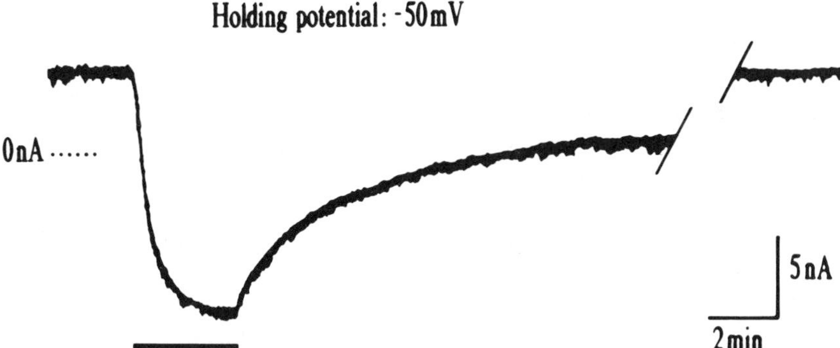

Fig. 13.2. Effect of YGGFMRFamide on muscle membrane conductance. An increase in holding current in a muscle fibre, voltage-clamped at a depolarized level, is elicited by application of 2×10^{-7} M peptide (bar). During wash-out the trace is interrupted for 20 min. $(K^+)_0$ = 10 mM; $(Ca^{2+})_0$ = 2 mM; $(Mg^{2+})_0$ = 12 mM; K^+-equilibrium potential −72 mV.

strongly in the presence of peptide since the counteraction by the polarizing K^+ resting current is diminished. The maximal reduction of the resting conductance could be as much as −50%, such as with a saturating dose of YGGFMRFamide.

Proctolin was the most potent of the peptides reducing the membrane resting conductance, giving pronounced effects ($\Delta G_{in} \sim -40\%$) already at 2×10^{-10} M. LMS caused a conductance decrease ($\Delta G_{in} \sim -15\%$) as did all other members of the FM(L)RFamide family. The potency of the FMRFamide–peptides did not differ from that of the homologous FLRFamides; YGGFM(L)RFamide ($\Delta G_{in} \sim -30\%$ at 10^{-7} M) is about 100 times more potent than FM(L)RFamide. LSKII gave a poor effect ($\Delta G_{in} \sim -5\%$), LSK none at all. While CARP was weakly active ($\Delta G_{in} \sim -10\%$). YGGFMRF, NPY, CCK8, AKH and CCAP were inactive.

13.5. **Comparison between pre- and postsynaptic peptide receptors**

The structural requirements for a peptide to be active seem to be largely similar at the pre- and postsynaptic (i.e. non-synaptic) side. This is illustrated (Table 13.1) by a ranking of the peptides with regard to pre- and postsynaptic effectiveness. In the case of FM(L)RFamide–peptides the C-terminal amidation is highly relevant. There are, however, at least two important differences. Firstly, proctolin may be presynaptically inactive (or else act on a receptor of a much lower affinity than that of the postsynaptic receptor) while postsynaptically it is extremely potent. Secondly, the postsynaptic effect of LMS is not opposite to that of other FM(L)RFamide peptides as is the case presynaptically.

In summary, the various peptide effects in the neuromuscular preparation suggest the presence of at least four different receptors. Postsynaptically there may be two receptors which distinguish between structurally very different peptides (proctolin vs. FM(L)RFamide–peptides) but initiate apparently identical physiological responses though the intermediate (second messenger?) pathways may differ. Presynaptically the two putative receptors must discriminate between structurally similar peptides (all from the FM(L)RFamide family) and initiate opposite physiological responses.

In this context it may be of interest that, according to combined HPLC and bioassay analysis of extracts from locust nervous tissue, FM(L)RF-amide probably does not occur in the locust (Schiebe *et al.*, 1989) and that all FM(L)RFamide-related peptides so far sequenced from insect and e.g., lobster nervous systems are N-terminally elongated by at least four amino acids (e.g. Trimmer *et al.*, 1987; Nambu *et al.*, 1988). This should allow sufficient structural diversity to provide selectivity for interactions of the various peptides with different FM(L)RFamide–peptide receptors.

13.6. **Peptide receptors remain to be explored**

There is – for obvious reasons – a large divergence between the number of neuropeptides isolated from various animal species and our knowledge about peptide receptors. As few agents which block peptide–receptor interactions are available, one way to characterize the putative receptors is to test still more structurally related peptides and to assess structure–activity relationships for narrowly defined physiological functions. In that respect the work presented here, though not strictly quantitative, is a step forward from the earlier investigations where effects on a complex parameter, i.e. neurally evoked contraction, were compared (Walther *et al.*, 1984; Evans and Myers, 1986b).

Receptor binding studies and computer modelling of the three-dimensional peptide configuration would obviously give further necessary information. The elucidation of the cellular processes initiated via peptide–receptor binding, as well as being of intrinsic interest, will provide additional criteria for receptor classification. Molecular genetics may ultimately yield the desired information, as it is currently doing for ionic channels. For example in *Drosophila* impressive advances with K^+ channels have been obtained (e.g. Schwarz *et al.*, 1988; Butler *et al.*, 1989). In this animal rather little seems to be known about the cellular effects of neuropeptides while, for example, a proctolinergic innervation of skeletal muscles (Anderson *et al.*, 1988) and the presence of FM(L)RFamide-peptides in the CNS (Nambu *et al.*, 1988) have been demonstrated.

13.7. **Functional aspects of neuromuscular modulations**

The potentiating effect on neurally evoked contractions, which results from the peptide-mediated reduction of K^+ resting conductance, has been outlined above. Conversely it would not be surprising if the potentiation of transmitter release could eventually be traced back to a peptide effect on an ionic conductance, perhaps similar to the one found in the muscle membrane. Of the various possibilities modulation of the Na^+ conductance is unlikely, since the presynaptic effect of YGGFMRFamide is the same regardless of whether tetrodotoxin is present or not (Schiebe and Walther, 1988).

Consideration of the possible utility of such neuropeptidergic mechanisms to the living insect presents an unsatisfactory perspective: Why, for example, have such complex modulatory systems for the jumping muscle? We know hardly anything about their levels of activation in the living animal. Clearly more work in that direction, taking account of developmental, behavioural and environmental aspects, is required. This would provide a more satisfying frame for the work at the single-cell level.

Acknowledgements

This work was supported by grants from the Deutsche Forschungsgemeinschaft (Schi 151/2; Wa 223/3; postdoctoral fellowship to K.E.Z.). We thank Dr K. Voigt for continuous support and Dr G. McGregor for critical reading of the manuscript.

References

Adams, M. E. and O'Shea, M. (1983). Peptide cotransmitter at a neuromuscular junction. *Science*, **221**, 286–89.

Anderson, M. S., Halpern, M. E. and Keshishian, H. (1988). Identification of the neuropeptide transmitter proctolin in *Drosophila* larvae: characterization of muscle fiber-specific neuromuscular endings. *J. Neurosci.*, **8**, 242–55.

Ashcroft, F. M. and Stanfield, P. R. (1982). Calcium and potassium currents in muscle fibres of an insect (*Carausius morosus*). *J. Physiol.*, **323**, 93–115.

Bishop, C. A., O'Shea, M. and Miller, R. J. (1981). Neuropeptide proctolin (H-Arg-Tyr-Leu-Pro-Thr-OH): immunological detection and neuronal localization in insect nervous system. *Proc. Natl. Acad. Sci. USA*, **78**, 5899–902.

Bishop, C. A., Wine, J. J., Nagy, F. and O'Shea, M. R. (1987). Physiological consequences of a peptide cotransmitter in a crayfish nerve-muscle preparation. *J. Neurosci.*, **7**, 1769–79.

Butler, A., Wei, A., Baker, K. and Salkoff, L. (1989). A family of putative potassium channel genes in *Drosophila. Science*, **243**, 943–7.

Cuthbert, B. A. and Evans, P. D. (1989). A comparison of the effects of FMRFamide-like peptides on locust heart and skeletal muscle. *J. Exp. Biol.*, **144**, 395–415.

Del Castillo, J. and Katz, B. (1954). Quantal components of the endplate potential. *J. Physiol.*, **124**, 560–73.

Evans, P. D. and Myers, C. M. (1986a). Peptidergic and aminergic modulation of insect skeletal muscle. *J. Exp. Biol.*, **124**, 143–76.

Evans, P. D. and Myers, C. M. (1986b). The modulatory action of FMRFamide and related peptides on locust skeletal muscle. *J. Exp. Biol.*, **126**, 403–22.

Evans, P. D. and O'Shea, M. (1977). An octopaminergic neurone modulates neuromuscular transmission in the locust. *Nature (Lond.)*, **270**, 257–59.

Hirata, T., Kubota, I., Takabatake, I., Kawahara, A., Shimamoto, N. and Muneoka, Y. (1987). Catch-relaxing peptide isolated from *Mytilus* pedal ganglia. *Brain Res.*, **422**, 374–6.

Hirata, T., Kubota, I., Imada, M., Muneoka, Y. and Kobayashi, M. (1989). Effects of catch-relaxing peptide on molluscan muscle. *Comp. Biochem. Physiol.*, **92C**, 283–288

Holman, G. M., Cook, B. J. and Nachman, R. J. (1986). Isolation, primary structure and synthesis of leucomyosuppressin, an insect neuropeptide that inhibits spontaneous contractions of the cockroach hindgut. *Comp. Biochem. Physiol.*, **85C**, 329–33.

Hoyle, G. (1975). Evidence that insect dorsal unpaired median (DUM) neurons are octopaminergic. *J. Exp. Zool.*, **193**, 425–31.

Hoyle, G. (1978). Distributions of nerve and muscle fibre types in locust jumping muscle. *J. Exp. Biol.*, **73**, 205–33.

Kiss, T. (1988). Catch-relaxing peptide (CARP) decreases the Ca-permeability of snail neuronal membrane. *Experientia*, **44**, 998–1000.

Krouse, M. E., Bishop, C. A. and Wine, J. J. (1988). Proctolin-sensitive Ca^{2+}

channels in isolated crayfish tonic flexor muscles. *Neurosci. Abstr.*, **14**, part I, p. 279, 111.5.

May, T. E., Brown, B. E. and Clements, A. N. (1979). Experimental studies upon a bundle of tonic fibres in the locust extensor tibialis muscle. *J. Insect Physiol.*, **25**, 169–81.

Murck, H., Zittlau, K. E. and Walther, C. (1989). Inactivation of a K^+ leak-conductance in locust skeletal muscle by two neuropeptides: involvement of G-proteins. In: Elsner, N. and Singer, W. (eds), *Dynamics and Plasticity in Neuronal Systems*. Thieme; Stuttgart and New York, p. 433.

Myers, C. M. and Evans, P. D. (1985a). An FMRFamide antiserum differentiates between populations of antigens in the ventral nervous system of the locust, *Schistocerca gregaria. Cell Tiss. Res.*, **242**, 109–14.

Myers, C. M. and Evans, P. D. (1985b). The distribution of bovine pancreatic polypeptide/FMRFamide-like immunoreactivity in the ventral nervous system of the locust. *J. Comp. Neurol.*, **234**, 1–16.

Nachman, R. J., Holman, G. M., Haddon, W. F. and Ling, N. (1986a) Leucosulfakinin, a sulfated insect neuropeptide with homology to gastrin and cholecystokinin. *Science*, **234**, 71–3.

Nachman, R. J., Holman, G. M., Cook, B. J., Haddon, W. F. and Ling, N. (1986b). Leucosulfakinin-II, a blocked sulfated insect neuropeptide with homology to cholecystokinin and gastrin. *Biochem. Biophys. Res. Commun*, **140**, 357–64.

Nachman, R. J., Holman, G. M. and Haddon, W. F. (1988). Aspects of gastrin/CCK-like insect leucosulfakinins and FMRF-amide. *Peptides*, **9**(1), 137–43.

Nambu, J. R., Murphy-Erdosh, C., Andrews, P. C., Feistner, G. J. and Scheller, R. H. (1988). Isolation and characterization of a *Drosophila* neuropeptide gene. *Neuron*, **1**, 55–61.

O'Shea, M. (1985). Are skeletal motoneurons in arthropods peptidergic? In: Selverston, A. I. (ed.), *Model Neural Networks and Behavior*. Plenum Press, New York, pp. 401–13.

O'Shea, M. and Adams, M. (1986). Proctolin: from 'gut factor' to model neuropeptide. *Adv. Insect Physiol.*, **19**, 1–28.

Robb, S., Packman, L. C. and Evans, P. D. (1989). Isolation, primary structure and bioactivity of SchistoFLRF-amide, a FMRF-amide-like neuropeptide from the locust, *Schistocerca gregaria. Biochem. Biophys. Res. Commun.*, **160**, 850–6.

Schiebe, M., Orchard, I., Watts, R., Lange, A. B. and Atwood, H. L. (1989). Characterization and partial purification of different factors with contraction potentiating activities from neurohaemal organs of the locust. *J. Comp. Neurol.*, **19**, 305–12.

Schiebe, M. and Walther, C. (1988). Pre- and postsynaptic actions of FMRF-NH_2-like peptides in insect neuromuscular synapse. *Symp. Biol. Hungarica*, **36**, 365–76.

Schiebe, M. and Walther, C. (1989) FMRFamide-like peptides in the locust. Immunohistochemical and HPLC-investigations. *Pestic. Sci.*, **25**, 97–103.

Schooneveld, H., Tesser, G. I., Veenstra, J. A. and Romberg-Privee, H. M. (1983). Adipokinetic hormone and AKH-like peptide demonstrated in the corpora cardiaca and nervous system of *Locusta migratoria* by immunocytochemistry. *Cell Tiss.* Res., **230**, 67–76.

Schwarz, T. L., Tempel, B. L., Papazian, D. M., Jan, Y. N. and Jan, L. Y. (1988). Multiple potassium-channel components are produced by alternative splicing at the *shaker* locus in *Drosophila. Nature*, **331**, 137–42.

Stangier, J., Hilbich, C., Beyreuther, K. and Keller, R. (1987) Unusual cardioactive peptide (CCAP) from pericardial organs of the shore crab *Carcinus maenas*. *Proc. Natl. Acad. Sci. USA*, **84**, 575–9.

Stangier, J., Hilbich, C. and Keller, R. (1989). Occurrence of crustacean cardioactive peptide (CCAP) in the nervous system of an insect, *Locusta migratoria*. *J. Comp. Physiol.*, **159**, 5–12.

Stone, J. V., Mordue, W., Batley, K. E. and Morris, H. E. (1976). Structure of locust adipokinetic hormone, a neurohormone which regulates lipid utilization during flight. *Nature (Lond.)*, **263**, 207–11.

Trimmer, B. A., Kobierski, L. A. and Kravitz, E. A. (1987) Purification and characterization of FMRFamidelike immunoreactive substances from the lobster nervous system: isolation and sequence analysis of two closely related peptides. *J. Comp. Neurol.*, **266**, 16–26.

Usherwood, P. N. R. (1978). Amino acids as neurotransmitters. *Adv. Comp. Physiol. Biochem.*, **7**, 227–309.

Walther, C. and Schäfer, S. (1988). FMRFamide-like immunoreactivity in the metathoracic ganglion of the locust (*Schistocerca gregaria*). *Cell Tiss. Res.*, **253**, 489–91.

Walther, Chr. and Schiebe, M. (1987). FMRF-NH_2-like factor from neurohaemal organ modulates neuromuscular transmission in the locust. *Neurosci. Lett.*, **77**, 209–14.

Walther, C., Schiebe, M. and Voigt, K. H. (1984). Synaptic and non-synaptic effects of molluscan cardioexcitatory neuropeptides on locust skeletal muscle. *Neurosci. Lett.*, **45**, 99–104.

Walther, C. and Zittlau, K. E. (1988) Do YGGFMRFamide and proctolin act on the same potassium conductance in locust skeletal muscle? *J. Physiol.*, **410**, 32P.

Worden, M. K., Witten, J. L. and O'Shea, M. (1985). Proctolin is a cotransmitter for the SETi motoneuron. *Neurosci. Abstr.*, **11**, 327.

Yamamoto, D., Ishikawa, S., Holman, G. M. and Nachman, R. J. (1988). Leucomyosuppressin, a novel insect neuropeptide, inhibits evoked transmitter release at the mealworm neuromuscular junction. *Neurosci. Lett.*, **95**, 137–42.

14 *David B. Morton and James W. Truman*

The molecular basis for steroid regulation of eclosion hormone action in the tobacco hornworm, *Manduca sexta*

14.1. **Introduction**

The responsiveness of a target tissue to a neurotransmitter or neurohormone can vary dramatically over the lifespan of an animal. The time-course of this plasticity can range from milliseconds to years, and an understanding of the causes and mechanisms of such changes is fundamental to an integrated picture of how the nervous system functions. Many studies have focused on short-term changes in CNS responsiveness such as facilitation, habituation, etc. (Zucker, 1989). Relatively fewer studies, however, have been directed to longer-term changes in CNS plasticity. One set of these longer-term changes involves the effects of steroids on the CNS. There are a number of neuropeptide-triggered behaviours, the responsiveness of which is dependent on prior exposure to steroid hormones. For example, the action of luteinizing hormone-releasing hormone (LHRH) to trigger lordosis behaviour in female rats requires that the CNS be 'primed' by exposure to the steroid hormone, oestrogen (Pfaff, 1973). In the case of oxytocin-triggered maternal behaviour in rats, the peptide is effective only when the CNS has been previously exposed to oestrogen (Fahrbach *et al.*, 1985). Ecdysis behaviour in the tobacco hornworm, *Manduca sexta*, is another example in which steroids and peptides must interact to bring about a behavioural response. This chapter outlines some of the studies which we have carried out in an attempt to understand the mechanisms by which these types of hormones might interact to alter CNS function.

14.2. **Eclosion hormone and ecdysis in *Manduca sexta***

The interaction of the steroid hormone, 20-hydroxyecdysone (20-HE), and the neuropeptide, eclosion hormone (EH) on the CNS has been studied in most detail at pupal ecdysis in *Manduca sexta*. The last step in the moulting of all insects is ecdysis, the shedding of the cuticle of the previous instar. This stereotyped behaviour is triggered by the direct action of EH on the CNS (Truman, 1978; Weeks and Truman, 1984).

EH is a sequenced 62 amino acid neuropeptide (Marti *et al.*, 1987). It is synthesized in two pairs of brain neurons, whose axons project the length of the ventral CNS to the proctodeal nerves, from which EH is then released into the blood to trigger ecdysis (Truman and Copenhaver, 1989).

Animals injected with EH 4 h before their normal time of ecdysis show premature ecdysis 1–2 h after injection. Those treated with EH about 20 h earlier, however, show no behavioural response. Behavioural sensitivity to EH first appears about 8 h before ecdysis, and then abruptly disappears once the behaviour has occurred. The animals will not respond to EH again until shortly before the next ecdysis. What is responsible for this narrow window of sensitivity, and what are the biochemical mechanisms underlying it?

Various behavioural experiments have shown that ecdysteroids are responsible for the onset of behavioural sensitivity to EH (Truman *et al.*, 1983). Analysis of ecdysteroid titres in the haemolymph of *Manduca* shows that 2–3 days before pupal ecdysis there is a large peak of ecdysteroids, which is responsible for the synthesis of new pupal cuticle and for coordinating the moult (Bollenbacher *et al.*, 1981; Riddiford, 1985). About 36 h before ecdysis the levels of ecdysteroids begin to fall, reaching a minimum level about 12 h before ecdysis. It is this decline in the levels of steroid which stimulates the release of EH and renders the CNS capable of responding to it (Truman *et al.*, 1983).

14.3. **Action of EH**

Before the action of ecdysteroids can be understood one has to know the sequence of events which underlie the action of EH. Presumably, the effects of the steroid involve one or more of the steps in the biochemical cascade that mediates EH action.

Figure 14.1 summarizes the events which underlie the action of EH on the CNS. By analogy with other neuropeptides, EH presumably acts via a cell surface receptor. The location, within the CNS, of the cells containing EH receptors is unknown. The action of EH on these receptors is to stimulate an increase in the levels of cGMP within the CNS (Morton and Truman, 1985). There are now a number of pathways described by which ligands regulate intracellular levels of cGMP. In vertebrate smooth muscle the peptide, atrial natriuretic factor (ANF), acts to elevate cGMP levels via a transmembrane protein which has both ANF receptor and guanylate cyclase regions located on a single polypeptide (Paul *et al.*, 1987). Acetylcholine elevates cGMP in the CNS via muscarinic receptors. These receptors are linked via a G-protein to phospholipase C which, when activated, liberates inositol tris-phosphate and diacylglycerol. The diacylglycerol is further metabolized to arachidonic acid, which is then thought to stimulate the guanylate cyclase (Snider *et al.*, 1984). Glutamate appears to elevate cGMP in the brain by an even more circuitous

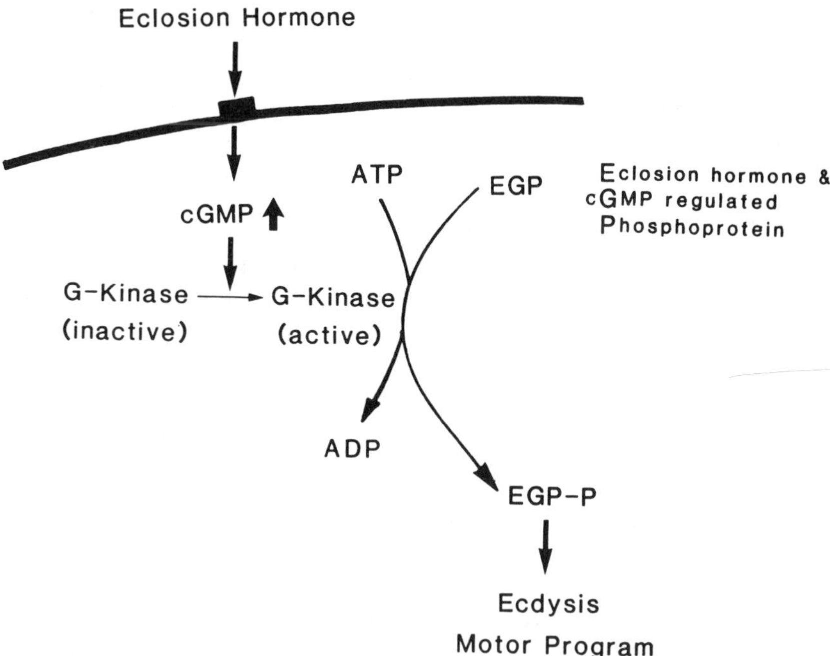

Fig. 14.1. Summary of the mechanism of action of EH on the CNS of *Manduca sexta*.

route. Activation of glutamate receptors on cerebellar cells stimulates the breakdown of L-arginine, causing the production of nitric oxide. The nitric oxide can then diffuse out of the cell and into neighbouring neurons and glia, where it directly activates guanylate cyclase (Garthwaite *et al.*, 1988). All of these mechanisms activate guanylate cyclase, but alterations in cGMP levels could also be achieved via regulation of the phosphodiesterase in an analogous situation to vertebrate rod cells (Stryer, 1986). In the case of *Manduca* we do not yet know the pathway through which the reception of EH is eventually translated into an increase in cGMP.

The action of cyclic nucleotides is generally mediated by the activation of a cyclic nucleotide-dependent protein kinase which catalyses the phosphorylation of one or more specific phosphoproteins (Drummond, 1984). EH appears to act in the same manner. We have confirmed the presence of a cGMP-dependent protein kinase in the CNS of *Manduca* (Morton and Truman, 1986). Furthermore, we have shown that EH and cGMP stimulate the phosphorylation of two 54 kD proteins which we have called the EGPs (*E*closion hormone and c*G*MP regulated *P*hosphoproteins; Morton and Truman, 1986, 1988a). The EGPs are loosely associated with the membrane fraction from the CNS, but it is not known how their phosphorylation leads to the activation of the ecdysis central pattern generator. Although our knowledge of the biochemical pathway that mediates the action of EH is incomplete, we can nevertheless examine the effects of ecdysteroids on some of the components of this pathway.

14.4. **Steroid regulation of EH action**

Truman *et al.* (1983) showed that the ability of EH to trigger ecdysis is regulated by the ecdysteroids. The CNS must be exposed to ecdysteroids, but then the steroid levels must decline before the CNS becomes responsive to EH. In general, ecdysteroids, like other steroid hormones, exert their effects via changes in gene expression (O'Connor, 1985). Studies with the protein synthesis inhibitor, cycloheximide, indicate that acquisition of EH sensitivity in *Manduca* requires protein synthesis (Morton and Truman, unpublished). Presumably, the fall in ecdysteroids prior to ecdysis stimulates *de novo* synthesis of a protein or proteins, the presence of which enables EH to trigger ecdysis.

Any or all of the proteins shown in Fig. 14.1 could be regulated by ecdysone, and the absence of any of these would render the CNS behaviourally unresponsive to EH. Our strategy to discover which step(s) is (are) regulated by ecdysteroids was to start at the receptor and work down the cascade.

The simplest way to monitor the presence or absence of functional receptors was to test the ability of the CNS to show an increase in cGMP levels in response to EH exposure. The normal increases in cGMP after EH exposure would indicate the presence of functional receptors, as well as all of the steps between the receptor and elevation of cGMP. The lack of an increase, however, would not show at what step the block occurred.

Figure 14.2B shows that the CNS does not always respond to EH with an increase in cGMP levels. It first becomes capable of showing an EH-stimulated cGMP increase about 36 h before ecdysis, at about the time the ecdysteroid levels begin to fall. Do ecdysteroids regulate this ability of the CNS to show an EH-stimulated cGMP increase?

Ecdysteroids are secreted from the prothoracic glands which are located in the thorax. One can deprive the abdominal CNS of ecdysteroids by tying a ligature between the thorax and the abdomen and removing the head and thorax. If this is done before the peak of ecdysteroids which initiates pupal development the resultant isolated abdomens survive for well over a week, as arrested larval abdomens and show no signs of pupal development. Ecdysteroids can then be replaced by infusing the isolated abdomens with 20-HE.

Animals were ligated just before the large peak of ecdysteroids (4 days before ecdysis) and then injected with EH 4 days later, 4 h before control

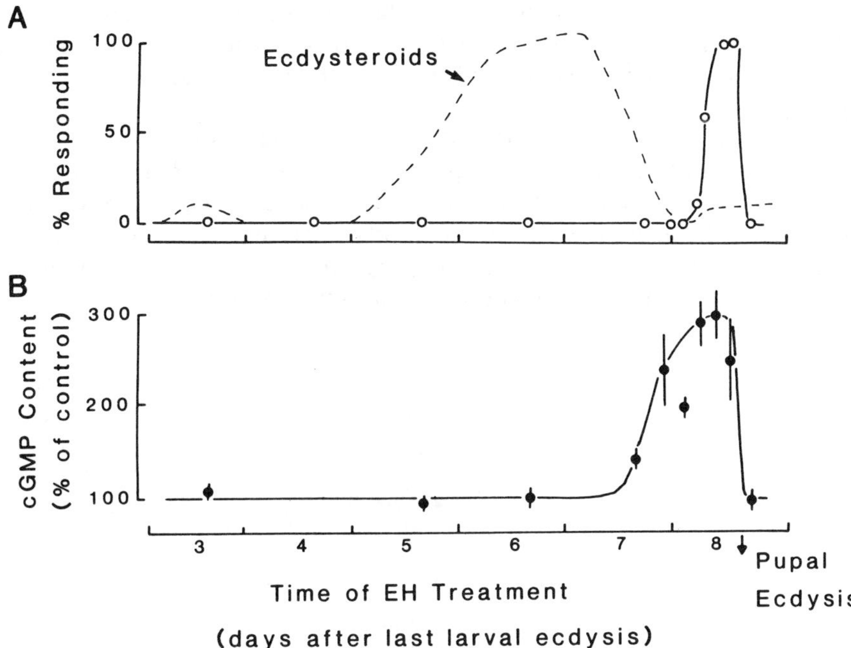

Fig. 14.2. Timing of development of EH sensitivity in *Manduca sexta*. **A**: Development of behavioral sensitivity; the dashed line gives approximate ecdysteroid titres. **B**: Development of the ability of EH to stimulate an increase in cGMP in the CNS. Each point represents the mean ± SEM of at least four determinations. (From Morton and Truman, 1985.)

(unligated) animals ecdysed. The CNS of these ligated animals showed no increase in cGMP levels in response to EH, whereas control animals showed the usual increase. Another group of ligated animals was infused with 20-HE for 12 h, beginning a day after ligation. These animals subsequently showed the formation of a pupal cuticle and they were injected with EH at various times after the end of the infusion period. By 48 h after the end of the infusion, the CNS of these preparations became capable of showing an EH-stimulated increase in cGMP (Morton and Truman, 1985). This indicates that the ability of EH to elevate cGMP levels is regulated by ecdysteroids, but it is not known if this represents the appearance of receptors or some other step between EH receptors and the elevation of cGMP. Preliminary data indicate that the ecdysteroid effects probably do not involve the guanylate cyclase, since activity levels of this enzyme do not change during the larval to pupal transformation.

Although ecdysteroids clearly have an effect on the early steps that lead from EH reception to the generation of the second messenger signal, these effects by themselves are not sufficient to render the CNS responsive to EH. Figure 14.2 shows that EH first elevates cGMP levels at least 24 h before it is able to trigger ecdysis behaviour. Therefore, there must be a second regulated event that is distal to the receptor/second messenger system.

Assays of the levels of cGMP-dependent protein kinase in the CNS showed no differences when assayed at 24 h and 4 h before ecdysis (Morton and Truman, 1986). Although these data do not preclude changes in the activity of the kinase in subcompartments of the CNS, for example in the EH target cells, they do not indicate that the protein kinase is a site for the regulation of behavioural sensitivity.

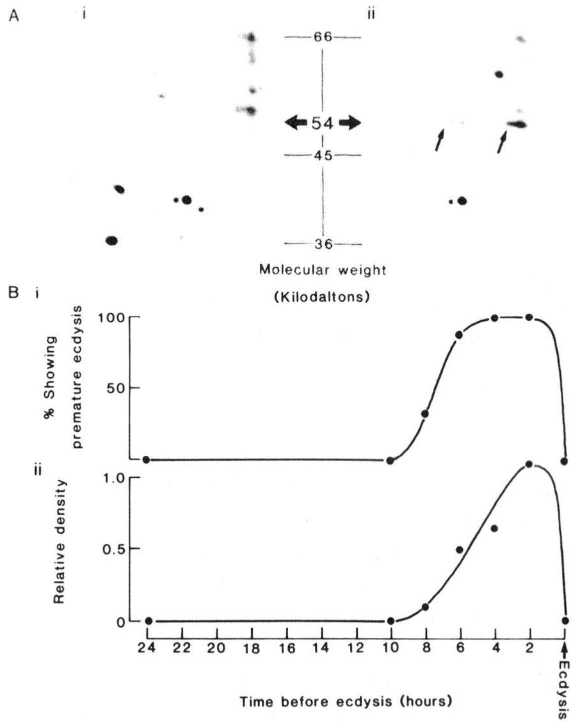

Fig. 14.3. **A**: Fluorograms showing phosphorylation of endogenous proteins in the CNS of *Manduca sexta*. The CNS from animals 24 h (i) and 4 h (ii) before ecdysis were homogenized, phosphorylated with [^{32}P]ATP in the presence of cGMP and the labelled proteins were separated by 2D SDS polyacrylamide gel electrophoresis. The gels were then exposed to X-ray films. The two proteins marked with arrows in (ii) are the EGPs. **B**: The relationship between the development of behavioural sensitivity to EH and the appearance of the EGPs. (i) The ability of EH to trigger ecdysis; (ii) the relative density of the EGPs. The relative density of the EGPs was measured relative to a 40 kD control phosphoprotein marked with an asterisk (From Morton and Truman, 1986.)

Next we turned our attention to the EGPs. Figure 14.3A shows fluorograms from phosphorylated CNS proteins, separated by 2D SDS-PAGE, from animals 24 h and 4 h before ecdysis. The EGPs (arrowed) are clearly visible on the fluorogram from the −4 h animals, but not from the −24 h animals. A more detailed analysis of the time of appearance of the EGPs showed a marked correlation with the time of acquisition of behavioural sensitivity to EH (Fig. 14.3B). Ten hours before ecdysis the animals were unresponsive to EH, and there was no trace of the EGPs on the fluorogram. By −8 h, however, a few animals had become responsive and a low level of the EGPs first appeared on the fluorograms. The level of EGPs then gradually increased as more animals became responsive (Figure 14.3B) and as response latencies decreased (Truman *et al.*, 1983).

The EGPs are too scarce to be detected directly. Consequently, their absence on fluorograms could be due to the absence of the proteins themselves, or due to their inability to accept ^{32}P-labelled phosphate. Hence, their appearance at 8 h before ecdysis could be due to *de novo* synthesis of the EGPs or to a change in their ability to be phosphorylated. We favour the former hypothesis, because the injection of the protein synthesis inhibitor cycloheximide at 10 h before ecdysis prevents both the subsequent appearance of the EGPs and the ability of the CNS to respond to EH. Later injections do not then affect either. Thus, protein synthesis is necessary for the appearance of the EGPs, but as yet we have not directly demonstrated that the EGPs themselves are synthesized, starting at 8–10 h before ecdysis.

Is the appearance of the EGPs regulated by ecdysteroids? To answer this question we used similar manipulations to those described above for examining the regulation of the cGMP system. These experiments showed that the EGPs do not appear in the CNS of isolated abdomens. If, however, the abdomens received a 12 h infusion of 20-HE, then the EGPs appear in the CNS about 48 h after the termination of the infusion (Morton and Truman, 1988b). Therefore, ecdysteroids are required for the appearance of the EGPs in the CNS.

Since the appearance of responsiveness to EH requires the appearance and then withdrawal of ecdysteroids (Truman *et al.*, 1983) we tested whether the appearance of the EGPs had the same requirements. This is indeed the case, and can be demonstrated in two ways. At 24 h before ecdysis the EGPs are absent. When the CNS was removed at this time and maintained in culture for 24 h, in the absence of ecdysteroids, the EGPs were then evident in the CNS by the end of the culture period. By contrast, the EGPs failed to appear in similar nervous systems that were maintained in culture in the presence of ecdysteroids (Morton and Truman, 1988b). The minimum level of 20-HE required to block the appearance of the EGPs was 0·05 μg/ml, which corresponds to the minimum level of ecdysteroids seen *in vivo* (Truman *et al.*, 1983).

The normal decline in ecdysteroids can also be interrupted *in vivo* by injecting animals with 20-HE at various times during this decline. As development proceeds, however, there comes a time when such injections are no longer effective in delaying the appearance of EH responsiveness. It is presumed that at this time the normal steroid decline has triggered some event that then leads (in a steroid-independant fashion) to the onset of EH responsiveness. For pupal ecdysis the onset of EH responsiveness can no longer be blocked by 20-HE injection after 12–14 h before ecdysis. The end of the steroid-sensitive period for the appearance of the EGPs is also at about this time (Morton and Truman, 1988b). The fact that the steroid requirements for the appearance of the EGPs and for the onset of EH sensitivity share similar time-courses supports the hypothesis that the ecdysteroid control over responsiveness is mediated through its effects on the EGPs. The cycloheximide results alluded to above suggest that the regulation is exerted at the level of the *de novo* synthesis of these phosphoproteins.

14.5. **Summary**

The above results have led us to the following hypothesis for the mechanism of steroid regulation of EH action. This model is summarized in Fig. 14.4 and shows two steps in the acquisition of responsiveness. During the intermoult the CNS contains only some of the elements of the EH cascade. Guanylate cyclase (GC) and the cGMP-dependent protein kinase (gPK_i) are present but the EGPs and possibly the EH receptors are absent. The CNS is unresponsive to EH at this point. At about 36 h before ecdysis, ecdysteroids, presumably via gene expression and protein synthesis, render the CNS partially responsive to EH, so that the peptide will elevate cGMP levels but not trigger ecdysis. This step is presumably accomplished by the synthesis or activation of EH receptors or some element of the transduction pathway. Ecdysteroids have to be present to attain this first step, and this ability is acquired before the steroid titres start to decline. About 36 h before ecdysis the level of ecdysteroids begins to fall, reaching a minimum level at about 12 h before ecdysis. This low level initiates the events that render the CNS able to respond to EH. These events presumably involve the production of the mRNA for the EGPs at about −12 h, followed by the synthesis of the EGPs themselves at −8 h. The pathway is complete and EH can then trigger ecdysis via the cascade of elevating cGMP, activation of the cGMP-dependent protein kinase and phosphorylation of the EGPs.

These two levels of regulation are independant of each other, as nervous tissue from − 24 h animals will show an EH-stimulated increase in cGMP, yet the appearance of the EGPs can still be blocked by preventing the decline in ecdysteroid levels. Furthermore, as the EH-stimulated

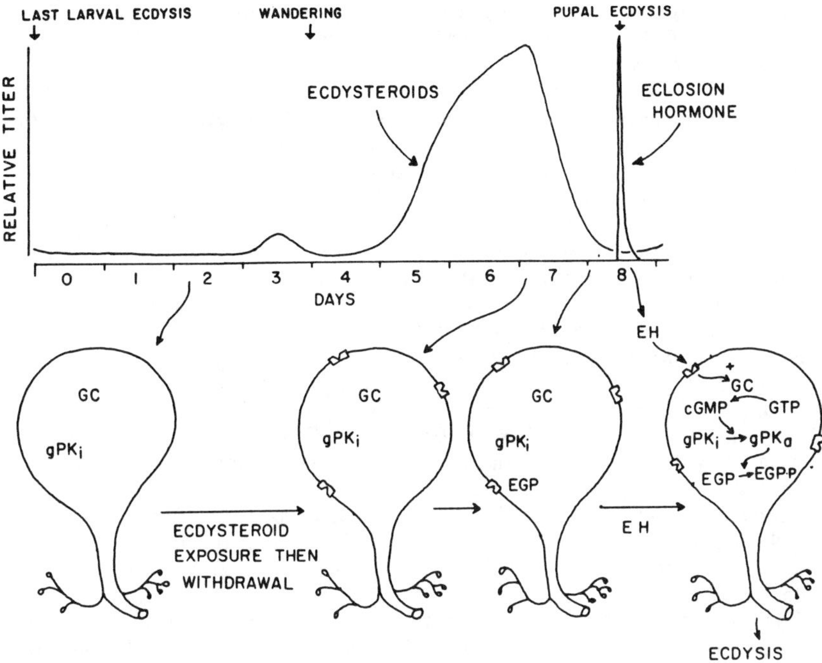

Fig. 14.4. Model for the mechansim of steroid regulation of EH sensitivity in the CNS of *Manduca sexta*.

cGMP increase can first be seen before the level of ecdysteroids fall, but the appearance of the EGPs requires an ecdysteroid decline, the mechanism of steroid regulated gene activation would appear to be different for the two steps.

There are a number of other systems where the responsiveness of the target tissue to a neurotransmitter or neurohormone is regulated by the up- or down-regulation of receptors or the underlying G-proteins (Hollenberg, 1985; Ros *et al.*, 1988). The regulation of EH responsiveness has this level of regulation as well as another level, distal to the second messenger system. It is possible that, on closer investigation, other systems will show similar multiple levels of regulation.

References

Bollenbacher, W. E., Smith, W. L., Goodman, W. and Gilbert, L. I. (1981). Ecdysteroid titer during larval–pupal–adult development of the tobacco hornworm, *Manduca sexta. Gen. Comp. Endocrinol.*, **44**, 302–6.

Drummond, G. I. (1984). *Cyclic Nucleotides in the Nervous System.* Raven Press, New York.

Fahrbach, S. E., Morrell, J. I. and Pfaff, D. W. (1985). Role of oxytocin in the onset of estrogen facilitated maternal behavior. In: Amico, J. A. and Robertson, A. G. (eds), *Oxytocin: Clinical and Laboratory Studies*, Elsevier, New York, pp. 372–88.

Garthwaite, J., Charles, S. L. and Chess-Williams, R. (1988). Endothelium-derived relaxing factor release on activation of NMDA receptors suggests role as intercellular messenger in the brain. *Nature*, **336**, 385–8.

Hollenberg, M. D. (1985). Examples of homospecific and heterospecific receptor regulation. *Trends Pharmacol. Sci.*, **6**, 242–5.

Marti, T., Takio, K., Walsh, K. A., Terzi, G. and Truman, J. W. (1987). Microanalysis of the amino acid sequence of the eclosion hormone from the tobacco hornworm, *Manduca sexta*. *FEBS Lett.*, **219**, 415–18.

Morton, D. B. and Truman, J. W. (1985). Steroid regulation of the peptide-mediated increase in cGMP in the nervous system of the hawkmoth, *Manduca sexta*. *J. Comp. Physiol. A*, **157**, 423–32.

Morton, D. B. and Truman, J. W. (1986). Substrate phosphoprotein availability regulates eclosion hormone sensitivity in an insect CNS. *Nature*, **323**, 264–7.

Morton, D. B. and Truman, J. W. (1988a). The EGPs: the eclosion hormone and cyclic GMP-regulated phosphoproteins. I. Appearance and partial characterization in the CNS of *Manduca sexta*. *J. Neurosci.*, **8**, 1326–37.

Morton, D. B. and Truman, J. W. (1988b). The EGPs: the eclosion hormone and cyclic GMP-regulated phosphoproteins. II. Regulation of appearance by the steroid hormone 20-hydroxyecdysone in *Manduca sexta*. *J. Neurosci.*, **8**, 1338–45.

O'Connor, J. D. (1985). Ecdysteroid action at the molecular level. In: Kerkut, G. A. and Gilbert, L. A. (eds), *Comprehensive Insect Physiology, Biochemistry and Pharmacology*, vol. 8: *Endocrinology II*, Pergamon, New York, pp. 85–98.

Paul, A. K., Marala, R. B., Jaiswal, R. K. and Sharma, R. K. (1987). Coexistence of guanylate cyclase and atrial natriuretic factor receptor in a 180-kD protein. *Science*, **235**, 1224–6.

Pfaff, D. W. (1973). Luteinizing hormone-releasing factor potentiates lordosis behavior in hypophysectomized ovariectomized female rats. *Science*, **182**, 1148–9.

Riddiford, L. M. (1985). Hormone action at the cellular level. In: Kerkut, G. A. and Gilbert, L. A. (eds) *Comprehensive Insect Physiology, Biochemistry and Pharmacology*, vol. 8: *Endocrinology II*, Pergamon, New York, pp. 37–84.

Ros, M., Northup, J. K. and Malbon, C. C. (1988). Steady-state levels of G-proteins and beta-adrenergic receptors in rat fat cells. Permissive effects of thyroid hormones. *J. Biol. Chem.*, **263**, 4362–8.

Snider, R. M., McKinney, M., Forray, C. and Richelson, E. (1984). Neurotransmitter receptors mediate cyclic GMP formation by involvement of arachidonic acid and lipoxygenase. *Proc. Natl. Acad. Sci.*, **81**, 3905–9.

Stryer, L. (1986). The cyclic GMP cascade of vision. *Ann. Rev. Neurosci.*, **9**, 87–119.

Truman, J. W. (1978). Hormonal release of stereotyped motor programmes from the isolated nervous system of the *Cecropia* silkmoth. *J. Exp. Biol.*, **74**, 151–73.

Truman, J. W., Rountree, D. B., Reiss, S. E. and Schwartz, L. M. (1983). Ecdysteroids regulate the release and action of eclosion hormone in the tobacco hornworm, *Manduca sexta* (L). *J. Insect Physiol.*, **29**, 895–900.

Truman, J. W. and Copenhaver, P. F. (1989). The larval eclosion hormone neurones in *Manduca sexta*: identification of the brain – proctodeal neurosecretory system. *J. Exp. Biol.*, **147**, 457–70.

Weeks, J. C. and Truman, J. W. (1984). Neural organization of peptide-activated ecdysis behavior during the metamorphosis of *Manduca sexta*. I. Conservation of the peristalsis motor pattern at the larval-pupal transformation. *J. Comp. Physiol. A*, **155**, 407–22.

Zucker, R. S. (1989). Short-term synaptic plasticity. *Ann. Rev. Neurosci.*, **29**, 13–31.

15 *Heinrich Dircksen*

Neurosecretory endings in the pericardial organs of the shore crab, *Carcinus maenas L.*, and their identification by neuropeptide immunocytochemistry

The pericardial organs (PO) are nerve plexuses and trunks in the lateral pericardial cavity of the heart of decapod crustaceans. In brachyurans these plexuses spread across the orifices of the branchiocardiac veins. They are anastomoses of at least seven segmental nerves mainly composed of peripheral projections of neurons originating in the dorsal parts of the thoracic ganglia mass. Further projections of some of these neurons form plexuses on the repiratory muscles commonly named the anterior ramifications complex (AR) arising from a branch of the first segmental nerve (Maynard, 1961). Since Alexandrowicz (1953) discovered the neurosecretory nature of the PO, they have been known as the neurohaemal release sites of several cardioactive factors. These include the biogenic amines dopamine, octopamine and serotonin, and cardioacceleratory neuropeptides (see for review: Cooke and Sullivan, 1982). Following the studies of Maynard and Welsh (1959) who presented first evidence for the existence of cardioactive peptides, preliminary isolations and characterizations were carried out by Belamarich (1963), Belamarich and Terwilliger (1966) and Terwilliger *et al.* (1970), suggesting at least two small bioactive peptides. Sullivan (1979) first suggested one of them to be proctolin, RYLPT, a neuropeptide known from insects (Brown and Starratt, 1975; Starratt and Brown, 1975). This has been biochemically confirmed by Schwarz *et al.* (1984) in the astacuran *Homarus americanus*, and by Stangier *et al.* (1986) in the brachyuran *Carcinus maenas* L. In both studies, proctolin was demonstrated in axons and neurosecretory terminals in the PO by means of light microscopic immunocytochemistry.

In the PO of *Carcinus maenas* L., a novel crustacean cardio-active peptide (CCAP) with the cyclic nonapeptide structure PFCNAFTGC-NH_2 has recently been structurally elucidated and demonstrated by immunocytochemistry (Stangier *et al.*, 1987; Dircksen and Keller, 1988). Immunocytochemical evidence for the existence of further neuropeptides related to the FMRF- or FLRFamide family and Leu-enkephalin has been provided by Dircksen *et al.* (1987) and Dircksen and Keller (1988) in the PO of this species.

Different neurosecretory neurons with origin in the thoracic ganglia of *Carcinus maenas* L. are immunoreactive with antisera against these peptides. The neurons project via the segmental nerves to neurosecretory endings in the PO and the AR. By immunocytochemical labelling of whole-mount preparations, semithin and ultrathin sections of PO at light and electron-microscopic level, neuropeptide contents could be attributed to at least four out of six morphologically distinguishable types of endings in the PO. Dircksen and Keller (1988) described two distinct types of neurosecretory granules immunoreactive with antisera against CCAP and FMRFamide. A further FMRFamide- and a proctolin-immunoreactive type of granule different from the others has now been identified. However, the relation of the Leu-enkephalin-like immunoreactivity to a distinct granule type is still uncertain, because the antigenicity of this peptide is destroyed by fixatives containing osmic acid.

Preabsorption controls and immunodot-blotting assays of HPLC-separated crude PO extracts were performed to confirm the immunocytochemical results. Single immunopositive peak fractions have been found for CCAP (Dircksen and Keller, 1988) and proctolin at the retention times of the authentic peptides. For FMRFamide-like immunoreactive peptides, retention times different from authentic FMRFamide were obtained. The latter results point to the existence of at least two different peptides related to FMRF- or FLRFamides or its elongated octapeptide analogues, SDRNFLRFamide and TNRNFLRFamide, that have recently been isolated and characterized by Trimmer *et al.* (1987) from the PO of *Homarus americanus*. At present the biochemical nature of the FMRFamide-like and the Leu-enkephalin-like peptides in the PO of *Carcinus maenas* is under investigation. These neuropeptides might, together with proctolin and CCAP, participate in the complex control of heartbeat and respiration in the shore crab.

References

Alexandrowicz, J. S. (1953). Nervous organs in the pericardial cavity of the decapod Crustacea. *J. Mar. Biol. Assoc.*, UK, **31**, 563–80.

Belamarich, F. A. (1963). Biologically active peptides from the pericardial organs of the crab *Cancer borealis*. *Biol. Bull.*, **124**, 9–16.

Belamarich, F. A. and Terwilliger, R. (1966). Isolation and identification of cardio-excitor hormone from the pericardial organs of *Cancer borealis*. *Am. Zool.* **6**, 101–6.

Brown, B. E. and Starratt, A. N. (1975). Isolation of proctolin, a myotropic peptide from *Periplaneta americana*. *J. Insect Physiol.*, **21**, 1879–81.

Cooke, I. M. and Sullivan, R. E. (1982), Hormones and neurosecretion. In: Bliss, D. E. and Mantel, L. H. (eds), *The Biology of Crustacea*, vol 3. Academic Press, New York, pp. 205–90.

Dircksen, H. and Keller, R. (1988). Immunocytochemical localization of CCAP, a novel crustacean cardioactive peptide, in the nervous system of the shore crab, *Carcinus maenas* L. *Cell Tiss. Res.*, **254**, 347–60

Dircksen, H., Stangier, J. and Keller, R. (1987). Proctolin-like, FMRFamide-like, and Leu-enkephalin-like immunoreactivity in the pericardial organs of brachyuran crustaceans. Abstract of the 13th Conf. Europ. Comp. Endocrinologists, Belgrade, Yugoslavia. *Gen. Comp. Endocrinol.*, **66**, 42.

Maynard, D. M. (1961). Thoracic neurosecretory structures in Brachyura. I. Gross anatomy. *Biol. Bull.*, **121**, 316–29.

Maynard, D. M. and Welsh, J. H. (1959). Neurohormones of the pericardial organs of brachyuran Crustacea. *J. Physiol.*, **149**, 215–27.

Schwarz, T. L., Lee, G. M. H., Siwicki, K. K., Standaert, D. G. and Kravitz, E. A. (1984). Proctolin in the lobster: the distribution, release, and chemical characterization of a likely neurohormone. *J. Neurosci.*, **4**, 1300–11.

Stangier, J., Dircksen, H. and Keller, R. (1986). Identification and immunocytochemical localization of proctolin in the pericardial organs of the shore crab, *Carcinus maenas*. *Peptides*, **7**, 67–72.

Stangier, J., Hilbich, C., Beyreuther, K. and Keller, R. (1987). Unusual cardioactive peptide (CCAP) from pericardial organs of the shore crab *Carcinus maenas*. *Proc. Natl. Acad. Sci. USA*, **84**, 575–9.

Starratt, A. N. and Brown, B. E. (1975). Structure of the pentapeptide proctolin, a proposed neurotransmitter in insects. *Life Sci.*, **17**, 1253–6.

Sullivan, R. E. (1979). A proctolin-like peptide in crab pericardial organs. *J. Exp. Zool.*, **210**, 543–52.

Terwilliger, R. C., Terwilliger, N. B., Clay, G. A. and Belamarich, F. A. (1970). The subcellular localization of a cardioexcitatory peptide in the pericardial organs of the crab, *Cancer borealis*. *Gen. Comp. Endocrinol.* **15**, 70–9.

Trimmer, B. A., Kobierski, C. A., and Kravitz, E. A. (1987). Purification and characterization of immunoreactive substances from the lobster nervous system: Isolation and sequence analysis of two closely related peptides. *J. Comp. Neurol.*, **266**, 16–26.

16 *Joachim Stangier*

Biological effects of crustacean cardioactive peptide (CCAP), a putative neurohormone/neurotransmitter from crustacean pericardial organs

16.1. Introduction

The crustacean cardioactive peptide (CCAP) was originally isolated from the pericardial organs of the shore crab, *Carcinus maenas* (Stangier *et al.*, 1987). The pericardial organs (PO) are large neurohaemal release sites situated in lateral position to the heart and in front of the openings of the branchiocardiac veins (Alexandrowicz, 1953). The PO are therefore exposed to the haemolymph stream entering the pericardial cavity through the openings of the branchiocardiac veins. Thus, neurohormones released from the PO will be transported to the heart and possibly other more distant target organs (Alexandrowicz and Carlisle, 1953). The presence of cardioactive substances was demonstrated by application of extracts from pericardial organs to the heart of crustaceans. The active substances exerted a powerful chronotropic and inotropic effect on the heartbeat (for review see Cooke and Sullivan, 1982).

Studies on the nature of the cardioactive substances led to the identification of the biogenic amines 5-hydroxytryptamine, octopamine and dopamine (Florey and Rathmayer, 1978; Sullivan, 1978). Although the presence of cardioactive peptides in POs was already postulated by Alexandrowicz in 1953, it was more than two decades before the first cardioactive peptide was identified. In 1979 Sullivan reported on the isolation of a peptide from PO of *Cardisoma* which was very similar or identical to the insect pentapeptide proctolin (Sullivan, 1979; Starratt and Brown, 1975). In several subsequent studies, proctolin was shown to exist in the nervous system of the lobster *Homarus americanus* and in the shore crab, *Carcinus maenas* (Schwarz *et al.*, 1984; Stangier *et al.*, 1986). Since proctolin was found to be highly abundant in the PO it is thought to have a neurohormonal role in crustaceans.

Recently, a second cardioactive peptide was isolated from the PO of *Carcinus maenas* (Stangier *et al.*, 1987). CCAP (crustacean cardioactive peptide) is an amidated nonapeptide with a disulphide bridge:

Pro-Phe-Cys-Asn-Ala-Phe-Thr-Gly-Cys-NH_2 (disulphide bridge between the two Cys residues)

Table 16.1. Amount of CCAP immunoreactive material (pmol CCAPir/PO) present in PO of four crustacean species (n = 10, ±SEM)

Liocarcinus puber	24·41 ± 9·4
Carcinus maenas	17·36 ± 3·0
Hyas araneus	14·0 ± 0·3
Cancer pagurus	2·57 ± 0·6

By radioimmunoassay investigations, CCAP was shown to occur in the PO of all crustacean species investigated so far (Table 16.1). In addition, very recently the presence of CCAP in the nervous system of an insect, *Locusta migratoria*, was demonstrated (Stangier *et al.*, 1989). Thus it is likely that CCAP is widely distributed throughout the phylum Arthropoda.

The storage and release of the peptide from isolated pericardial organs in response to high-potassium saline (Stangier *et al.*, 1988) led to the assumption that the nonapeptide acts as a neurohormone involved in the control of heart activity. The threshold concentration for an accelerating effect of CCAP on the semi-isolated heart of *Carcinus* was estimated to be 10^{-10} M CCAP. Taking into account a blood volume of about 3 ml in the crabs and an amount of 40 pmol CCAP/2PO, it was calculated that the release of 1% of peptide stored in the PO would be sufficient to raise the level of circulating CCAP above threshold concentration (Stangier *et al.*, 1988).

The heart was considered the primary target organ of circulating CCAP until recent findings indicated that CCAP is possibly involved in control of hindgut motility and in control of ventilation (see Section 16.5). Additionally, radioimmunoassay and immunocytochemical investigations revealed the presence of CCAP in nearly every part of the nervous system. The existence of a complex system of CCAP-immunoreactive neurons points to a role as a neurotransmitter/modulator (Dircksen and Keller, 1988).

In this chapter a summary of the hitherto known biological effects of CCAP will be given. As far as data are available, the mode of action of CCAP will be compared with that of proctolin.

16.2. **The effect of CCAP and proctolin on the isolated heart of the shore crab, *Carcinus maenas***

The isolated heart of the shore crab mounted in a cylindrical chamber of 1·2 ml volume was continuously supplied with crab saline at a temperature of 12°C. Spontaneous contractions of the isolated heart were recorded with a force transducer. For application of test samples the

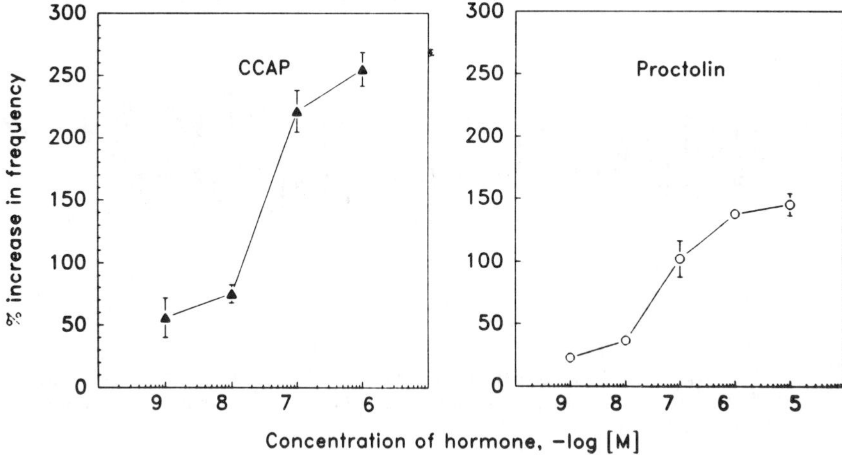

Fig. 16.1. Dose–response relationship of the inotropic effect of CCAP (left) and proctolin on the isolated heart of the shore crab, *Carcinus maenas*. Minimum number of tests per data point: 6, error bar: ± SEM.

chamber was depleted and immediately refilled with hormone solution made up in crab saline. After application of CCAP or proctolin, the force and frequency of the heartbeat increased in a dose dependent manner (Fig. 16.1). The increase in frequency amounted to 230% at a concentration of 10^{-6} M CCAP, whereas the response to proctolin did not exceed 150% at the same dose. It was observed that proctolin had a stronger effect on the force of contraction than CCAP. Although a complete dose–response relationship concerning the inotropic effects of both peptides remains to be established, it seems reasonable to conclude that CCAP is a more chronotropic and proctolin a predominantly inotropic agent.

16.3. **Effect of CCAP on the heart beat of the crayfish *Orconectes limosus***

Recent radioimmunological investigations revealed the occurrence of CCAP in the entire nervous system of the crayfish (Stangier, in preparation). Evidence pointing to a neurohormonal role of the peptide came from the detection of about 50 fmol CCAP per ml of haemolymph. The source of circulating CCAP in the crayfish remains to be elucidated, but preliminary results indicate the presence of the peptide in the pericardial organs and in the abdominal ganglia.

In order to investigate the possible role of the neurohormone in the control of crayfish heart activity, CCAP was administered to the isolated heart of *Orconectes limosus*. Similar to its action upon the heart of the

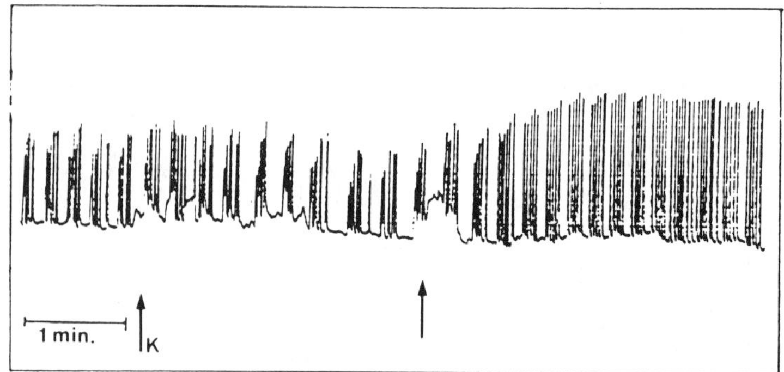

Fig. 16.2. Representative record of the response of the isolated crayfish heart to CCAP. Hormone (10^{-6} M) application is marked by an arrow. ↑K: control, application of saline.

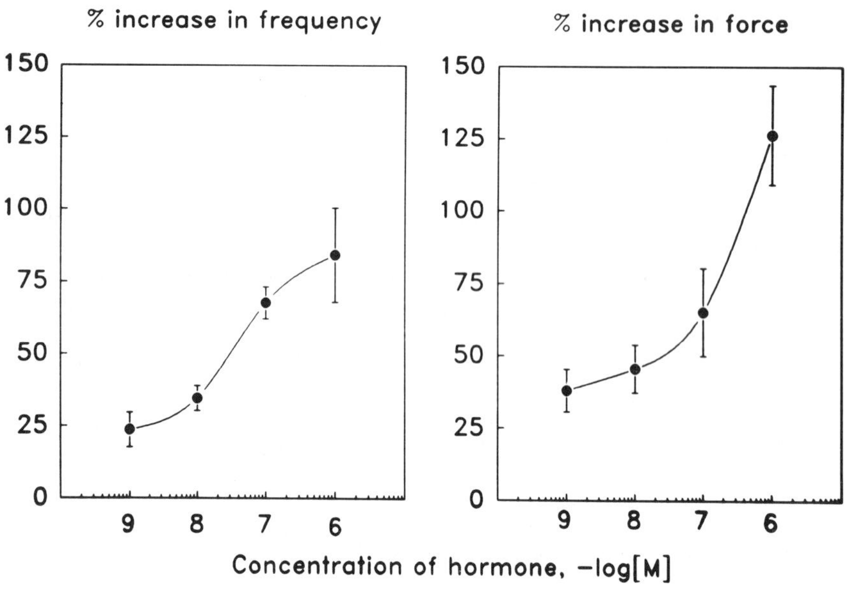

Fig. 16.3. Dose–response relationship of the chronotropic (left) and inotropic (right) effect of CCAP on the isolated heart of the crayfish, *Orconectes limosus*. $n = 6$, ± SEM.

shore crab, CCAP raised the frequency and force of contraction (Fig. 16.2).

However, it was evident from the dose–response relationship that, in comparison to the response of the crab heart, the peptide is less effective in the crayfish. The maximum response to 10^{-6} M CCAP was an 80%

increase in frequency and a 125% increase in force of contraction. It remains to be elucidated whether this effect is of physiological significance (Fig. 16.3).

16.4. **Effect of CCAP upon spontaneous contractions of the crayfish hindgut**

The isolated hindgut of crayfish is sensitive to a variety of stimulatory agents. Florey (1954) reported that acetylcholine and epinephrine increased hindgut motility. In a more recent study, Sylvia and Holman (1983) isolated two hindgut stimulatory peptides from the crayfish *Procambarus clarkii*. Until then, no information relating to the structure of these peptides was available. As mentioned above, CCAP is abundantly represented in the abdominal ganglia of *Orconectes*. Radioimmunoassay investigations revealed the presence of approximately 10 pmol CCAP in the sixth, terminal abdominal ganglion (Stangier, in preparation). Since the rectum is innervated by posterior intestinal nerves arising from the terminal ganglion (Winlow and Laverack, 1972), it seems reasonable to suppose that CCAP represents one of the peptides involved in the control of hindgut motility. Actually, CCAP enhances the force and frequency of spontaneous contractions and initiates contractions of inactive hindgut preparations (Fig. 16.4).

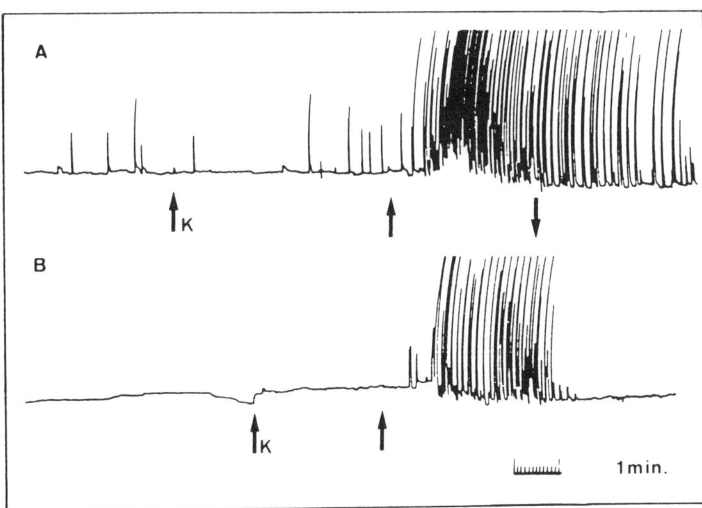

Fig. 16.4. Records of isolated hindgut contractions of the crayfish *Orconectes limosus*. **A**: Application of 10^{-6} M CCAP (↑) led to an increase of force and frequency of spontaneous contractions. Application of CCAP to quiescent preparations (**B**) initiated contractions. ↑K: control application of saline, ↓: Wash with saline.

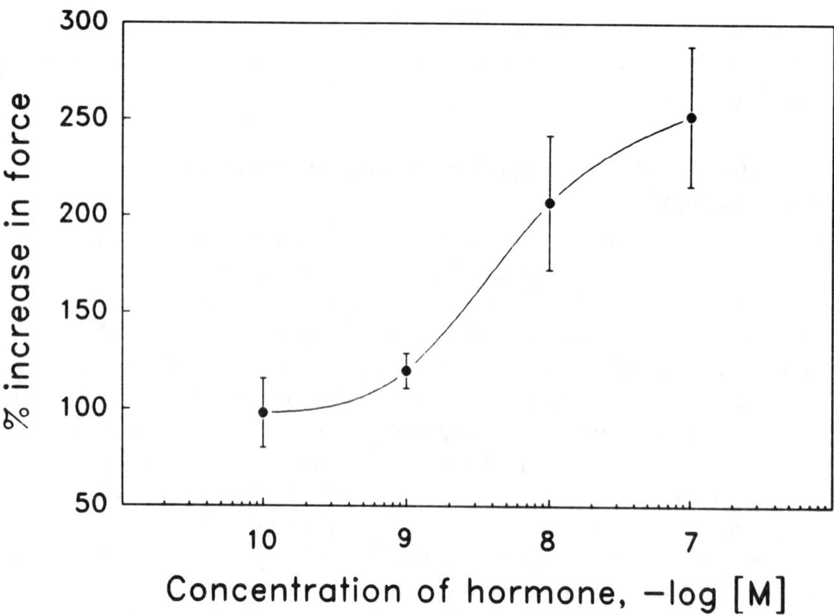

Fig. 16.5. Dose–response relationship of the inotropic effect of CCAP on the isolated hindgut of the crayfish, *Orconectes limosus*. $n = 6$, ± SEM.

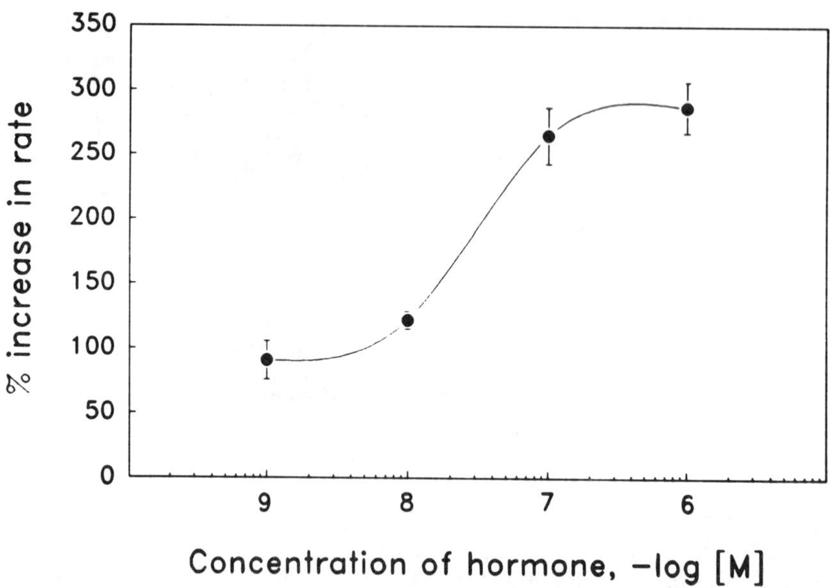

Fig. 16.6. Chronotropic response of the isolated hindgut of an insect, *Locusta migratoria*, to CCAP. $n = 10$, ± SEM.

CCAP is active at a low concentration (10^{-10} M) and led to a 250% increase in frequency at a concentration of 10^{-7} M CCAP (Fig. 16.5). Recently, CCAP was shown to be present in the nervous system of the migratory locust, *Locusta migratoria* (Stangier *et al.*, 1989). Its distribution in the nervous system was studied by radioimmunoassay, and subsequent isolation and sequence analysis of the immunoreactive peptide confirmed its identity with crustacean CCAP. Interestingly, the locust hindgut was found to respond to CCAP in a way very similar to the crayfish gut. CCAP accelerated the frequency of spontaneous hindgut contractions at low concentrations (Fig. 16.6) and induced contractions in hindgut preparations that were not spontaneously active (Stangier *et al.*, 1989).

16.5. Cardiac and ventilatory responses to CCAP by the edible crab, *Cancer pagurus*

Crabs respond to handling and surgery with stress-like increase in heart rate and ventilatory rate (Wilkens *et al.*, 1985). The persistent effect of the ventilatory response often exceeds 2 h, suggesting a neurohormonal basis. Exogenously supplied PO extract, biogenic amines and proctolin can mimic these manifestations of stress (Wilkens *et al.*, 1985). In order to investigate whether CCAP is also involved in stress responses, the neurohormone was injected into intact edible crabs, *Cancer pagurus*. Heart rate and ventilation were monitored simultaneously by photophlethysmography, a non-invasive technique based on emission and reflection of infrared light (Depledge, 1984). A bolus injection of CCAP resulting in a calculated systemic concentration of 10^{-7} M led to a long-lasting increase (300%) in ventilation, whereas no or only very small effects on heart rate were observed (Fig. 16.7). In this regard CCAP resembles the action of proctolin. The pentapeptide was found to increase the rate of ventilation when injected into *Carcinus maenas*, but failed to induce changes in heart rate (Wilkens *et al.*, 1985).

16.6. Conclusions

CCAP, the crustacean cardioactive peptide, was originally considered to be a cardioregulatory neurohormone. The storage of large amounts of CCAP in the PO, the demonstration of *in vitro* release, the occurrence of CCAP in the haemolymph and its strong cardioactive effect, even at low concentration, substantiated the hypothesis that CCAP is released from PO to act on the heart. Interestingly, *in vivo* administration of CCAP did not cause a change in heart rate, a finding that is in contrast to the results obtained with isolated or semi-isolated hearts. However, photoplethysmography is only capable of measuring the rate of heart beat. Changes in force of contraction or modulation of the blood flow by

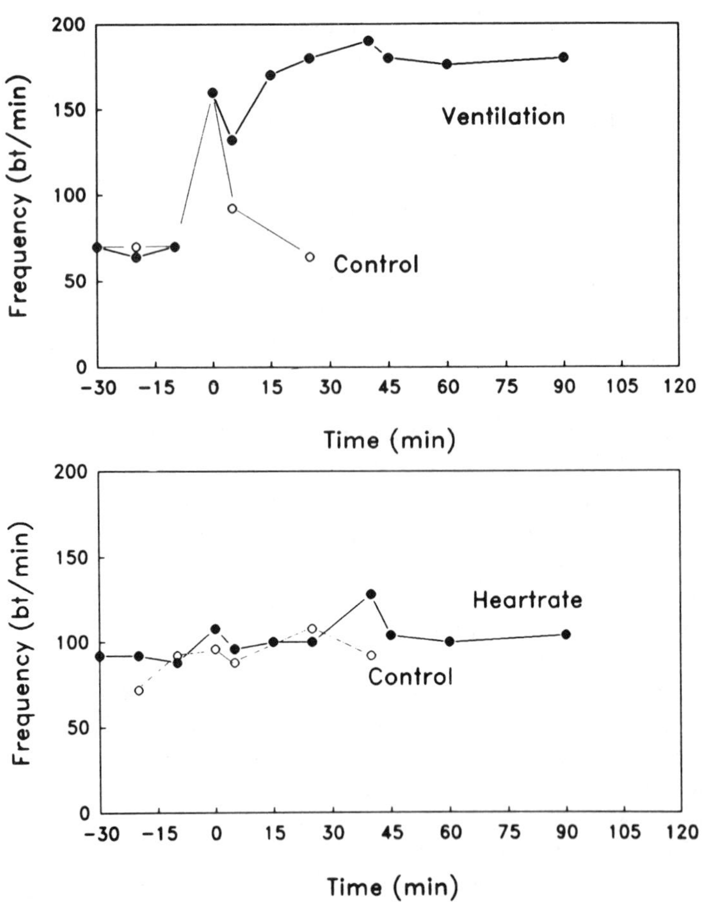

Fig. 16.7. Responses of ventilation (upper panel) and heart rate (lower panel) of *Cancer pagurus* to an injection of CCAP. The calculated systemic concentration of hormone was 10^{-7} M.

influencing the arterial valves (Kuramoto and Ebara, 1984) are not detectable.

The occurrence of a complex system of CCAPir neurons in parts of the nervous system not related to neurohaemal organs (Dircksen and Keller, 1988) pointed to a functional role as a neurotransmitter/modulator, albeit a neurotransmitter function remains to be proven. The strong accelerating effect of CCAP on spontaneous macruran hindgut contraction led to the assumption that the gut is an additional target organ of CCAP. It is supposed that CCAP plays a role in the control of gut motility, the transport of faeces and in defaecation. Whether this is based on a neurotransmitter or neurohormonal mode of action is currently under investigation. Immunocytochemical studies on the intestinal nerve, for instance, will help to reveal the possible neurotransmitter function of CCAP.

More than 80% of the total amount of CCAP present in the nervous system of the shore crab is localized in the pericardial organs (Stangier *et al.*, 1988), thus the PO represent the main source of CCAP. However, it cannot be ruled out that additional release sites contribute to the level of circulating hormone. The abundance of CCAP in the abdominal ganglia (Stangier, unpublished) points to release sites possibly localized in the ganglionic sheet similar to FMRFamide-like neurosecretory endings recently identified in the lobster, *Homarus americanus* (Kobierski *et al.*, 1987).

CCAP from peripheral neurohaemal areas is likely to act on target organs other than the heart. It is known from earlier investigations (Berlind, 1976) that PO extract increases the frequency of scaphognathite beating. As was demonstrated by *in vivo* administration, CCAP caused a persistent increase in frequency, and might therefore represent one of the active substances involved in regulation of ventilation. Future work will have to focus on the mechanism employed. Possible modes of action include the modulation of the central pattern generator producing the motor output to the muscles driving the scaphognathite, or a direct effect of CCAP on the ventilatory musculature. The latter possibility seems to be feasible because Dircksen and Keller (1988) identified CCAP-like immuno-reactive neurosecretory endings in the 'anterior ramifications', neurohaemal structures situated in close proximity to the ventilatory muscles.

The physiological function of CCAP relative to that of proctolin is of particular interest. Both peptides occur in insects and crustaceans, and they have similar effects on crustacean heart and hindgut and on the insect hindgut. The specific role each peptide has to play will be the main point addressed in future studies.

References

Alexandrowicz, J. S. (1953). Nervous organs in the pericardial cavity of the decapod Crustacea. *J. Marine Biol. Assoc.*, **34**, 563–80.

Alexandrowicz, J. S. and Carlisle, D. B. (1953). Some experiments on the function of the pericardial organs in Crustacea. *J. Marine Biol. Assoc.*, **32**, 175–92.

Berlind, A. (1976). Neurohemal organ extracts effect the ventilation oscillator in crustaceans. *J. Exp. Zool.*, **195**, 165–70.

Cooke, I. M. and Sullivan, R. E. (1982). Hormones and neurosecretion. In: Bliss, D. (ed.), *Biology of Crustacea*, Vol. 3. Academic Press, New York, pp. 205–90.

Depledge, M. H. (1984) Photophlethysmography – A non-invasive technique for monitoring heart beat and ventilation rate in decapod crustaceans. *Comp. Biochem. Physiol.*, **77A**(2), 369–71.

Dircksen, H. and Keller, R. (1988). Immunocytochemical localization of CCAP, a novel crustacean cardioactive peptide, in the nervous system of the shore crab, *Carcinus maenas* (L.). *Cell Tiss. Res.*, **245**, 347–60.

Florey, E. (1954). Über die Wirklung von Acetylcholin, Adrenalin, Noradrenalin, Factor I und anderen Substanzen auf den isolierten Enddarm des Flusskrebses *Cambarus clarkii. Z. Vergleich. Physiol.*, **36**, 1–8.

Florey, E. and Rathmayer, M. (1978). The effects of temperature, anoxia and sensory stimulation on the heart rate of unrestrained crabs. *Comp. Biochem. Physiol.*, **61C**, 229–37.

Kobierski, L. A., Beltz, B. S., Trimmer, B. A. and Kravitz, E. A. (1987) FMRFamide like peptides of *Homarus americanus*: distribution, immunocytochemical mapping and ultrastructural localization of terminal varicosities. *J. Comp. Neurol.*, 266, 1–15.

Kuramoto, T. and Ebara, A. (1984). Neurohormonal modulation of the cardiac outflow through the cardioarterial valves in the lobster. *J. Exp. Biol.*, **3**, 123–30.

Schwarz, T. L., Lee, G. M., Siwicki, K. K., Standaert, D. G. and Kravitz, E. A. (1984). Proctolin in the lobster: the distribution, release and chemical characterization of a likely neurohormone. *J. Neurosci.*, **4**(5), 1300–11.

Stangier, J., Dircksen, H. and Keller, R. (1986). Identification and localization of proctolin in pericardial organs of the shore crab, *Carcinus maenas*. *Peptides*, **7**, 67–72.

Stangier, J., Hilbich, C., Beyreuther, K. and Keller, R. (1987). Unusual cardioactive peptide (CCAP) from pericardial organs of the shore crab, *Carcinus maenas*. *Proc. Natl. Acad. Sci. USA*, **84**, 575–9.

Stangier, J., Hilbich, C., Dircksen, H. and Keller, R. (1988). Distribution of a novel cardioactive neuropeptide (CCAP) in the nervous system of the shore crab, *Carcinus maenas*. *Peptides*, **9**, 795–800.

Stangier, J., Hilbich, C. and Keller, R. (1989). Occurrence of crustacean cardioactive peptide (CCAP) in the nervous system of an insect, *Locusta migratoria*. *J. Comp. Physiol. B*, **159**; 5–11.

Starratt, A. N. and Brown, A. E. (1975). Structure of the pentapeptide proctolin, a proposed neurotransmitter in insects. *Life Sci.* **7**, 1253–6.

Sullivan, R. E. (1978). Stimulus coupled ^{3}H-serotonin release from identified neurosecretory fibers in the spiny lobster, *Panulirus interruptus*. *Life Sci.*, **22**, 1429–38.

Sullivan, R. E. (1979). A proctolin like peptide in the crab pericardial organs. *J. Exp. Zool.*, **210**, 543–52.

Sylvia, V. L. and Holman, G. M. (1983). Isolation and high performance liquid chromatographic purification of a myotropic peptide from the hindgut of the crayfish, *Procambarus clarkii*. *J. Chromatogr.*, **261**, 158–62.

Wilkens, J. L., Mercier, A. J. and Evans, J. (1985). Cardiac and ventilatory responses to stress and to neurohormonal modulators by the shore crab, *Carcinus maenas*. *Comp. Biochem. Physiol.*, **82C**(2), 337–43.

Winlow, W. and Laverack, M. S. (1972). The control of hindgut motility in the lobster, *Homarus gammarus*. *Mar. Behav. Physiol.*, **1**, 1–27.

17 *François Van Herp and Janine L. Kallen*

Neuropeptides and neurotransmitters in the X-organ sinus gland complex, an important neuroendocrine integration centre in the eyestalk of Crustacea

17.1. Introduction

The neuroendocrine system in Crustacea is complex, and elements of it are found throughout the central nervous system. Neurons with a putative neuroendocrine function occur in all ganglia, but three regions show remarkable clustering of them. These are the optic ganglia which contain the x-organ sinus gland complex, the pericardial organs and the postcommissural organs. Until now, the x-organ sinus gland complex has received most attention. It represents the most important source of neuroendocrine products which are involved in the regulation of a variety of biological processes such as moulting, reproduction, pigment migration, osmoregulation and carbohydrate metabolism. These neurohormonal functions have been described in mainly decapod Crustacea and have been reviewed in recent years by Cooke and Sullivan (1982), Keller (1983), Kleinholz (1985), Quackenbush (1986), and Webster and Keller (1987). Andrew (1983) and Chaigneau (1983) have recently reviewed the morphology of neurohaemal organs in Crustacea.

Immunocytochemical application of several antisera against biologically active peptides from vertebrates and invertebrates resulted in the detection of a more extensive (neuro)peptidergic system than formerly recognized in Crustacea. Van Herp (1988) has summarized these immunocytochemical studies carried out during recent years. Although there is at present little information concerning the role of 'vertebrate-like' and 'invertebrate-like' peptides, it is clear that the histological distributions of the peptidergic material reflect the versatility of the central nervous system in Crustacea for neurohormonal as well as neuromodulatory functions. This view is also supported by the occurrence of aminergic centres, illustrating their neurotransmittory as well as their neurohormonal functions. Definitive evidence for synthesis and storage of different biogenic amines in the nervous system of Crustacea is reported by Sullivan (1978), Kravitz *et al.* (1980), Elofsson (1983), Bellon-Humbert and Van Herp (1988) and Kallen (1988). With respect to the neuroendocrine functions of the optic ganglia in the eyestalk of decapod Crustacea, the organization of both systems postulates their importance for the regulation of synthesis, transport and release of typical crustacean neuropeptides such as the crustacean hyperglycaemic hormone (CHH), the red pigment-concentrating hormone (RPCH), the pigment-dispersing hormone (PDH), the moult-inhibiting hormone (MIH) and the gonad-inhibiting hormone (GIH), all synthesized in the x-organ of the medulla terminalis (MTGX). Immunocytochemical localization of these 'native' crustacean neuropeptides has been made possible recently by the availability of specific antisera for CHH (Van Herp and Van Buggenum, 1979; Jaros and Keller, 1979), for RPCH (Bellon-Humbert *et al.*, 1986; Mangerich *et al.*, 1986; Schooneveld *et al.*, 1987), for PDH (Mangerich *et al.*, 1987), for MIH (Dircksen *et al.*, 1988) and for GIH (Kallen and Meusy, 1989).

In this chapter the functional significance of the x-organ sinus gland complex in the eyestalk of decapod Crustacea is illustrated by reviewing our knowledge of the crustacean hyperglycaemic hormone-(CHH) producing system.

17.2. **Morphology of the CHH-producing system**

About 10 years ago it became possible to identify the CHH-producing cells in the eyestalk of decapod Crustacea by immunocytochemical detection using a polyclonal antiserum against purified CHH of the crayfish, *Astacus leptodactylus* (Van Herp and Van Buggenum, 1979) and an antiserum against CHH of the crab, *Carcinus maenas* (Jaros and Keller, 1979).

In the eyestalk of the crayfish the CHH-producing cells form part of the medulla terminalis x-organ (MTGX) where they are found as a distinct group, consisting of about 35 cells, situated at the rostrodistal part, lateroventrally on the MTGX. Their axons form a tract which transverses the medulla terminalis neuropil and runs to the sinus gland. Here the axons split up into numerous axon terminals (Figs 17.1–17.4). By light and fluorescence microscopy the CHH material is visible as granular material randomly distributed throughout the cytoplasm. Electron microscopy reveals that the immunopositive reaction occurs in Golgi sacculi and in the CHH-containing granules present in the cell body, the axon and the axon terminals. In the sinus gland the immunoreaction is restricted to one granule type (Gorgels-Kallen and Van Herp, 1981).

A comparative study of the CHH system in the eyestalk of several decapod Crustacea (Gorgels-Kallen *et al.*, 1982) has indicated that the CHH-producing cells of all investigated species are found in a region similar to that described for the crayfish. From these results one can conclude that the sites of production, storage and release of CHH are comparable for the infra-orders of Astacidea, Brachyura, Caridea and Palinura.

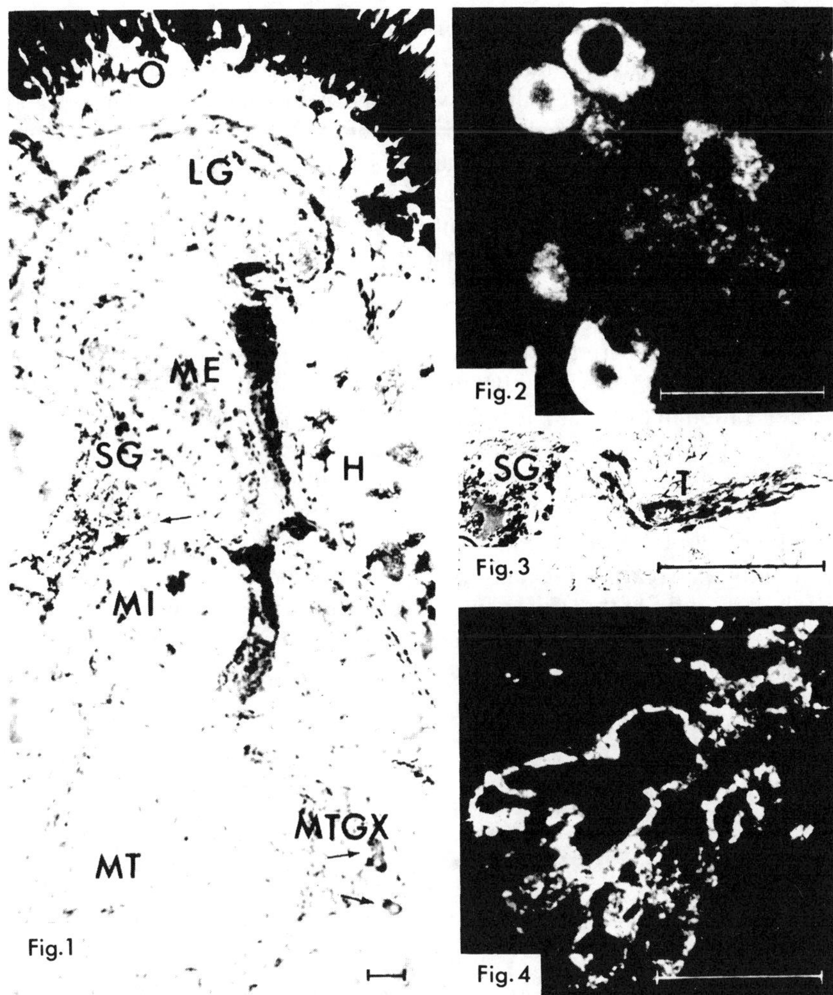

Fig. 17.1. Longitudinal section of the eyestalk of a crayfish, showing CHH immunopositive (PAP) reaction in cells of the MTGX and in the sinus gland (arrows). Post-staining of the nuclei with hemalum Mayer. Bar: 100 μm. (Abbreviations: H = haemolymph; LG = lamina ganglionaris; ME = medulla externa; MI = medulla interna; MT = medulla terminalis; MTGX = medulla terminalis ganglionic x-organ; O = ommatidia; SG = sinus gland).

Fig. 17.2. Immunofluorescent staining of CHH material in the perikarya of the CHH system in the MTGX, showing the difference in intensity. Bar: 100 μm.

Fig. 17.3. Immunocytochemical (PAP) staining of CHH in the tract (T), descending from the MTGX and leading to the sinus gland (SG). Bar: 100 μm.

Fig. 17.4. Immunofluorescence of CHH material in the sinus gland. Bar: 100 μm.

17.3. Cytophysiology of the CHH-producing system

The secretory stages of individual CHH-producing cells in the eyestalk of the crayfish can be determined by means of immunocytochemistry and morphometry (Gorgels-Kallen and Voorter, 1984). Immunostaining demonstrates striking differences in staining intensity among individual CHH cells. Electron microscopic investigations reveal that these differences are correlated with differences in numerical density of the neurosecretory granules in the cytoplasm, and that they reflect differences in synthetic activity among the CHH cells. Morphometric analyses at the light- and electron-microscopic level indicate that three categories of immunopositive cells represent different stages in the CHH synthesizing process of the cells. When these data are related to the 24 h rhythmicity of the glucose concentration in the haemolymph, the secretory dynamics of the synthesizing cell group can be determined in the course of the day/night cycle (Gorgels-Kallen and Voorter, 1985). The synthetic activity of the CHH cells receives a stimulus 2 h before the beginning of the dark period, resulting in a pronounced transfer of CHH granules into the axons. These CHH granules reach the axon terminals in the sinus gland at the onset of the dark period. At that time a burst of exocytotic activity occurs, causing a strong release of CHH into the haemolymph. Four hours later this CHH release results in hyperglycaemia. The same process, though with less

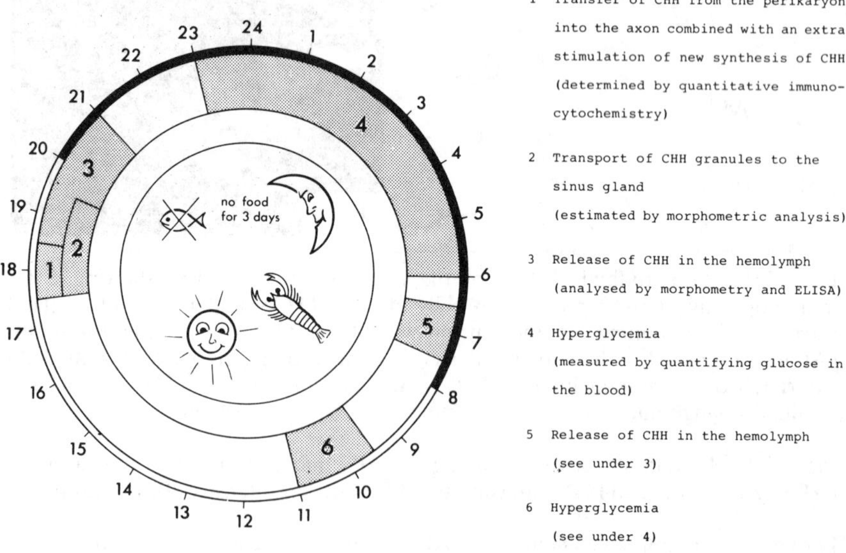

Fig. 17.5. Diagram illustrating the secretory dynamics of the CHH-producing cells and the hyperglycaemic effect in the blood of the crayfish during a day/night cycle of 12 h L/12 h D.

intensity, is repeated and causes a second smaller glucose peak at the beginning of the light period (Fig. 17.5). The rhythmicity in the CHH level in the blood, measured by a double-antibody sandwich ELISA for CHH, reflects the releasing activity of the CHH-producing system (Kallen and Abrahamse, in preparation).

The secretory dynamics of the CHH system in the crayfish are firmly reduced during the moulting period of the animal. In addition, examination of the 24 h blood glucose rhythmicity reveals the complete absence of a nocturnal peak (Kallen, 1985).

An immunocytochemical study of the CHH system during the larval and postlarval life cycle of the crayfish indicates that the system appears immediately after hatching and is already active during both life periods. The number of active CHH cells increases during postlarval life (Gorgels-Kallen and Mey, 1985).

By studying the CHH-producing system in the prawn *Palaemon serratus* a relationship is found with the seasonal as well as the moulting cycle of those animals (Van Herp *et al.*, 1984). In conclusion, the aforementioned results suggest that several rhythmic phenomena may play a role in the regulation of the activity of the CHH-producing system and related glycaemia in Crustacea.

17.4. **Innervation/regulation of the CHH-producing system**

Investigations on the effects of constant darkness, light/dark phase shift, covered eyes, eyestalks and rostral regions, as well as optic tract sectioning on the entrainment of the daily rhythmicity in secretory activity of the CHH cells and blood glucose variations, indicate an endogenous circadian rhythm entrained by a light/dark schedule, at one side, and suppose that the biological clock must be located within the optic lobes (Kallen *et al.*, 1988). Neither the retinas nor the caudal receptors represent the main modulating receptor for blood glucose level and synthetic activity of the CHH cells. Therefore one can postulate that the eyestalks possess the major receptors for the entraining light stimulus and also contain the oscillatory centre for the rhythmicity in blood glucose concentration and CHH cell activity.

By injection of the fluorescent dye Lucifer Yellow into individual CHH cells the shape of the cells can be traced. A highly fluorescent perikaryon gives rise to an axon that can be followed by the fluorescent label to the neurohaemal region, the sinus gland. The proximal part of that axon sends out extensive branches into the neuropil of the medulla terminalis. Ultrastructural investigations reveal synaptic input on these CHH axonal ramifications (Gorgels-Kallen, 1985). In the direct neighbourhood of these CHH axonal ramifications a considerable number of neuronal processes is found. They are closely packed with two types of inclusions: one type is moderately dense and surrounded by a halo; the second type is an electron-lucent vesicle. In addition, these processes show synaptic contacts with the CHH axonal ramifications. All these synapses have the following features: (a) the adjacent CHH ramifications and neuronal processes are separated by a 20 nm synaptic cleft; (b) the neuronal processes contain synaptic vesicles in presynaptic regions; (c) the CHH processes contain a distinct synaptic density contiguous with the synaptic cleft. Moreover, one presynaptic process may contact more than one CHH branch, and the morphology of the synapses suggests that the CHH branches are postsynaptic (Figs 17.6 and 17.7).

Previous studies on synaptic input on neurosecretory axons in Crustacea suggest aminergic neuromodulators in the presynaptic axons (Shivers, 1967; Andrew and Saleuddin, 1978). However, the view that information transmitted by neurons takes place only by means of non-peptidergic substances has been superseded, and recent studies reveal also neuromodulatory functions for peptides (Van Deynen, 1986; Van Herp, 1988). In Crustacea, immunocytochemical studies of the eyestalk by means of antisera against several neuropeptides reveal distinct reactions in neuroendocrine cells and axons. In crayfish, for example, positive reactions are found with antisera against FMRF-amide, α-MSH, vasotocin, gastrin, CCK, oxytocin, secretin, glucagon and GIP. The morphology of these peptidergic systems suggests that one part is neurohormonal and the other part neurotransmitter-like or neuromodulatory, but that all of them may play a role in the neuroendocrine integration of the medulla terminalis neuropil (Van Deynen *et al.*, 1985). It is striking that the distribution of these described peptidergic systems shows considerable similarities to that obtained for monoamines, as studied by Elofsson and Klemm (1972), Elofsson *et al.* (1977) and Strolenberg (1979). It becomes clear that both systems are in very close contact and are even intermingled.

Detailed information on the chemical nature of the substance(s) in the aforementioned neuronal processes in synaptic contact with CHH axonal ramifications is important because they probably represent the only neuronal input to the CHH system. As physiological experiments dealing with the effects of injections of several neurotransmitters and neuropeptides on glycaemia in Crustacea have indicated that serotonin, and on a lower scale also dopamine, provoke hyperglycaemia (Keller and Beyer, 1968; Strolenberg and Van Herp, 1977; Martin, 1978; Kallen, 1988) and that Met-enkephalin decreases the blood glucose level as well as the nocturnal peak (Kallen, 1988), their relationship with the CHH cell system is being studied, using immunocytochemical techniques. Recently, it became possible to demonstrate immunopositive reactions for anti-serotonin, anti-dopamine and anti-Met-enkephalin, together with anti-CHH using double-immunostaining methods at the light microscopic level (Kallen *et al.*, in preparation). Figure 17.8 summarizes these preliminary

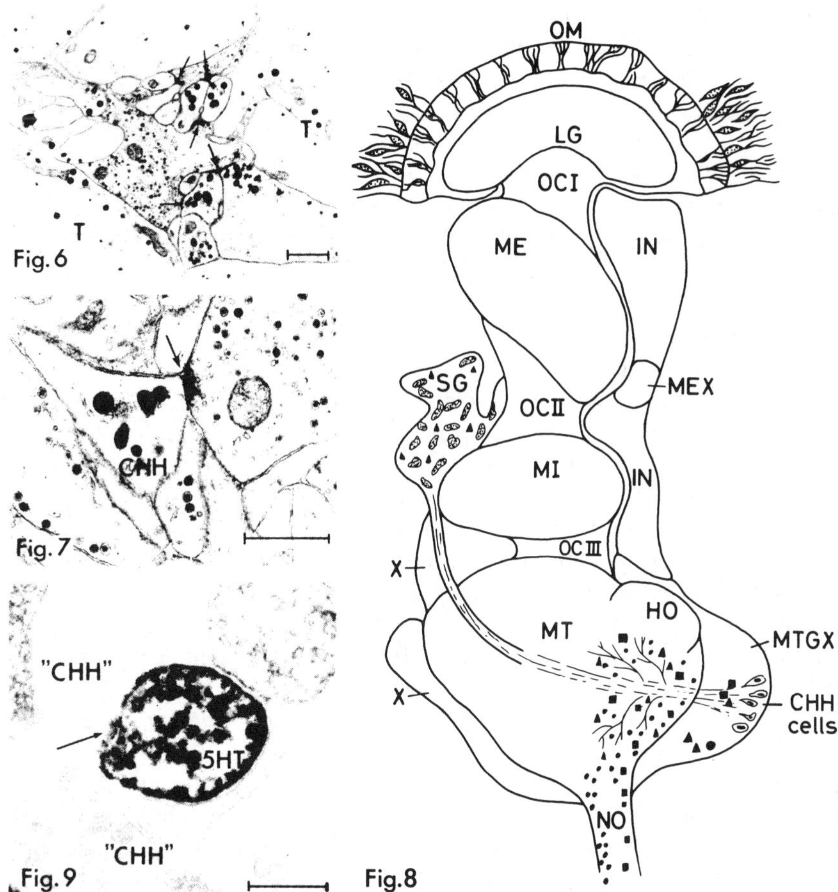

Fig. 17.6. Survey of a presynaptic axon making synaptic contact with more than one CHH process (arrows) and a main CHH axon (large arrow). Bar: 1 μm, (T = tract).

Fig. 17.7. Ultrastructure of synaptic contact on a CHH axon branch (arrow). Bar: 500 nm.

Fig. 17.8. Putative relationships between serotonin, dopamine, Met-enkephalin and the CHH-producing system in the neuropile region of the MT (● = serotonin; ■ = dopamine; ▲ = Met-enkephalin).

Fig. 17.9. Electron micrograph demonstrating a serotoninergic terminal (5HT), forming a synaptic junction (arrow) with a presumptive CHH axonal ramification ('CHH'). Bar: 500 nm. Immunostaining is carried out on 'pre-embedded' tissue sections.

results. Furthermore, immunocytochemistry applied at the ultrastructural level, and directed at the medulla terminalis neuropil region – where neuronal processes and CHH axonal ramifications are abundant – demonstrates serotoninergic synaptic input on 'CHH' axon branches (Fig. 17.9). Comparable studies for dopamine and Met-enkephalin are in progress. Summarizing the results obtained so far, one can postulate that serotonin and dopamine, as well as Met-enkephalin, are good candidates as regulatory factors for the activity of the CHH-producing system.

17.5. **Conclusion**

In recent years modern methods such as pulse chase labelling of neuropeptide precursors, microanalytical chemistry for protein sequencing, molecular biological techniques for studying DNA and mRNA, *in vitro* culture methods for crustacean neuroendocrine cells, as well as electrophysiological methods, have been introduced by several research teams interested in crustacean neuroendocrinology, and their results will contribute to elucidate the complexity of the presented CHH-producing system and to relate it to physiological processes. Furthermore, recent studies have also demonstrated that the 'native' crustacean neuropeptides responsible for moult inhibition (MIH), red pigment concentration (RPCH), pigment dispersion (PDH) and gonad inhibition (GIH) are synthesized in the same regions of the x-organ sinus gland complex of the medulla terminalis. From this point of view crustacean neuroendocrine research dealing with topics such as synthesis, storage and release of 'native' neuropeptides; seasonal-, diurnal- and activity-related changes of neuropeptide production, storage, release and action; and functional versatility of neuroendocrine cells in the optic ganglia have good prospects in the near future.

References

Andrew, R. D. (1983). Neurosecretory pathways supplying the neurohemal organs in Crustacea. In: Gupta, A. P. (ed.), *Neurohemal Organs of Arthropods*. Charles C. Thomas, Springfield, IL, pp. 90–117.

Andrew, R. D. and Saleuddin, A. S. M. (1978). Structure and innervation of a crustacean neurosecretory cell. *Can. J. Zool.*, **56**, 423–30.

Bellon-Humbert, C. and Van Herp, F. (1988). Localization of serotonin-like immunoreactivity in the eyestalk of the prawn *Palaemon serratus* (Crustacea, Decapoda, Natantia). *J. Morph.*, **196**, 307–20.

Bellon-Humbert, C., Van Herp, F. and Schooneveld, H. (1986). Immunocytochemical study of the red pigment concentrating material in the eyestalk of the prawn *Palaemon serratus*, Pennant, using rabbit antisera against the insect adipokinetic hormone. *Biol. Bull.*, **171**, 647–59.

Chaigneau, J. (1983). Neurohemal organs in Crustacea. In: Gupta, A. P. (ed.),

Neurohemal Organs of Arthropods. Charles C. Thomas, Springfield, IL, pp. 53–89.

Cooke, I. M. and Sullivan, R. E. (1982). Hormones and neurosecretion. In: Bliss, D. E., Atwood, H. L. and Sandeman, D. C. (eds), *The Biology of Crustacea*, vol. 3, Academic Press, New York, pp. 205–90.

Dircksen, H., Webster, S. G. and Keller, R. (1988). Immunocytochemical demonstration of the neurosecretory systems containing putative moult-inhibiting hormone and hyperglycemic hormone in the eyestalk of brachyuran crustaceans. *Cell Tiss. Res.*, **251**, 3–12.

Elofsson, R. (1983). 5-HT-like immunoreactivity in the central nervous system of the crayfish, *Pacifastacus leniusculus*. *Cell Tiss. Res.*, **232**, 221–36.

Elofsson, R. and Klemm, N. (1972). Monoamine-containing neurons in the optic ganglia of crustaceans and insects. *Z. Zellforsch.*, **133**, 475–99.

Elofsson, R., Nässel, D. and Myhrberg, H. (1977). A catecholaminergic neuron connecting the first two optic neuropiles (lamina ganglionaris and medulla externa) of the crayfish, *Pacifastacus leniusculus*. *Cell Tiss. Res.*, **182**, 287–97.

Gorgels-Kallen, J. L. (1985). Appearance and innervation of CHH-producing cells in the eyestalk of the crayfish *Astacus leptodactylus*, examined after tracing with Lucifer yellow. *Cell Tiss. Res.*, **240**, 385–91.

Gorgels-Kallen, J. L. and Mey, J. T. A. (1985). Immunocytochemical study of the hyperglycemic hormone (CHH)-producing system in the eyestalk of the crayfish *Astacus leptodactylus*, during larval and postlarval development. *J. Morph.*, **183**, 155–63.

Gorgels-Kallen, J. L. and Van Herp, F. (1981). Localization of the crustacean hyperglycemic hormone (CHH) in the X-organ sinus gland complex in the eyestalk of the crayfish *Astacus leptodactylus* (Nordmann, 1842). *J. Morph.*, **50**, 347–55.

Gorgels-Kallen, J. L. and Voorter, C. E. M. (1984). Secretory stages of individual CHH-producing cells in the eyestalk of the crayfish *Astacus leptodactylus*, determined by means of immunocytochemistry. *Cell Tiss. Res.*, **237**, 291–8.

Gorgels-Kallen, J. L. and Voorter, C. E. M. (1985). The secretory dynamics of the CHH-producing cell group in the eyestalk of the crayfish *Astacus leptodactylus*, in the course of a day/night cycle. *Cell Tiss. Res.*, **241**, 361–6.

Gorgels-Kallen, J. L., Van Herp, F. and Leuven, R. S. E. W. (1982). A comparative immunocytochemical investigation of the crustacean hyperglycemic hormone (CHH) in the eyestalk of some decapod Crustacea. *J. Morph.*, **174**, 161–8.

Jaros, P. P. and Keller, R. (1979). Immunocytochemical identification of the hyperglycemic hormone-producing cells in the eyestalk of *Carcinus maenas*. *Cell Tiss. Res.*, **204**, 379–85.

Kallen, J. L. (1985). The hyperglycemic hormone-producing system in the eyestalk of the crayfish *Astacus leptodactylus*. Thesis, Faculty of Sciences, Catholic University of Nijmegen, The Netherlands.

Kallen, J. L. (1988). Quelques aspects de la régulation du système neuroendocrine produisant la CHH et de la relation entre le rythme circadien et la glycémie. In: Le Gal, Y. and Van Wormhoudt, A. (eds), *Aspects récents de la biologie des Crustacés. Actes de Colloques*, 8. Ifremer, Brest, pp. 105–7.

Kallen, J. L. and Meusy, J. J. (1989). Do the neurohormones VIH (vitellogenesis inhibiting hormone) and CHH (crustacean hyperglycemic hormone) of crustaceans have a common precursor? Immunolocalization of VIH and CHH in the X-organ sinus gland complex of the lobster *Homarus americanus*. *Invert. Repr. Dev.*, **16**, 43–52.

Kallen, J. L., Rigiani, N. R. and Trompenaars, H. J. A. J. (1988). Aspects of entrainment of CHH cell activity and hemolymph glucose levels in crayfish. *Biol. Bull.*, **175**, 137–43.

Keller, R. (1983). Biochemistry and specificity of the neurohemal hormones in Crustacea. In: Gupta, A. P. (ed.), *Neurohemal Organs of Arthropods*, Charles C. Thomas, Springfield, IL, pp. 118–48.

Keller, R. and Beyer, J. (1968). Zur hyperglykämischen Wirkung von Serotonin und Augenstielextract beim Flusskrebs *Orconectes limosus*. *Z. Vergl. Physiol.*, **59**, 78–85.

Kleinholz, L. H. (1985). Biochemistry of crustacean hormones. In: Bliss, D. E. and Mantel, L. H. (eds), *The Biology of Crustacea*, vol. 5, Academic Press, New York, pp. 464–522.

Kravitz, E. A., Glusman, S., Harris-Warrick, R. M., Livingstone, M. S., Schwartz, T. and Goy, M. F. (1980). Amines and a peptide as neurohormones in lobsters: Actions on neuromuscular preparations and preliminary behavioural studies. *J. Exp. Biol.*, **89**, 159–75.

Mangerich, S., Keller, R. and Dircksen, H. (1986). Immunocytochemical identification of structures containing putative red pigment-concentrating hormone in two species of decapod crustaceans. *Cell Tiss. Res.*, **245**, 377–86.

Mangerich, S., Keller, R., Dircksen, H., Ranga Rao, K. and Riehm, J. P. (1987). Immunocytochemical localization of pigment-dispersing hormone (PDH) and its coexistence with FMRF-amide immunoreactive material in the eyestalk of the decapod crustaceans *Carcinus maenas* and *Orconectes limosus*. *Cell Tiss. Res.*, **250**, 365–75.

Martin, G. (1978). Action de la sérotonine sur la glycémie et sur la libération des neurosécrétions contenues dans la glande du sinus de *Porcellio dilatatus* Brandt (Crustacé, Isopode, Oniscoide). *C. R. Soc. Biol.*, **172**, 304–8.

Quackenbush, L. S. (1986). Crustacean endocrinology, a review. *Can. J. Fish. Aquat. Sci.*, **43**, 2271–82.

Schooneveld, H., Van Herp, F. and Van Minnen, J. (1987). Demonstration of substances immunologically related to the identified arthropod neuropeptides AKH/RPCH in the CNS of several invertebrate species. *Brain Res.*, **406**, 224–32.

Shivers, R. R. (1967). Fine structure of crayfish optic ganglia. *Univ. Kansas Sci. Bull.*, **XLVII**, 677–733.

Strolenberg, G. E. C. M. (1979). Functional aspects of sinus gland in the neurosecretory system of the crayfish *Astacus leptodactylus*. An ultrastructural approach. Thesis, Faculty of Sciences, Catholic University of Nijmegen, The Netherlands.

Strolenberg, G. E. C. M. and Van Herp, F. (1977). Mise en évidence du phénomène d'exocytose dans la glande du sinus d'*Astacus leptodactylus* (Nordmann) sous l'influence d'injection de sérotonine. *C. R. Acad. Sci. D*, **284**, 57–60.

Sullivan, R. E. (1978). Structure and function of spiny lobster ligamental nerve plexuses: Evidence for synthesis and storage of biogenic amines and their secretion as neurohormones. *J. Neurobiol.*, **8**, 581–605.

Van Deynen, J. E. M. (1986). Structural and biochemical investigations into the neuroendocrine system of the optic ganglia of decapod Crustacea. Thesis, Faculty of Sciences, Catholic University of Nijmegen, The Netherlands.

Van Deynen, J. E., Vek, F. and Van Herp, F. (1985). An immunocytochemical study of the optic ganglia of the crayfish *Astacus leptodactylus* (Nordmann, 1842) with antisera against biologically active peptides of vertebrates and invertebrates. *Cell Tiss. Res.*, **240**, 175–83.

Van Herp, F. (1988). La neuroendocrinologie des Crustacés: Evaluation des progrès obtenus ces dernières années. In: Le Gal, Y. and Van Wormhoudt, A. (eds), *Aspects récents de la biologie des Crustacés. Actes de Colloques*, 8. Ifremer, Brest, pp. 85–99.

Van Herp, F. and Van Buggenum, H. J. M. (1979). Immunocytochemical localization of hyperglycemic hormone (HGH) in the neurosecretory system of the eyestalk of the crayfish *Astacus leptodactylus*. *Experientia*, **35**, 1527–28.

Van Herp, F., Van Wormhoudt, A., Van Venrooy, W. A. J. and Bellon-Humbert, C. (1984). Immunocytochemical study of crustacean hyperglycemic hormone (CHH) in the eyestalks of the prawn *Palaemon serratus* (Pennant) and some other Palaemonidae, in relation to variations in blood glucose level. *J. Morph.*, **182**, 85–94.

Webster, S. G. and Keller, R. (1987). Physiology and biochemistry of crustacean neurohormonal peptides. In: Thorndyke, M. and Goldsworthy, G. (eds), *Neurohormones in Invertebrates*. Cambridge University Press, London, pp. 173–95.

18 *Heinz Penzlin and Eckehard Baumann*

Neurohormone D: function and inactivation

18.1. Introduction

The first neuropeptide of an invertebrate to be identified was the red pigment-concentrating hormone (RPCH) from the shrimp *Pandalus* in 1972 (Fernlund and Josefsson, 1972). It is an octapeptide which controls the centripetal migration of the pigment in the erythrophores. Four years later, in 1976, another peptide from the corpora cardiaca of the locust was identified. It regulates lipid utilization during prolonged flight by provoking the release of diacylglycerol from the fat body; it was named 'adipokinetic hormone' (AKH) by Mayer and Candy in 1969. The amino acid sequence of this peptide turned out to be very similar to the RPCH. Meanwhile, 11 members of this peptide family have been identified (Fig. 18.1). It must be expected that their number will continue to increase in the future.

All members of this family are C- and N-terminally blocked, N-terminally by pyroglutamic acid and C-terminally by acid amide. In every case we find in position 4 phenylalanine (F), and in position 8 tryptophan (W). Positions 2, 3, and 5 are also very uniform: leucine (L) or valine (V) in position 2, asparagine (N) or threonine (T) in position 3, serine (S) or threonine (T) in position 5. In all these cases only one alteration of a single nucleotide base pair is necessary to come from the codon for the first amino acid to the codon for the second one. When present, glycine (G) is in position 9 and threonine (T) in position 10. Until now members of this peptide family have been found only in arthropods, although certain similarities exist to bradykinin, on the one hand, and to glucagon/secretin, on the other (Penzlin, 1989).

The physiological effects of these peptides may concern lipid metabolism

Abbreviations: DDT: 2,2-bis(*p*-chlorophenyl)-1,1,1,-tri-chloroethane; DDVP (=Dichlorvos): *O*,*O*-dimethyl-2,2-dichlorovinylphosphate; dieldrine: 1,2,3,4,10,10-hexachloro-exo-6,7-epoxy-1,4,4a,5,6,7,8,8a-octahydro-1,4-endo,exo-5,8-dimethanonaphthalene; malathion: *O*,*O*-dimethyl-S-1,2-di(carboethoxy)ethyl phosphorodithioate; EDTA: ethylenediamine tetra-acetic acid; AEBSF: 4-β-aminoethylbenzenesulphonylfluoride; PMSF: phenylmethanesulphonyl fluoride; TLCK: tosyl-lysine chloromethyl ketone; TPCK: tosyl-phenylalanyl chloromethyl ketone.

Peptide	Sequence	Occurrence
Bradykinin	R–P–P–G–F–S–P–F–R	Mammalia
RPCH	pQ–L–N–F–S–P–G–W•NH_2	Pandalus (Crustacea)
AKH II-L	pQ–L–N–F–S–A–G–W•NH_2	Locusta
AKH II-S	pQ–L–N–F–S–T–G–W•NH_2	Schistocerca
AKH-G = Ro II	pQ–V–N–F–S–T–G–W•NH_2	Gryllus, Romalea
Hypertrehalosemic h.	pQ–V–N–F–S–P–G–W–G–T•NH_2	Blaberus, Nauphoeta
Neurohormone D(MI)	pQ–V–N–F–S–P–N–W•NH_2	Periplaneta
Ro I	pQ–V–N–F–T–P–N–W–G–T•NH_2	Romalea
AKH I	pQ–L–N–F–T–P–N–W–G–T•NH_2	Locusta, Schistocerca
Hypertrehalosemic h.II	pQ–L–T–F–T–P–N–W–G–T•NH_2	Carausius
M II	pQ–L–T–F–T–P–N–W•NH_2	Periplaneta
H-AKH = M-AKH	pQ–L–T–F–T–S–S–W–G•NH_2	Heliothis, Manduca
Glucagon	H–S–Q–G–T–F–T–S–D–Y–S–··· (29)	Mammalia
Secretin	H–S–D–G–T–F–T–S–E–L–S–··· (27)	Mammalia
Leucopyrokinin	pQ–T–S–F–T–P–R–L•NH_2	Periplaneta

Fig. 18.1. Members of the AKH family of neuropeptides and their sequence homologies with other neuropeptides. (From Penzlin, 1989.)

and transport, carbohydrate metabolism, protein synthesis and the activity of certain muscles (Orchard, 1987).

18.2. **Neurohormone D: its biological functions**

Neurohormone D (MI, Periplanetin CC-1) discovered by Hans Unger in 1957, occurs in cockroaches and is an octapeptide showing homologies with RPCH and AKH. It increases the beat frequency of the dorsal blood vessel *in vitro* and *in situ* in a dose-dependent manner. Its distribution seems to be restricted to roaches. It is rapidly inactivated by intact Malpighian tubules and also by their homogenates. The peptide is actively removed from the haemolymph followed by enzymatic degradation. The process of inactivation by the Malpighian tubules can be inhibited by insecticides (dieldrine, malathion, DDVP). Also in intact animals poisoning with insecticides results in a decrease of the capacity of Malpighian tubules to inactivate neurohormone D.

From the corpora cardiaca of the cockroach, *Periplaneta americana*, two peptides of the RPCA/AKH-family were isolated. Analysis of structure revealed the identity of one of these peptides, periplanetin CC-1 or the so-called myoactive factor MI, with the sequence of neurohormone D (Baumann and Penzlin, 1984; Scarborough *et al.*, 1984; Witten *et al.*, 1984), first described and separated by Unger in Jena from aqueous extracts of the brain–corpora cardiaca complex by paper chromatography

(Unger, 1957). Neurohormone D was later isolated from 5000 pairs of corpora cardiaca of *Periplaneta americana* by Baumann and Gersch in 1982. Approximately 20 μg of the pure product containing less than 10% contamination with other peptides or amino acids were obtained after purification by a four-step separation procedure, and the amino acid composition was determined (Baumann and Gersch, 1982).

A variety of physiological processes is regulated by neurohormone D in the cockroach. Besides its hyperglycaemic effect (Scarborough *et al.*, 1984; Gäde, 1985) the peptide seems primarily to act on the nervous system and on muscles. Both the contractile dorsal blood vessel (heart) and the dilator muscle of the antennal heart in the head are highly sensitive to neurohormone D, suggesting the possibility of a hormonal control of haemolymph circulation in cockroaches (Gersch *et al.*, 1982).

The increase in heart rate is caused by both a direct effect of neurohormone D on the myocardium (Hertel, 1975) and stimulation of the frequency of discharge of neurosecretory neurones in the lateral cardiac nerve cord (Richter and Gersch, 1974). Neurohormone D shows a clear dose–response relationship in the semi-isolated heart preparations as well as in *in vitro* tests (Fig. 18.2).

A comparison between the dose–response curves of the acceleration

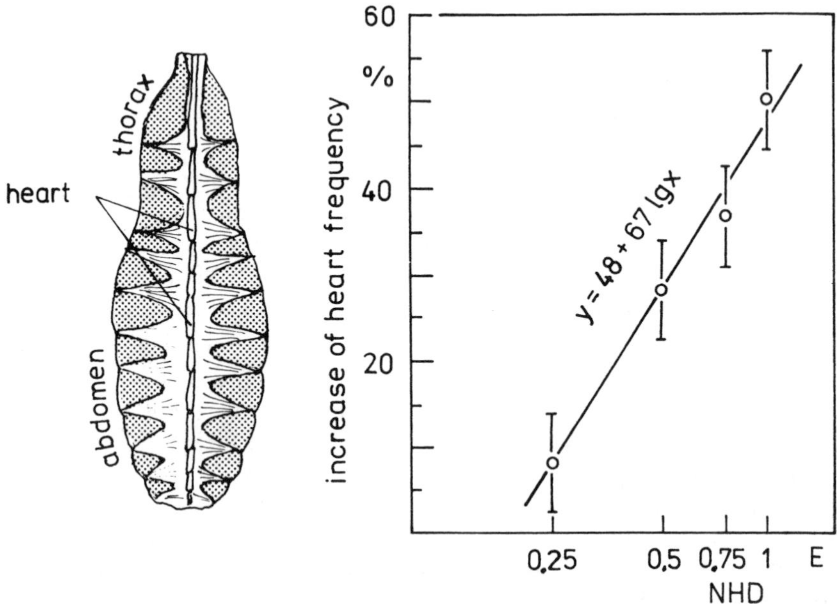

Fig. 18.2. Increase of the beat frequency of semi-isolated hearts of *Periplaneta americana* in response to different neurohormone D (NHD) doses. (From Gersch *et al.*, 1982.)

effect in semi-isolated heart preparations of ethanol extracts of the brain, the suboesophageal ganglion and the sixth abdominal ganglion of *Periplaneta americana* with the 'standard curve' of the corpora cardiaca extract ('four-point bioassay' – Gaddum, 1953) revealed that, in the cases of brain and sixth abdominal ganglion, the slopes of the curves were identical. In contrast, the slope of the dose–response curve of the suboesophageal ganglion extract differed significantly (Baumann *et al.*, 1984). It was therefore concluded that the factor in the suboesophageal ganglion differs from neurohormone D. In lower concentrations it failed to produce any increase in beat frequency in semi-isolated heart preparations. Gel chromatography showed that only in the case of brain extracts, but not in the case of sixth abdominal ganglion, the heart-accelerating activity was eluted coinciding with the elution volume of neurohormone D. However, only very small amounts of heart-accelerating material could be extracted from the brain. Excised corpora cardiaca from *Leucophaea maderae* and *Nauphoeta cinerea* contained as much heart-accelerating activity as the corpora cardiaca of *Periplaneta americana*. The slope of the dose–response curves showed no significant differences. Since the amino acid composition of the peptide fractions purified by reverse-phase chromatography were also very similar it was suggested that neurohormone D is stored in corpora cardiaca of all three cockroach species (Baumann *et al.*, 1984). In contrast, heart-accelerating substances similar to neurohormone D could not be detected in extracts of whole heads of *Bombyx mori* or in sinus gland extracts of *Orconectes limosus* (Baumann *et al.*, 1984). So it seems that neurohormone D in its occurrence is restricted to *Blattaria*, and in these animals to the corpora cardiaca and – in distinctly smaller amounts – to the cerebral ganglion.

18.3. **Neurohormone D: its inactivation**

In contrast to the long-lasting effect on semi-isolated heart preparations, the acceleration of heart rate in intact cockroaches after injection of neurohormone D rapidly decreases and returns to the previously recorded level within 20–40 min (Gersch *et al.*, 1982) (Fig. 18.3). This decrease of heart-accelerating effect of neurohormone D injected into intact animals can be the result of elimination of the peptide from the haemolymph or the desensitization of receptors. A second injection of neurohormone D into cockroaches increases the heart-beating frequency to the same degree as recorded with the first injection, excluding any change of the sensitivity of receptors. The peptide must be removed from the haemolymph by excretion, or inactivated by enzymatic cleavage.

Since the Malpighian tubules in locusts are known to accumulate and destroy AKH (Mordue and Stone, 1978) a comparable mechanism for inactivation of neurohormone D has been proposed to exist in cock-

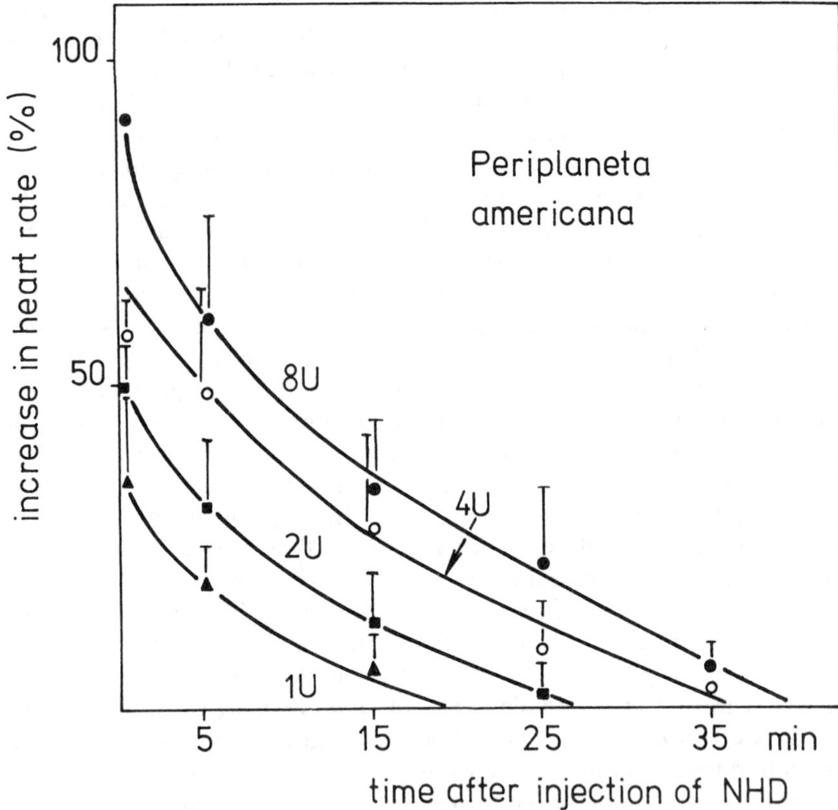

Fig. 18.3. Effect of injected neurohormone D in different doses (U = arbitrary unit) on the heart-beat rate of intact animals (n = 10). (From Gersch *et al.*, 1982.)

roaches. Indeed, incubation of neurohormone D solution for 30 min with Malpighian tubules dramatically decreases the heart-accelerating activity in contrast to other tissues of the cockroach, e.g. muscle, fat body, midgut, heart, brain, and haemolymph (Baumann and Penzlin, 1987) (Fig. 18.4).

As Gold and Fahrney demonstrated (1964), phosphosulphofluorides inhibit proteolytic activity by an irreversible reaction with a serine residue at the active centre. Malpighian tubules treated with phenylmethanesulphonyl fluoride (PMSF) for 20 min lost the ability to destroy neurohormone D to some degree. 4β-Aminoethylbenzenesulphonyl fluoride (AEBSF) was less effective than PMSF and chloromethyl ketones were only partially effective in higher concentrations. These results suggest the participation of serine proteases in the mechanism of inactivation of neurohormone D (Baumann and Penzlin, 1987).

Further experiments (Baumann and Penzlin, 1987) revealed that in the

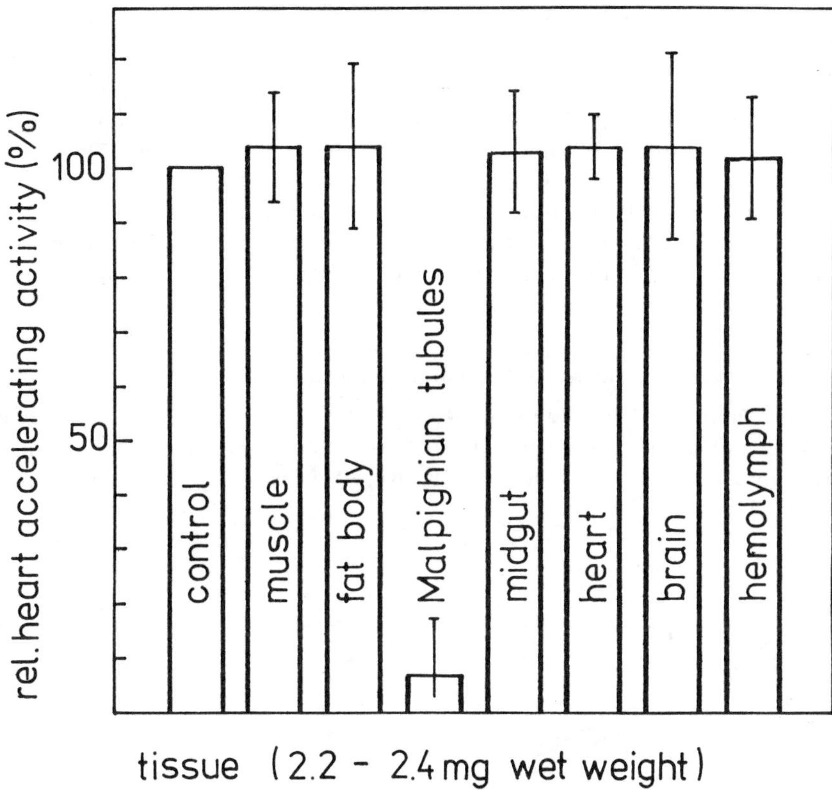

Fig. 18.4. Incubation of neurohormone D for 30 min with different tissues of the cockroach. Only Malpighian tubules dramatically decrease the heart-accelerating activity of neurohormone D solution.

presence of the chelating reagent EDTA the neurohormone-D-destroying activity of intact Malpighian tubules was diminished. This effect could be reversed by several divalent cations including Ca^{2+}, Co^{2+}, and Mn^{2+}, indicating an involvement of metal-activated enzymes or metalloendopeptidases.

The cells of Malpighian tubules may secrete the proteases into the surrounding saline or take up the peptide from the solution and digest it internally. The inactivation of neurohormone D by intact Malpighian tubules depends on oxidative metabolism. Under nitrogen, or in the presence of cyanide ions, the tubules lose their inactivating capacity, which means that the heart-accelerating effect of the neurohormone D solution remains. Oxygen supply or washing out the cyanide restores the tubules' activity to inactivate the peptide. From these experiments we conclude that the inactivation of neurohormone D by Malpighian tubules depends on an active transport mechanism supported by an oxidative metabolism. Perhaps the peptide is then degraded in the tubule cells by serine proteases. This is confirmed by results using homogenized Malpighian tubules. Only the supernatant after centrifugation was able to destroy the heart-accelerating activity of a neurohormone D solution. The homogenate is more effective in comparison with intact tubules, as shown by the time-course of inactivation. The difference in inactivation rate indicates that the uptake of neurohormone D by the cells is most probably the rate-limiting step in the sequence of events (Baumann and Penzlin, 1987).

18.4. **Insecticides and inactivation of neurohormone D**

Acetylcholinesterase, like the neurohormone-D-degrading enzyme, needs a serine in its active centre for catalytic activity. Furthermore, besides esterolytic activity, acetylcholinesterase also exhibits trypsin-like and

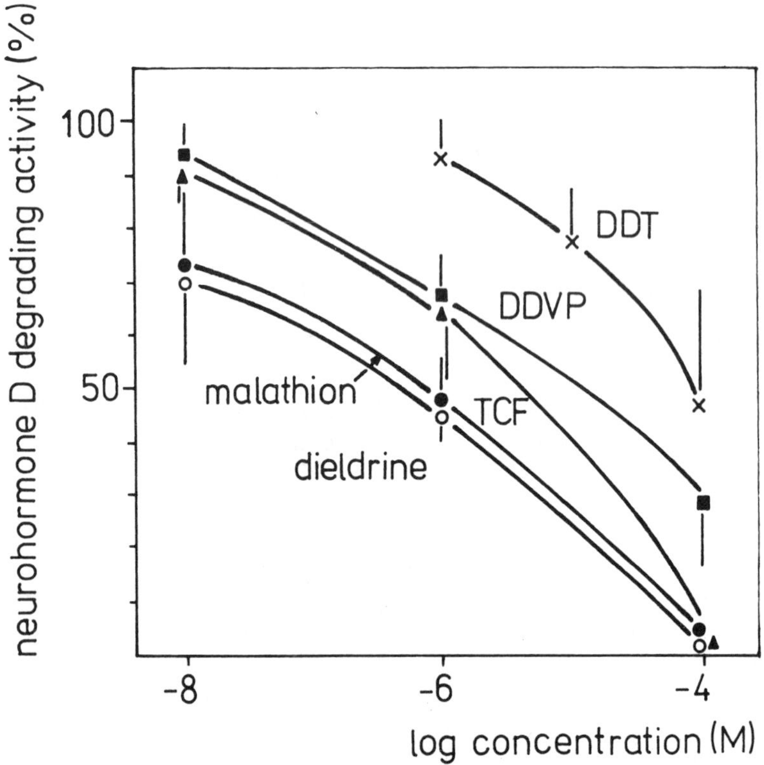

Fig. 18.5. Inhibition of neurohormone D inactivation by preincubation of a homogenate of Malpighian tubules with several insecticides.

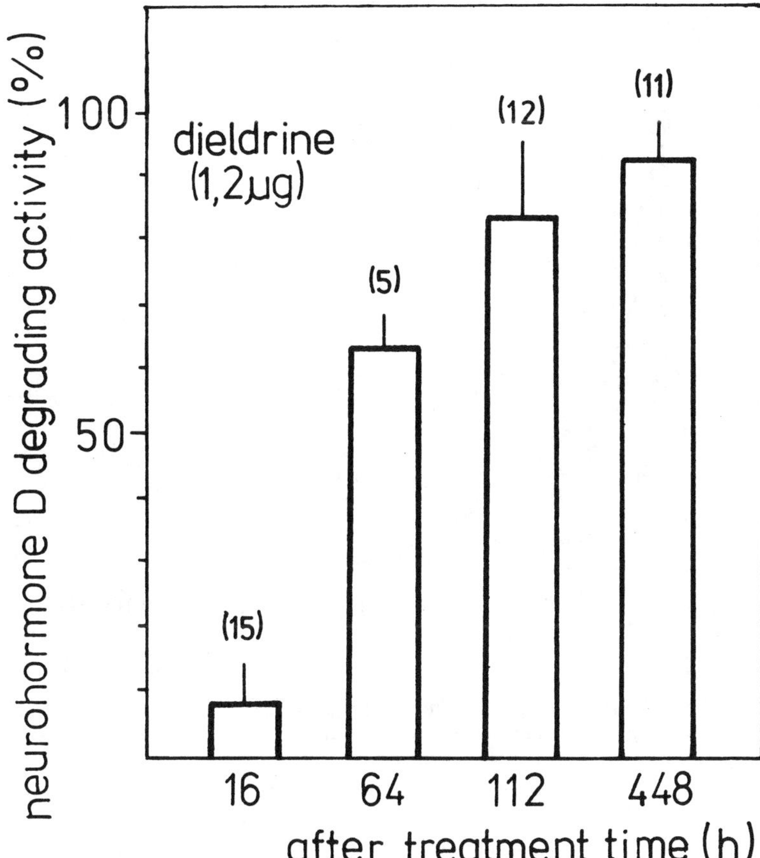

Fig. 18.6. Restoration of neurohormone D inactivating capacity of Malpighian tubules in poisoned cockroaches. In each case 15 animals were poisoned by topical administration of dieldrine (1.2 μg). Figures in parentheses show number of surviving animals.

metalloexopeptidase-like activities (Small *et al.*, 1987). Thus, it is possible that neurohormone-D-degrading enzyme may be inhibited by insecticides in a comparable way as is acetylcholinesterase.

Malpighian tubules were homogenized and insoluble fragments were centrifuged off. The supernatant was preincubated for 15 min with insecticides. Then incubation continued for ½ h after addition of neurohormone D. After that neurohomone D was separated from the buffer salts and the insecticides by passing through reversed-phase columns and tested in respect to its heart-accelerating capacity. Preincubation with malathion, dieldrine, or trichlorfon (TCF), in concentration of 10^{-4} M, completely inhibited the neurohormone-D-degrading enzyme (Fig. 18.5).

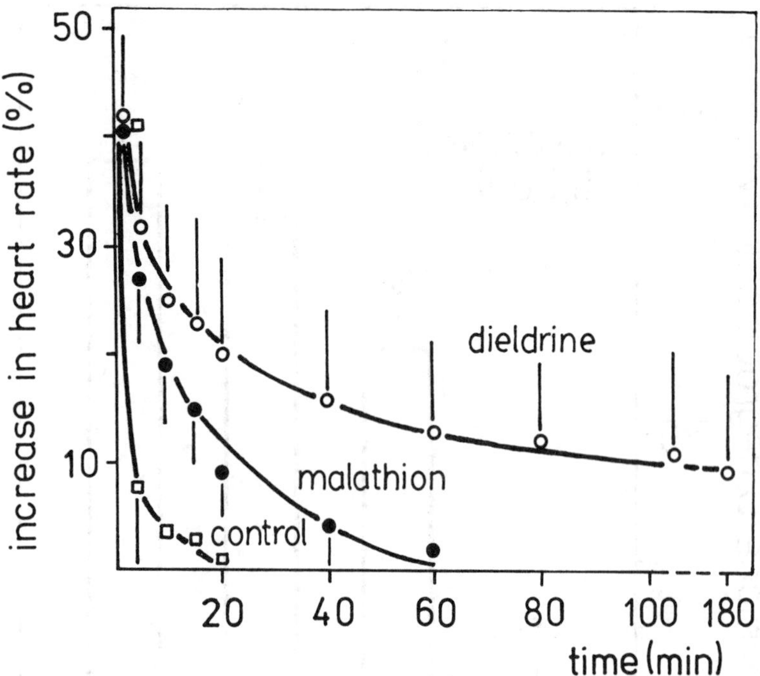

Fig. 18.7. The heart-accelerating effect of injected neurohormone D decreases in insecticide poisoned cockroaches (n = 25) more slowly than in untreated control animals.

In consequence the heart-accelerating effect of the neurohormone D solution remained unchanged after preincubation. This effect was dose-dependent (Baumann and Penzlin, 1990). Dichlorvos (DDVP) and especially DDT were less effective.

In intact animals the capacity of Malpighian tubules to inactivate neurohormone D decreases progressively with time of insecticide treatment. In each case 10 µl of the insecticide dieldrine or malathion dissolved in acetone or methyl alcohol were administered under the pronotum. After poisoning for 8 h the Malpighian tubules were completely inhibited. The capacity of Malpighian tubules to degrade neurohormone D is restored to its original level within 10–20 days after dieldrine application (Fig. 18.6). This is thought to be the result of a *de novo* enzyme synthesis.

If neurohormone-D-degrading activity is decreased in cockroaches poisoned with insecticide, and provided that the function of neurohormone D receptors is not impaired by the insecticide, injection of neurohormone D into these animals will result in prolonged acceleration of heart rate. This is indeed the case. The heart-beating frequency in cockroaches treated with sublethal doses of dieldrine or malathion needs some hours to return to the normal level (Fig. 18.7). In control animals this happens after only 20 min.

It seems to us that, despite the relatively low specificity of the reaction, the neurohormone-D-inactivating system of Malpighian tubules is an interesting physiological mechanism to search for both the action mechanism of insecticide and the development of new compounds highly and specifically toxic to insects.

References

Baumann, E. and Gersch, M. (1982). Purification and identification of neurohormone D, a heart accelerating peptide from the corpora cardiaca of the cockroach *Periplaneta americana. Insect Biochem.*, **12**, 7–14.

Baumann, E. and Penzlin, H. (1984). Sequence analysis of neurohormone D, a neuropeptide of an insect, *Periplaneta americana. Biomed. Biochim. Acta*, 43, K13–K16.

Baumann, E. and Penzlin, H. (1987). Inactivation of neurohormone D by Malpighian tubules in an insect, *Periplaneta americana. J. Comp. Physiol. B*, **157**, 511–17.

Baumann, E. and Penzlin, H. (1988). Influence of insecticides on the mechanism of inactivation of an insect neurohormone. In: Sehnal, F. *et al.* (eds), *Endocrinological Frontiers in Physiological Insect Ecology*. Wrocℓaw Technical University Press, Wrocℓaw, pp. 759–62.

Baumann, E. and Penzlin, H. (1990). Influence of insecticides on the inactivation of neurohormone D in *Periplaneta americana. Zool. Jb. Physiol.*, **94**, 1–3.

Baumann, E., Gersch, M. and Penzlin, H. (1984). The occurrence of neurohormone D in the nervous system of several arthropods. *Zool. Jb. Physiol.*, **88**, 213–23.

Fernlund, R. and Josefsson, L. (1972). Crustacean color-change hormone amino acid sequence and chemical synthesis. *Science*, **177**, 173–5.

Gaddum, J. H. (1953). Bioassay and mathematics. *Pharmacol. Rev.*, **5**, 87–134.

Gäde, G. (1985). Hypertrehalosaemic hormones and myoactive factors from cockroach corpus cardiacum are very likely identical. *Naturwissenschaften*, **72**, 95–6.

Gersch, M., Baumann, E., Birkenbeil, H., Pass, G. and Penzlin, H. (1982). Die Wirkungsweise des herzaktiven Peptids Neurohormon D von *Periplaneta americana. Zool. Jb. Physiol.*, **86**, 17–33.

Gold, A. M. and Fahrney, D. (1964). Sulfonyl fluorides as inhibitors of esterases. II. Formation and reactions of phenylmethane sulfonyl α-chymotrypsin. *Biochemistry*, **3**, 783–91.

Hertel, W. (1975). Untersuchungen zur Beeinflussung des Herzschlags von *Periplaneta americana* L. unter Bedingungen der Kurzzeitkultur von Herzsegmentpräparaten *in vitro. Zool. Jb. Physiol.*, **79**, 70–6.

Mayer, R. J. and Candy, D. J. (1969). Control of hemolymph lipid concentration during locust flight: an adipokinetic hormone from the corpora cardiaca. *J. Insect Physiol.*, **25**, 169–81.

Mordue, W. and Stone, J. V. (1978). Structure and metabolism of adipokinetic hormone. In: Gaillard, P. J. and Boer, H. H. (eds), *Comparative Endocrinology*. Elsevier/North Holland, Amsterdam, pp. 487–90.

Orchard, I. (1987). Adipokinetic hormones – an update. *J. Insect Physiol.*, **33**, 451–63.

Penzlin, H. (1989). Neuropeptides – occurrence and functions in insects. *Naturwissenschaften*, **76**, 243–52.

Richter, K. and Gersch, M. (1974). Untersuchungen über die Wirkung cholinerger, aminerger und peptiderger Substanzen am lateralen Herznerven und den funktionellen Zusammenhang dieser Wirkungen bei der Herzregulation von *Blaberus craniifer* Burm. (Insecta: Blattaria). *Zool. Jb. Physiol.*, **78**, 16–32.

Scarborough, R. M., Jamieson, G. C., Kalish, F., Kramer, S. J., McEnroe, G. A., Miller, C. A. and Schooley, D. A. (1984). Isolation and primary structure of two peptides with cardio-acceleratory and hyperglycaemic activity from the corpora cardiaca of *Periplaneta americana. Proc. Natl. Acad. Sci., USA*, **81**, 5575–9.

Small, D. H., Ismael, J. and Chubb, I. W. (1987). Acetylcholinesterase exhibits trypsin-like and metalloexopeptidase-like activity in cleaving model peptide. *Neuroscience*, **21**, 991–5.

Unger, H. (1957). Untersuchungen zur neurohormonalen Steuerung der Herztätigkeit bei Schaben. *Biol. Zbl.* 76, 204–25.

Witten, J. C., Schaffer, M. H., O'Shea, M., Cook, J. C., Hemling, M. E. and Rinehart, K. L. (1984). Structure of two cockroach neuropeptides assigned by fast atom bombardment mass spectrometry. *Biochem. Biophys. Res. Commun.*, **124**, 344–9.

19 *Jacques Proux, Bernard Fournier, Josiane Girardie and Martine Picquot*

Involvement of neuropeptides in insect water regulation: the African locust model. Comparison with other insect and vertebrate systems

19.1. Introduction

Water percentage of terrestrial insect ranges from 46% to 96% and is, as well as haemolymph composition, mostly regulated by food ingestion and excretion. Excretion is chiefly achieved by two organs: Malpighian tubules producing primary urine and rectum where selective reabsorptions occur. The involvement of neurohormones in the control of this excretory process has been demonstrated for roughly three decades (see Phillips, 1983). The relative importance of the Malpighian tubules and rectum vary according to the type of insect studied. Because of the quantitative variations of its hydrous diet, the African locust (*Locusta migratoria*) has been chosen by several laboratories to study the hormonal control of insect diuresis.

In this chapter we will discuss the most significant data obtained on this insect and then compare the data to those obtained on other insects as well as vertebrates. Finally, insects as models to study basic aspects of hormonal control of animal diuresis will be discussed.

19.2. The African locust model

The African locust is herbivorous, living in a dry environment where very hydrous food (fresh grass) is available for short periods at a time. This means it must either achieve an important fluid reabsorption to save water and ions or lose extra water. To match this diversity, several diuretic and antidiuretic hormones (DH and ADH) are present, acting selectively on Malpighian tubules and rectum.

19.2.1. *Origin and release of DH and ADH*

Two major sources of DH and ADH have been reported so far: the pars intercerebralis-corpora cardiaca (PI-CC) complex and the ventral nerve cord (VNC). The PI-CC complex is comparable to the mammal hypothalamo-hypophysial system. Neurohormones are synthesized in the PI (cell bodies) and transported down axons, stored in the nervous lobes of the CC (NCC) and then released in the haemolymph. In *Locusta* this complex is known to contain both DH and ADH (Cazal and Girardie, 1968). Morgan and Mordue (1983) extracted DH from the NCC, then ADH were found in both NCC and glandular lobes of the CC (GCC) by Proux *et al.* (1984), confirming previous reports (see Phillips, 1983). Later, Fournier and Girardie (1988) proved that the ADH of the NCC were the neuroparsins, two proteins produced by the Al-type neurosecretory cells of the PI (Girardie *et al.*, 1987).

The second major source of DH is the VNC. Histological techniques (Chalaye, 1965), *in vitro* tests (Cazal and Girardie, 1968) and surgery (Proux, 1978a, b) demonstrated diuretic factor(s) in the ganglia of the VNC. The discovery of the arginine-vasopressin-like insect diuretic hormone (AVP-like IDH) (Proux *et al.*, 1987) confirmed these previous results and enhanced this field of investigation. This neuropeptide, immunologically related to AVP, is synthesized by two A-type neurosecretory cells of the suboesophageal ganglion. It is a homodimer present with its monomeric form. Rémy *et al.* (1979) and Rémy and Girardie, 1980 performed an immunological exploration of the entire CNS and demonstrated immunoreactive tracts in the two suboesophageal perikarya that run forward towards the brain and backward towards the thoracic and abdominal ganglia. Collaterals branch from these tracts and give arborizations in the root of each segmental nerve of thoracic and abdominal ganglia. These arborizations end in swellings under the nerve sheaths (Girardie and Rémy, 1980), and represent the releasing sites of neurosecretion into the haemolymph (Proux and Rougon-Rapuzzi, 1980). Their abundance compensates for the scarcity of the producing structures (two perikarya). DH and ADH release occurs as a response to different environmental modifications or physiological events. Feeding is probably the main phenomenon able to disrupt fluid and ion balance.

In *Locusta*, Picquot and Proux (1987) demonstrated an increase in both haemolymphatic titre in AVP-like substances and diuresis 3–4 h after food was supplied. It is likely that this release from neurohaemal sites can be initiated by cell depolarization, since release of AVP-like substances is induced by bathing neurohaemal organs in high K^+ solution (Picquot, 1989). DH-related stretch or osmotic receptors have been reported in different insects (Maddrell and Gardiner, 1976; Nijhout, 1984; Nicholls, 1985) but never on *Locusta* where the release of AVP-like substances from the ganglia of the VNC seems to be more complex and under the control of releasing factors. Proux *et al.* (1980) and Proux and Girardie (1982) demonstrated the antagonistic control of PI and subocellar median neurosecretory cells (SOM-NSC) on this release. PI contains a hormonal factor inhibiting the release of AVP-like substances into the haemolymph, while SOM-NSC contain a substance which increases this release. Thus, several neurosecretory centres (PI, SOM-NSC, VNC) previously thought of as being separately involved in *Locusta* diuresis, are combined in a single regulatory system.

Feeding is not the only event capable of provoking changes in *Locusta* water and ion balance. Animals submitted to a 5 h flight exhibit a 5% increase in water content (Albrecht *et al.*, 1982). This increase may be due to the production of metabolic water, but the involvement of the ADH, contained in the GCC, may also occur. Indeed, Hérault *et al.* (1985) demonstrated that ADH release, as well as that of the adipokinetic hormone (AKH) (Lafon-Cazal and Morandini, 1982: Lange and Orchard, 1986), is controlled by octopamine. This control of the release of two hormones by the same aminergic system is physiologically justified, since it occurs during flight and might improve insect autonomy: AKH providing energy (ATP) and ADH recycling water and ions.

Changes in the relative humidity (RH) of the environment result in a disruption of water balance and then in subsequent hormonal regulation. This was clearly demonstrated in *Locusta*. As described below, the AVP-like IDH coexists with its monomeric form. The influence of hygrometric conditions on the metabolism of both AVP-like IDH and its monomeric form was recently established by Picquot and Proux (1989). Low RH and dehydrated food (i.e. a dry environment) decrease concentration and supply of dimer (AVP-like IDH), monomer being predominant in the nervous tissues. On the other hand, high RH and hydrated food (i.e. a damp environment) stimulate monomer and dimer production, the dimer becoming predominant in the tissue and being available in large amounts to increase diuresis. Monomer is probably synthesized then transformed to dimer according to physiological needs. This transformation seems to be due to an enzyme coexisting with monomer and dimer within the neurosecretory granules (Picquot, 1989).

19.2.2. *Chemical nature and structure of DH and ADH*

DH and ADH have been partially purified in several insects (reviewed by Phillips, 1983) and fully characterized in *Locusta* (Proux *et al.*, 1987; Girardie *et al.*, 1989) and *Manduca* (Kataoka *et al.*, 1989). Schooley *et al.* (1987) and Proux *et al.* (1987) isolated and characterized two AVP-like substances, a homodimer, termed AVP-like IDH, which acts by increasing Malpighian tubule excretion, and its monomer, devoid of stimulatory activity at the Malpighian tubule level. The AVP-like IDH is an antiparallel homodimer, i.e. two monomers linked by disulphide bonds. The monomer exhibits provocative sequence homologies (66% and 78% respectively) with the vertebrate neurohormones: AVP and arginine-vasotocin (AVT). Two *Locusta* ADH, the neuroparsins A and B, were characterized by Girardie *et al.* (1989). These two similar substances are thermostable sulphur-containing proteins consisting of two polypeptide chains linked by disulphide bonds. Neuroparsin B was found as a homodimer. Its monomer was sequenced and comprises 78 residues. Neuroparsin B and the incomplete N-terminal sequence of another *Locusta* cardiacum peptide (Mordue *et al.*, 1985) present unambiguous homologies. Both neuroparsin A and B are produced by the Al-type neurosecretory cells of the PI and stored in NCC. Fournier and Girardie (1988) demonstrated that neuroparsin A and B are the only ADH contained in the NCC since crude NCC extracts treated with anti-neuroparsin antibody prior to the assay lose their antidiuretic activity.

19.2.3. *Transduction of the DH and ADH signals*

Cyclic AMP is the most frequent second messenger of the animal kingdom, including insects (reviewed by Smith and Combest, 1985). The ability of cAMP to mimic DH action on *Locusta* was reported by Rafaeli *et al.* (1984). This result is questionable because cAMP is supposed not to enter cells (Lehninger, 1979) and could act externally, as a hormone (Devreotes, 1983). Using cAMP analogues, Morgan and Mordue (1985) gave more convincing proof, then Proux and Hérault (1988) brought definitive evidence of the involvement of cAMP on AVP-like IDH transduction when using synthetic hormone.

The involvement of cAMP as second messenger of *Locusta* ADH has also been reported by Hérault and Proux (1987), who stated that ADH from GCC crude extract acts via cAMP, but this result with crude GCC extract awaits confirmation. On the other hand, neuroparsins, the ADH from NCC, are not mediated by cAMP or cGMP (Fournier and Dubar, 1989) but by two other second messengers, diacylglycerol stimulating protein kinase C and Ca^{2+} (Fournier, 1989).

19.2.4. *Bioassays for* Locusta *Malpighian tubules and rectum*

Bioassays are necessary to study the hormonal modes of action and to follow hormones during the successive purification steps. The priorities will be different according to the chosen purpose. A bioassay devoted to this study of the hormonal action can be time-consuming and laborious; that used to screen the numerous fractions yielded by purification steps must be easy to handle, fast and quantitatively reliable.

Until recently, most research on the activity of Malpighian tubules was carried out using the bioassays developed by Ramsay (1954) for isolated tubules, and by Maddrell and Klunsuwan (1973) for semi-isolated tubules, both having advantages and drawbacks (Proux *et al.*, 1988). A new bioassay based on the excretory activity of whole Malpighian tubules *in vitro* was recently developed (Proux *et al.*, 1988). It permits the collection of large samples of primary urine, allowing easy measurement of ion concentration, precise determination of the rate of action of biological substances, and maintenance of the tubules for prolonged periods of time, but it is not suitable for running large numbers of tests in a single set of tubules due to the duration of the reversibility of biological effects. Proux *et al.* (1987) and Schooley *et al.* (1987) followed a different course. They used a radioimmunoassay which allowed far more rapid monitoring of fractions than would been possible with available bioassay procedures.

Among the tissues involved in insect diuresis, the locust rectum has obviously been the most studied, owing to two bioassays: the everted sac preparation (Goh and Phillips, 1978) and the voltage clamp method (Williams *et al.*, 1978). Everted rectal sacs reabsorb water over a 5–6 h period *in vitro*; this reabsorption is coupled to an active transport of ions (Cl^-, Na^+ and K^+; Goh and Phillips, 1978). Using everted sac preparations (Hérault *et al.*, 1985; Fournier *et al.*, 1987) *Locusta* GCC and NCC were found to contain two different AD factors (that of NCC being neuroparsins) and the duration of the assay was reduced (from 5 to 2 h).

19.2.5. *Turnover and degradation of DH and ADH*

Little is known about DH and ADH turnover in insects (Phillips, 1983), because of the difficulty in quantifying hormonal haemolymphatic concentrations. Picquot (1989) obtained some preliminary data on AVP-like IDH breakdown, and hence turnover. When synthetic AVP-like IDH is injected into locusts it disappears rather rapidly, and 3 h after the injection only 5% remains probably shared between still-intact AVP-like and degradative compounds. These degradative compounds, probably due to enzymes present in Malpighian tubules, may be present in urine (Proux, personal communication).

19.3. **Comparison with some other insects**

As stressed below, locusts are good models to study diuresis, due to their way of life when dry and damp periods succeed each other. But other insects, owing to special adaptations, can yield valuable information in the field of diuresis. Sucking insects, for example, which feed on blood at intervals, must lose extra water very fast, and are thereby good models to study DH. The most significant results have been obtained on the reduviid *Rhodnius prolixus* and the mosquito *Aedes aegypti*.

The pioneering work of Maddrell led to knowledge concerning *Rhodnius* DH system. A DH, localized in the ventral ganglionic mass (Maddrell, 1963, 1976), is released in the haemolymph (Maddrell, 1966) 15 s after feeding begins (Maddrell and Gardiner, 1976). This rapidity is made essential by the enormous payload due to the consumption of blood that impairs flight and reduces manoeuvrability. It is likely that DH release is initiated by cell depolarization and (Maddrell and Gardiner, 1976) the involvement of stretch receptors in the post-prandial release of DH. Such a model was rightly considered a good candidate for the study and subsequent characterization of DH and an *in vitro* bioassay of Malpighian tubules (Maddrell, 1976). Unfortunately, the *in vitro* instability of *Rhodnius* DH prevented its characterization (Aston and Hugues, 1980).

Another blood-sucking insect, *Aedes*, seems to be more promising. The brain is the site of diuretic activity shared by three DH (Petzel *et al.*, 1985). As in *Rhodnius*, diuresis begins before the blood meal has been completed (Beyenbach and Petzel, 1987). Whether release of DH after the blood meal is controlled by stretch receptors, as in *Rhodnius*, is still controversial. Current progress in isolating *Aedes* DH is largely due to the bioassay (Petzel *et al.*, 1985). They used the methods of Burg *et al.* (1966), and Ramsay (1954), to record transepithelial voltage and to measure fluid secretion on isolated Malpighian tubules. They demonstrated that three fractions eluted during chromatographic purification steps of brain extracts exhibited natriuretic or (and) diuretic activities. The major interest of this bioassay is its rapidity. Because effects on the voltage are immediate and reversible on washing, it was possible to scan up to 120 fractions from a single purification step using only a few Malpighian tubules.

In addition, see the DH work of Weltens and Van Kerkhove (1985) on *Formica* Malpighian tubules; Spring and Hazelton (1987) on *Acheta*; Bourême *et al.* (1989); Fournier (personal communication). Kataoka *et al.* (1989) on *Manduca*, Proux *et al.* (1987) and Girardie *et al* (1989) are the only ones to get a complete amino acid sequence (a DH made from 41 residues and completely different from AVP-like IDH).

19.4. **Comparison with vertebrates**

19.4.1. *Excretory processes*

The physiology of insect and vertebrate excretory organs can be explained by the same general principles. Haemolymph (or blood) solutes are present in an iso-osmotic primary urine produced by the Malpighian tubules (or the Malpighian bodies). Selective reabsorption of water, ions and metabolites occurs afterwards at the rectal (or tubular) level to give the final urine. But the pathways to reach this result are quite different. The kidneys of all vertebrates, except a few teleost fishes, work roughly on a 'filtration–reabsorption' principle, while the insect excretory system is based on tubular secretion followed by rectal reabsorption.

In the vertebrate process the primary urine is forced out by blood pressure and is filtered at a high rate. So all the blood compounds, except those of large molecular size (Ml > 80,000) are present. Then those of the filtered compounds which are useful are reabsorbed. This reabsorption mechanism consumes a lot of energy but has an advantage. When an animal encounters new substances to be excreted, it can do so with its existing secretory system, since this is able to excrete almost everything. This gives the animal greater possibilities to explore new environments.

The problem is quite different on the insect excretory system. It is slow-operating and the passive transfer of solutes from the haemolymph to the lumen of the tubules is usually restricted to sugars and amino acids. This system has an advantage in that a small amount of energy is involved in the reabsorption of valuable substances. The drawback is the necessity to develop specialized secretory mechanisms for each new potential harmful substance the insect may meet. Insects overcame this difficulty since their ability to excrete toxic substances exceeds that of vertebrates (Maddrell, 1977). In kidney tubules an organic anion transport able to handle two types of compounds, acylamides and sulphonates, by a common mechanism, is present. A similar system exists in insects, but it is much more efficient since the two compounds are handled by separate mechanisms and do not compete with each other (Maddrell, 1977). The significance of this double mechanism is obvious, since insect metabolism is able to detoxify many harmful compounds by excreting them through Malpighian tubules. These Malpighian tubules are also able to excrete alkaloids. Maddrell *et al.* (1974) demonstrated the ability of tubules to remove nicotine, atropine and morphine from the haemolymph at a high rate, and against concentration gradients. This ability to discard these alkaloids allows insects to settle in a wide variety of ecological niches since they are considered as a protective measure against plant-feeders.

19.4.2. *Hormonal regulation*

Most vertebrate and insect excretory systems are driven by DH and ADH. This hormonal diversity is probably a consequence of the complex way of life of these animals (variety of food sources, environmental conditions, etc) since, as previously shown, parasitic insects (*Rhodnius, Aedes*) developed more simple systems.

A comparison between the mammal and locust systems shows that they have both DH (atrial natriuretic factor and AVP-like IDH) and ADH (AVP, renin–angiotensin–aldosterone system and neuroparsins, GCC hormone).

19.4.3. *Hormonal relationships*

Grimm-Jorgensen and McKelvy (1975) demonstrated a thyrotrophin-releasing hormone-like substance in gastropod ganglia. It was the first evidence of the existence of an immunological relationship between vertebrate and invertebrate neuropeptides. Shortly afterwards an AVP-like substance, the future AVP-like IDH, was detected in *Clitumnus* and *Locusta* by Rémy *et al.* (1977, 1979). AVP and AVT (the ancestral molecule of AVP) for vertebrates, AVP-like IDH for insects and conopressin (Cruz *et al.*, 1987) an AVT-like substance for molluscs, are all members of the same interphyletic family. This structural evolution, which is occurring in the AVP family, is sometimes on a par with functional evolution. AVT, present in all vertebrates except mammals, plays a diuretic then an antidiuretic role, change occurring at the passage from aquatic to terrestrial life (Sawyer, 1972). On the other hand, although *Locusta* is a terrestrial animal, AVP-like IDH is a DH (maybe because ADH are available too). Hormones of the AVP family can act as neurotransmitter-neuromodulators too, as demonstrated for AVP (Dreifuss *et al.*, 1982) and conopressin (Cruz *et al.*, 1987). The *Manduca* DH characterized by Kataoka *et al.* (1989) also exhibits some structural analogies with vertebrate hormones involved in diuresis: sauvagine and urotensin-I, and this sauvagine/urotensin family is now represented in the classes Insecta–Amphibia–Pisces–Mammalia.

Insect DH and ADH are not necessarily related to vertebrate hormones, but can be members of intraphyletic families existing within the insect group. Thus neuroparsin-like substances were found in different insect orders (Bourême and Girardie, 1986; Bourême *et al.*, 1989; Tamarelle and Girardie, 1989).

19.5. **The future of the insect model**

Although 95% of the animal species belong to invertebrates, they attract only a small part of the research endeavour in biology. As stressed by Scharrer (1987), this restriction of the study of biological phenomena to a few mammalian orders, originates in a historical preference of biologists for the concept of homology (where structural relationships are privileged) over the prejudice of the concept of analogy (which dwells on the functional relatedness). The somewhat recent discovery of biological models in invertebrates (and especially insects) permitted demonstration of the interest of a comparative study of the functional relationships, and the concept of analogy is now of primary importance.

One of the consequences of this change in thinking was the discovery that these so-called 'inferior' animals offer numerous advantages in biology (relative simplicity, low cost, fast reproduction, lack of moral problems, etc.) and might often be considered as an alternative to 'superior' animals. In exchange, numerous discoveries made on mammals have been advantageous to invertebrate studies and have increased our knowledge. The present dramatic infatuation for structural relationships between invertebrate and vertebrate hormones, leading to the discovery of new hormones and of intra- and interphyletic families was made possible because physiologists using invertebrates have had the opportunity to use antibodies raised against mammals.

The discovery of the AVP-like IDH, as detailed above, is a good example of this use of complementary techniques. Thanks to an antibody raised against the mammalian AVP, this hormone has been synthesized and is now used to study the basic mechanism insect hydrous regulation, as well as for possible drug interest.

The study of insect diuresis is of prime interest at four levels: (1) comparison with vertebrate models leading to the advantages described above; (2) knowledge of insect physiology at the crucial level represented by maintenance of homeostasis; (3) a wider knowledge, extensible to animal and vegetable kingdoms, of maintenance of the size and composition of the water compartments; (4) use of Malpighian tubules and rectum to study basic mechanisms of diuresis (their major advantage when compared to the kidney being their structural simplicity).

Insect models also have some disadvantages. Their small size prevents the extraction of large amounts of biological material and this limitation is especially important when purifying and characterizing hormones. Luckily, the use of insects of economic interest, and availability in great numbers, such as the silkworm *Bombyx mori*, and above all the tremendous strides made in the field of biochemistry, allow us to overcome these difficulties. Consequently 38 structures of insect neuropeptides were determined in 1988, versus four in mid-1985 (Holman *et al.*, 1988).

References

Albrecht, F. O., Lauga, J., Lafon-Cazal, M., Baehr, J. C. and Casanova, D (1982). Influence du vol prolongé sur l'équilibre hydrique, la lipémie et l'hormone juvénile chez les solitaires verts et bruns de *Locusta migratoria migratorioides* (R. et F.) (Orthoptera: acridoidea) *C. R. Acad. Sci. Paris,* **295**, 737–40.

Aston, R. J. and Hugues, L. (1980). Diuretic hormone: Extraction and chemical properties; In: Miller, T. A. (ed.), *Neurohormonal Techniques in Insects.* Springer Verlag, New York, pp. 91–115.

Beyenbach, K. W. and Petzel, D. H. (1987). Diuresis in mosquitoes: role of a natriuretic factor. *NIPS*, **2**, 171–5.

Bourême, D. and Girardie, J. (1986). Présence de protéines à ponts disulfure dans les corpora cardiaca d'insectes. *C. R. Acad. Sci. Paris*, **302**, 655–60.

Bourême, D., Tamarelle, M. and Girardie, J. (1987). Production and characterisation of antibodies of neuroparsins A and B isolated from the corpora cardiaca of the locust. *Gen. Comp. Endocrinol.*, **67**, 167–77.

Bourême, D., Fournier, B., Matz, G. and Girardie, J. (1989). Immunological and functional cross-reactivities between locust neuroparsins and proteins from cockroach corpora cardiaca. *J. Insect. Physiol.*, **35**, 265–71.

Burg, M., Grantham, J., Abramow, M. and Orloff, J. (1966). Preparation and study of fragments of single rabbit nephrons. *Am. J. Physiol.*,**210**, 1293–8.

Cazal, M. and Girardie, A. (1968). Contrôle humoral de l'équilibre hydrique chez *Locusta migratoria migratorioides. J. Insect. Physiol.*, **14**, 655–68.

Chalaye, D. (1965). Recherches histochimiques et histophysiologiques sur la neurosécrétion dans la chaîne nerveuse ventrale du criquet migrateur *Locusta migratoria. C. R. Acad. Sci. Paris*, **260**, 7010–13.

Cruz, L. J., Santos, De, V., Zafarella, G. C., Ramilo, C. A., Zeikus, R., Gray, W. R. and Olivera, B. M. (1987). Invertebrate vasopressin/oxytocin homologs. *J. Biol. Chem.*, **262**, 15821–4.

Devreotes, P.N. (1983). Cyclic nucleotides and cell communication in dictyostelium discoideum; In: Greengard, P. and Robison, G. A. (eds), *Advances in Cyclic Nucleotides Research*. Raven Press, New York, pp. 55–93.

Dreifuss, J. J., Muhlethaler, M. and Gahwiller B. H. (1982). Electrophysiology of vasopressin in normal rats and in Brattleboro rats. *Ann. N. Y. Acad. Sci.*, **394**, 689–702.

Fournier, B. (1989). Neuroparsins induce phosphoinositide breakdown in the migratory locust rectal cells. *Comp. Biochem. Physiol.*, **95B**, 57–64.

Fournier, B. and Dubar, M. (1989). Relationship between neuroparsin-induced rectal fluid reabsorption and cyclic nucleotides in the migratory locust. *Comp. Biochem. Physiol.*, **94A**, 249–55.

Fournier, B. and Girardie, J. (1988). A new function for the locust neuroparsins: stimulation of water reabsorption. *J. Insect. Physiol.*, **34**, 309–13.

Fournier, B., Hérault, J.–P. and Proux, J. (1987). Study of an antidiuretic factor from the nervous lobes of the migratory locust corpora cardiaca. Improvement of an existing *in vitro* bioassay. *Gen. Comp. Endocrinol.*, **68**, 49–56.

Girardie, J. and Rémy, C. (1980). Particularités histo-cytologiques des prolongements distaux des 2 cellules à 'vasopressine-neurophysine-like' du criquet migrateur. *J. Physiol. Paris*, **76**, 265–71.

Girardie, J., Girardie, A., Huet, J. C. and Pernollet, J. C. (1989). Amino acid sequence of locust neuroparsins. *FEBS Lett.*, **245**, 4–8.

Girardie, J., Bourême, D., Couillaud, F., Tamarelle, M. and Girardie, A. (1987). Anti-juvenile effect of neuroparsin A, a neuroprotein isolated from locust corpora cardiaca. *Insect Biochem.*, **17**, 977–83.

Goh, S. and Phillips, J. E. (1978). Dependence of prolonged water absorption by *in vitro* locust rectum on ion transport. *J. Exp. Biol.*, **72**, 25–41.

Grimm-Jørgensen, Y. and McKelvy, J. (1975). Immunoreactive thyrotrophin releasing factor in gastropod circumoesophageal ganglia. *Nature*, **254**, 620.

Hérault, J.-P. and Proux, J. (1987). Cyclic AMP, the second messenger of an antidiuretic hormone from the glandular lobes of the migratory locust corpora cardiaca. *J. Insect Physiol.*, **33**, 487–91.

Hérault, J. –P., Girardie, J. and Proux, J. (1985). Separation and characterisation of antidiuretic factors from the corpora cardiaca of the migratory locust. *Int. J. Invert. Reprod. Develop.*, **1**, 325–35.

Holman, G.M., Wright, M. S. and Nachman, R. J. (1988). Insect neuropeptides: coming of age. *ISI Atlas Science: Animal Plant Sciences*, **1**, 129–36.

Kataoka, H., Troetshler, R. G., Li, J. P., Kramer, S. J., Carney R. L. and Schooley, D. A. (1989). Isolation and identification of a diuretic hormone from the tobacco hornworm, *Manduca sexta. Proc. Natl. Acad. Sci., USA*, **86**, 2976–80.

Lafon-Cazal, M. and Morandini, M, (1982). Contrôle neuroendocrine de la lipémie chez le criquet, *Locusta migratoria. J. Physiol. Paris*, **78**, 566–73.

Lange, A. B. and Orchard, I. (1986). Identified octopaminergic neurons modulate contractions of locust visceral muscle via adenosine 3', 5'-monophosphate (cyclic AMP). *Brain Res.*, **363**, 340–49.

Lehninger, A. L. (1979). *Biochemisrtry*. Worth Publishers, New York.

Maddrell, S. H. P. (1963). Excretion in the blood-sucking bug *Rhodnius prolixus* Stal. I: The control of diuresis. *J. Exp. Biol.*, **40**, 247–56.

Maddrell, S. H. P. (1966). Site of release of the diuretic hormone in *Rhodnius* – A new neurohaemal system in insects. *J. Exp. Biol.*, **45**, 499–508.

Maddrell, S. H. P. (1976). Functional design of the neurosecretory system controlling diuresis in *Rhodnius prolixus. Am. Zool.*, **16**, 131–9.

Maddrell, S. H. P. (1977). Insect Malpighian tubules. In: Gupta, B. L. *et al.* (eds), *Transport of Ions and Water in Animals*. Academic Press, London, New York, and San Francisco, pp. 541–69.

Maddrell, S. H. P. and Gardiner, B. O. C. (1976). Diuretic hormone in adult *Rhodnius prolixus*: total store and speed of release. *Physiol. Entomol*, **1**, 265–9.

Maddrell, S. H. P. and Klunsuwan, S. (1973). Fluid secretion by *in vitro* preparation of the Malpighian tubules of the desert locust *Schistocerca gregaria. J. Insect. Physiol.*, **19**, 1369–76.

Maddrell, S. H. P. Gardiner, B. O. C., Pilcher, D. E. M. and Reynolds, S. E. (1974). Active transport by insect Malpighian tubules of acidic dyes and of acylamides, *J. Exp. Biol.*, **61**, 357–78.

Mordue, W., Morgan, P. J. and Siegert, J. K. (1985). Strategies for the isolation of insect peptides. *Peptides*, **6**, 407–10.

Morgan, P. J. and Mordue, W.(1983). Separation and characteristics of diuretic hormone from the corpus cardiacum of *Locusta. Comp. Biochem. Physiol.*, **75B**, 75–80.

Morgan, P. J. and Mordue, W. (1985). Cyclic AMP and locust diuretic action. Hormone induced changes in cAMP levels offers a novel method for detecting biological activity of uncharacterized peptide. *Insect Biochem.*, **15**, 247–57.

Nicholls, S. P. (1985). Fluid secretion by the Malpighian tubules of the dragonfly *Libellula quadrimaculata. J. Exp. Biol.*, **116**, 53–67.

Nijhout, H. F. (1984). Abdominal stretch reception in *Dipetalogaster maximus* (Hemiptera: Reduviidae). *J. Insect Physiol.*, **30**, 629–33.

Petzel, D. H., Hagedorn, H. H. and Beyenbach, K. W. (1985). Preliminary isolation of mosquito natriuretic factor. *Am. Physiol. Soc.*, R379–R386.

Phillips, J. E. (1983). Endocrine control of salt and water balance: excretion. In: Downer R. G. H. and Laufer, H. (eds), *Endocrinology of Insects*, vol. 1. Liss, New York, pp. 411–25.

Picquot, M. (1989). La neurohormone diurétique, apparentée à la vasopressine du criquet migrateur: dualités structurale et fonctionnelle. Thesis, Université Bordeaux I.

Picquot, M. and Proux, J. (1987). Relationship between excretion of primary urine and haemolymphatic level of diuretic hormone along a circadian cycle in the migratory locust. *Physiol. Entomol.*, **12**, 455–60.

Picquot, M. and Proux, J. (1989). Influence of hydration on the concentration of two arginine-vasopressin-like neuropeptides coexisting in the ventral nerve cord of the migratory locust, *Locusta migratoria. J. Insect Physiol.*, **35**, 571–7.

Proux, J. (1978a). Ganglions de la chaîne nerveuse ventrale et équilibre hydrique chez le criquet migrateur. I. Ganglions abdominaux. *Arch. Biol Bruxelles*, **89**, 297–312.

Proux, J. (1978b). Ganglions de la chaîne nerveuse ventrale et équilibre hydrique chez le criquet migrateur. II. Ganglions sous-oesophagien et thoraciques. *Arch. Biol Bruxelles*, **89**, 313–28.

Proux, J. and Girardie, A. (1982). Neurosecretory regulation of the haemolymphatic titre of a vasopressin-like substance in the migratory locust. *Neurosci. Lett.*, **33**, 73–7.

Proux, J. and Hérault, J. P. (1988). Cyclic AMP, the second messenger of the newly characterized insect diuretic hormone: the AVP-like insect diuretic hormone. *Neuropeptides*, **12**, 7–12.

Proux, J. and Rougon-Rapuzzi, G. (1980). Evidence for vasopressin molecule in migratory locust. Radioimmunological measurements in different tissues: correlation with various states of hydration. *Gen Comp. Endocrinol.*, **47**, 378–83.

Proux, B., Proux, J. and Phillips, J. E. (1984). Antidiuretic action of corpus cardiacum (CTSH) on long term fluid absorption across locust recta *in vitro*. *J. Exp. Biol.*, **113**, 409–21.

Proux, J., Rougon-Rapuzzi, G. and Rémy, C. (1980). Influence de la pars intercerebralis et des cellules neurosécrétrices sous-ocellaires médianes sur la teneur en substance apparentée à la vasopressine dans la chaîne nerveuse ventrale et l'hémolymphe du criquet migrateur. Etude radioimmunologique et immunohistologique. *J. Physiol. Paris*; **76**, 277–82.

Proux, J., Picquot, M., Hérault, J.-P. and Fournier, B. (1988). Diuretic activity of a newly characterized neuropeptide: the AVP-like insect diuretic hormone. Use of a novel bioassay. *J. Insect Physiol.*, **34**, 919–27.

Proux, J., Miller, C. A., Li, G., Carney, R. L., Girardie, A., Delaage, M. and Schooley, D. (1987). Identification of an arginine vasopressin-like hormone from *Locusta migratoria*. *Biochem. Biophys. Res. Commun.*, **149**, 180–6.

Rafaeli, A., Pines, M., Stern, P. and Applebaum, S. W. (1984). Locust diuretic hormone-stimulated synthesis and excretion of cyclic-AMP: a novel Malpighian tubule bioassay. *Gen. Comp. Endocrinol.*, **54**, 35–42.

Ramsay, J. A. (1954). Active transport of water by the Malpighian tubules of the stick insect, *Dixippus morosus* (Orthoptera, Phasmidae). *J. Exp. Biol.*, **31**, 104–13.

Rémy, C. and Girardie, J. (1980). Anatomical organization of two vasopressin-neurophysin-like neurosecretory cells throughout the central nervous system of the migratory locust. *Gen. Comp. Endocrinol.*, **40**, 27–35.

Rémy, C. Girardie, J. and Dubois, M. P. (1977). Exploration immunocytologique des ganglions cerebroïdes et sous-oesophagien du phasme *Clitumnus extradentatus*. Existence d'une neurosécrétion apparentée à la vasopressine-neurophysine. *C. R. Acad. Sci. Paris*, **285**, 1495–7.

Rémy, C. Girardie, J. and Dubois, M. P. (1979). Vertebrate neuropeptide-like substance in the suboesophageal ganglion of two insects: *Locusta migratoria* R. et F. (Orthoptera) and *Bombyx mori* L. (Lepidoptera). Immunocytological investigation. *Gen. Comp. Endocrinol.*, **40**, 27–35.

Sawyer, W. H. (1972). Lungfishes and amphibians: endocrine adaptation and the transition from aquatic to terrestrial life. *Fed. Proc.* **31**, 1609–14.

Scharrer, B, (1987). Neurosecretion: beginnings and new directions in neuropeptide research. *Ann. Rev. Neurosci.*, **10**, 1–17.

Schooley, D. Miller, C. and Proux, J. (1987). Isolation of two arginine vasopressin-like factors from ganglia of *Locusta migratoria*. *Arch. Insect Biochem. Physiol.*, **5**, 157–66.

Smith, W. A. and Combest, W. L. (1985). Role of cyclic nucleotides in hormone action; In: Kerkut, C. A. and Gilbert, L. I. (eds), *Comprehensive Insect Physiology, Biochemistry and Pharmacology*. Pergamon Press, Oxford, vol. 8, pp.263–98.

Spring, J. H. and Hazelton, S. R. (1987). Excretion in the house cricket (*Acheta domesticus*): stimulation of diuresis by tissue homogenates. *J. Exp. Biol.*, **129**, 63–81.

Tamarelle, M. and Girardie, J. (1989). Immunohistological investigation of locust neuroparsin-like substances in several insects, in some invertebrates and vertebrates. *Histochemistry*, **91**, 431–5.

Weltens, R. and Van Kerkhove, E. (1985). Transport and electrophysiological phenomena in Malphighian tubules of *Formicidae*. *Arch. Int. Physiol. Biochem.*, **93**, P54–P55.

Williams, D., Phillips, J. E., Prince, W. T. and Meredith, J. (1978). The source of short-circuit current across locust rectum. *J. Exp. Biol.*, **77**, 107–22.

Part IV

Vertebrate studies

20 *Robin A. Barraco*

Purines: neuromodulators and metabolic regulators of bioreactivity in the nervous system

20.1. Emerging notions of chemical transmission in the nervous system

In recent years it has become increasingly clear that brain function cannot be understood simply in terms of 'hard-wiring' circuitry in the nervous system – neurons acting as relaying stations in a chemically transmitting telephone exchange. The integrative capacity of the nervous system must accrue from the enormous variability of the intracellular transmission process at any of its synapses. It is now apparent that not only the intercellular transmission process between central neurons may be extraordinarily varied but also the chemical vocabulary by which neurons communicate – the transmitter substance – is much more complex than originally envisioned by Dale. For example, certain neurons (e.g. amacrine cells) in the retina, which is part of the CNS, function almost entirely without the need for any propagated action potentials, suggesting that 'local' electrical and chemical circuits exist. At some of these retinal synapses, activation of the synapse may involve alteration or modulation of a continuous release of transmitter substances rather than the 'classic' view of the triggering of a release of a burst of transmitter. Similarly, evidence is mounting that some transmitters can be released from non-synaptic areas. For example, the release of dopamine has been observed from the dendrites and cell bodies of dopaminergic neurons in the substantia nigra following depolarization *in vitro* and in tissue slices (Osborne, 1981).

In a similar departure from the traditiional view of intercellular communication between neurons, there has been considerable recent evidence indicating the coexistence of putative transmitters in the same nerve cells (Milhorn and Hökfelt, 1988). It has thus become necessary to consider the possibility that neurons may release more than one transmitter at some or all of their terminals. For example, evidence for adrenergic-purinergic and cholinergic-purinergic dual function has been observed in microcultures of rat sympathetic neurons (Furshpan *et al.*, 1982). Similarly, there is evidence for a release of ATP from cholinergic and noradrenergic nerve terminals based on measurements of synaptosomal contents of this purine and its detection in the perfusate of stimulated nerves. ATP can also be detected in the vesicle fraction prepared from cholinergic and noradrenergic nerves. It has been shown that ATP is released together with acetylcholine from stimulated rat phrenic nerve terminals (Silinsky, 1975). The ATP is derived from the nerve terminal rather than the muscle as its release is not affected by blockade of the postsynaptic receptors with curare. In addition, Chan-Palay and colleagues (1978) have used a variety of histochemical techniques to study the coexistence of neurotransmitters such as the serotonin and substance P containing neurons in the rat brain. The coexistence of neuropeptides and amine-transmitter molecules has also been demonstrated in a number of other neurons (Hökfelt *et al.*, 1978). Cells of the rat superior cervical ganglion have been shown to contain norepinephrine and enkephalin-like material (Schultzberg *et al.*, 1979). Cholecystokinin-like immunoreactivity exists in some nigral dopamine containing cells (Milhorn and Hökfelt, 1988). Somatostatin and avian pancreatic polypeptide have been co-localized in human cortical neurons (Vincent *et al.*, 1982) and dynorphin (an endogenous opiate) and vasopressin coexist in certain cells of the hypothalamus (Watson *et al.*, 1982). More recently, direct evidence was shown for the concomitant release of noradrenaline, ATP and neuropeptide Y from sympathetic nerve supplying the guinea-pig vas deferens (Kasakov *et al.*, 1988).

Thus the histochemical evidence, along with other lines of investigation such as the release and uptake studies, suggest that transmitter-like molecules may coexist in certain vertebrate neurons. However, the fact that a neuron contains some amount of several transmitter-type molecules, does not prove that each substance is being released as a transmitter. There is also the possibility that some neurons release more than one type of transmitter-like substance from the same terminal, but that only one of these would function as the primary neurotransmitter with the others serving to modify or modulate the action of the primary transmitter.

20.2. **Intercellular mediators: neurotransmitters and neuromodulators**

The concept of neuromodulation is still in the process of development. The term 'modulator' has recently appeared in the literature with increasing frequency to characterize the actions of putative transmitters that do not seem to fit within the traditional concepts of neuronal interaction. Concurrently it has been suggested that the substances which elicit these uncoventional responses (purines, monoamines, peptides) should be considered as modulators or neuroregulators.

The basic criteria for identification of neurotransmitters have been described by a number of authors. The essential features of a neurotransmitter can be summarized as follows: the substance should be synthesized and stored in the neuron; it should be released by the terminal during nerve activity; and it should interact with specific receptors on the postsynaptic membrane to cause a metabolic or membrane potential change in the postsynaptic cell.

In contrast to neurotransmitters which interact with specific receptors, a neuromodulator might exert its effects on intercellular mediation through a variety of mechanisms. For example, it might affect neurotransmitter synthesis, release, receptor interactions, reuptake or metabolism. Further, neuromodulators may be released from neurons, glial cells, or secretory cells, to amplify or dampen local synaptic activity by altering the effect of the primary neurotransmitters. Finally, neuromodulators need not be generally associated with synaptic vesicles and may be synthesized on demand, as are the prostaglandins (Osborne, 1981).

Florey (1960) introduced the term 'modulator', and proposed a distinction between transmitter substance and modulator substance based on a model of release rather than mechanism of action. Transmitters would be released only by orthodromic nerve impulses while the modulator substance would be released in a continuous or intermitted secretion process. Furthermore while a transmitter substance would always be released from nerve endings, a modulator could be released at any specialized or even unspecialized region of a cell. Some definitions of modulators (Osborne, 1981) distinguish them from transmitters as being compounds which are important in general communication between nerve cells, but which operate in a 'hormone-like' fashion rather than transsynaptically. Barchas *et al.* (1978) have defined a number of criteria to characterize a neuromodulator, and of these the most pertinent is that the substance is not acting as a neurotransmitter in that it does not act transsynaptically and does not generate excitatory or inhibitory postsynaptic potentials. Barker (1978) has defined neuromodulation as a form of gain control over transmission at conventional synapses by alteration of synaptically activated, voltage-independent membrane conductances which might or might not be restricted to contiguous cells. Dismukes (1981) abandoned the use of the word modulator and recommended instead the term 'nonsynaptic transmitter'.

All of these classifications are based, to some extent, on the proximity of the postsynaptic structure to the releasing terminal, and tend to be ambiguous or to lead to inconsistencies. An alternative classification proposed by Phillis (1978) for intercellular mediators might be to categorize them as neurotransmitters if they mediate transmission at rapidly transmitting junctions by increasing specific, voltage-independent, ionic conductances (nicotinic actions of acetylcholine, amino acid transmitters) or as modulators if they initiate longer-lasting alterations in neuronal excitability or transmitter release which are not associated with increases in membrane conductance (muscarinic actions of acetylcholine, adenosine, monamines, peptides). Renshaw cells and sympathetic ganglion cells provide examples of neurons on which acetylcholine would have both neurotransmitter and modulator actions at nicotinic and muscarinic synapses respectively.

20.3. Regulation of neuronal sensitivity

The primary function of the nervous system is the transmission of information via electrochemical signals. In the simplest circuit a nerve impulse travels via the axon to a synaptic terminal where a transmitter substance is released, eliciting a postsynaptic response. The possible mechanisms for regulation of neuronal response in a simple circuit might include (1) changes in the efficacy of presynaptic mechanisms for excitation-secretion coupling and thus changes in the levels of neurotransmitter released; and (2) changes in the efficacy of the interaction of the neurotransmitter with the receptor (by, for example, affecting receptor sensitivity) (Daly, 1980). For the regulation of neuronal sensitivity in more complex systems a multiplicity of factors may be superimposed upon this basic design. For example, a neuron can have an intrinsic tonic or spontaneous pacemaker activity which can be influenced by a variety of local inputs or inputs from other neurons. Furthermore, the regulation of neuronal sensitivity for complex circuitry can involve a number of other modulatory mechanisms such as (1) direct excitatory or inhibitory inputs from other neurons, collaterals, or interneurons; (2) a variety of presynaptic inhibitory and excitatory regulatory mechanisms; and (3) local feedback modulation of neuronal activity (Daly, 1980). An example of local feedback modulation of neuronal basal activity is seen with adenosine, an agent generally inhibitory to central neurons; high electrical activity in central neurons leads to release of endogenous adenosine which produces local feedback inhibition of neuronal activity. Thus the alteration of any one or several of these factors – presynaptic excitation – secretion coupling, neurotransmitter release and receptor interactions, intrinsic pacemaker activity of neurons, local feedback modulation of neuronal basal activity, and input from other neuronal elements – can significantly affect neuronal sensitivity and, ultimately, bioreactivity.

In the mammalian nervous system the important cellular messengers are neurotransmitters and neuromodulators, cyclic nucleotides and their derivatives, phosphoinositides and calcium. Some of these second-messenger systems share common sites, G proteins, coupling neurotransmitter receptors with second-messenger generating enzymes (Worley *et al.*, 1987). Generally, neurotransmitters make it possible for cells to communicate with one another; cyclic nucleotides (or their purine derivatives) and calcium carry messages from one part of a cell to another. Thus the modulation of neuronal sensitivity can involve many time-scales since these three cellular messengers have complementary roles with respect to time as well as distance. For neurotransmitters and neuromodulators the response time and the duration of action can range from milliseconds to hours, involving diverse cellular events such as changes in membrane potential membrane fluidity and conductance (minutes), changes in the density of ionic channels (hours), and effects on steady-state levels or activities of modulatory factors or biosynthetic enzymes (days) (Daly, 1980). For cyclic AMP the time-scale is from seconds to minutes, and may involve changes in enzyme levels (adenylate cyclase, phosphodiesterase), membrane phosphorylation via cAMP-dependent protein kinases (seconds), and changes in membrane conductances (minutes). For calcium the response time and duration of action is in the millisecond range and may involve stimulation of potassium conductances, membrane phosphorylation via calcium-dependent protein kinases and, ultimately, neurotransmitter release. Thus the pivotal role of calcium as a molecular switch in regulation neuronal sensitivity, especially in the millisecond time-frame, has considerable relevance for the regulation of bioreactivity and, ultimately, for the enhancement of neural information processing.

20.4. **Calcium, calmodulin, and protein phosphorylation**

A variety of mechanisms have emerged, some intracellular and others occurring at the cell membrane, that are possibly relevant to the control of neuronal sensitivity, and it appears that most of these involve calcium. It is evident that Ca^{2+} plays a major role in the function of the nervous system, especially in the process of neurotransmitter release. Moreover, it has recently been shown that most of the effects of Ca^{2+} are mediated through a homologous class of Ca^{2+}-binding proteins, of which calmodulin appears to be the most ubiquitous and versatile (DeLorenzo, 1982). Calmodulin is a major Ca^{2+}-binding protein in the brain that may regulate many of the neuronal functions of Ca^{2+} by modulating the effects of calcium on several important enzyme systems. For example, calmodulin has been shown to regulate the effects of Ca^{2+} (1) on synaptic protein phosphorylation, which is stimulated by depolarization-dependent Ca^{2+} influx, (2) neurotransmitter release and synaptic vesicle–synaptic membrane interactions, and (3) phosphodiesterase and adenylate cyclase activity (DeLorenzo, 1982). Some of these regulatory effects of calcium on neuronal activity are intracellular while other events occur at the cell membrane. For example, activation of adenylate cyclase occurs at the membrane, but cAMP is generated intracellularly where it may activate membrane-bound or cytosol protein kinases. Likewise, neuronal depolarization can cause the influx of calcium ions, resulting in the activation of certain intracellular protein kinases. Calcium ions are also capable of activating intracellular guanylate cyclases. It follows that cyclic adenosine monophosphate, cGMP, and calcium-dependent protein phosphorylation could lead to long-term regulation of neuronal activity (Daly, 1980).

Protein phosphorylation represents one potential enzymatic means whereby the responsiveness of neurons can be regulated. The phosphorylation and dephosphorylation of neuronal proteins are dynamic processes that are controlled by protein kinases and phosphoprotein phosphatases. Cyclic adenosine monophosphate can regulate protein phosphorylation·through cAMP-dependent protein kinases and, in fact, most of the actions of cAMP appear to be linked to the phosphorylation of cellular proteins (Greengard, 1978). In addition, it appears that calcium ions and G proteins may also modulate neuronal activity through effects on protein phosphorylation via calcium-dependent and cGMP-dependent protein kinases. The regulation of neuronal sensitivity by protein phosphorylation can occur at the presynaptic nerve terminal or at the postsynaptic membrane. A similar cascade of events appears to occur for receptor-mediated phosphoinositide turnover which involves G proteins, inositol 1,4,5-triphosphate (IP_3), calcium ions and protein kinases (Worley *et al.*, 1987). However, the nature of the substrates for these protein kinases and their physiological significance is not yet clearly understood.

Moreover, the duration of the after-hyperpolarization following an action potential determines the maximum spike-firing frequency of a neuron (Busis and Weight, 1976). Consequently, the molecular mechanisms involved in controlling depolarization-evoked Ca^{2+} uptake and the resultant calcium-sensitive potassium conductances may be important for ultimately understanding the regulation of neuronal excitability and the bioreactivity of neuronal circuits. Thus not only does calcium-dependent phosphorylation at synaptic terminals probably play a role in the control of neurotransmitter release, but calcium ions also appear to play an important role in regulating the recovery time of a neuron after an action potential – the spike after-hyperpolarization (Daly, 1980). The duration of the after-hyperpolarization following an action potential determines neuronal excitability, which limits the spike-firing frequency of a neuron. This recovery period of a neuron has obvious relevance in the regulation of bioreactivity, and calcium appears to play a pivotal role. It is of great interest, therefore, that the depressant actions of adenosine on neuronal activity, and its modulatory effects on neurotransmitter release, may be mediated by its effects on the stimulus-induced calcium uptake and entry into the nerve terminal (Wu *et al.*, 1982). Ultimately, by regulating calcium fluxes, adenosine could modulate not only neurotransmitter release, but also neuronal recovery time through calcium-sensitive potassium conductances associated with spike after-hyperpolarization and also the activity of calcium-dependent protein kinases.

20.5. **Adenosine receptors, calcium and neurotransmitter release**

It has recently been proposed that adenosine, similar to many neuropeptides, is a homeostatic neuromodulator, and as such may represent the prototypic bioactive agent of this type (Williams, 1989). It is well established, for example, that adenosine and its analogues inhibit the calcium-dependent release of numerous neurotransmitter substances, including monoamines, from a variety of vertebrate central synapses, and peripheral nerve terminals (Ribeiro and Sebastiao, 1986; Fredholm and Dunwiddie, 1988). Many of the modulatory effects of adenosine on transmitter release appear to be mediated by extracellular receptors, negatively coupled (A_1) and positively coupled (A_2) to adenylate cyclase (Daly, 1985; Phillis and Barraco, 1985). Responses at both receptors can be competitively antagonized by methylxanthines such as theophylline (Daly, 1985). Moreover, recent work from ligand binding studies has suggested the existence of A_2 receptor subtypes (Bruns, 1988). Adenosine receptor activation can occur at both pre- and postsynaptic sites (Williams, 1989) and, as a result, the effects of adenosine may be indirect, inhibiting the release of a variety of conventional neurotransmitters from presynaptic terminals (Fredholm and Dunwiddie, 1988) or direct, by affecting postsynaptic processes (Rubio *et al.*, 1988). On the other hand, the site of action of adenosine, the role of cyclic AMP in mediating its effects and the mechanism of its release have not been clearly established (Phillis and Barraco, 1985; Williams, 1989).

The role of adenosine and its receptors in the regulation of presynaptic mechanisms of neurotransmitter release is also suggested by studies of its effects of cAMP-dependent protein phosphorylation in presynaptic terminals. In the brain the elevation of intracellular cAMP levels has been associated with the action of a number of neurotransmitters and neuromodulators on membrane receptors. The intracellular effect of cAMP is to activate cAMP-dependent protein kinases which phosphorylate specific substrate proteins (Dolphin and Greengard, 1981). One of these substrates, Protein I, is a synapse-specific neuronal phosphoprotein which is present throughout the CNS. This protein appears to be present primarily in nerve terminals and is associated with synaptic vesicles. In cell-free preparations this protein is a substrate for both cAMP-dependent and Ca^{2+}-dependent protein kinases. In one study it was shown that adenosine increased the phosphorylation of Protein I (Dolphin and Greengard, 1981). Thus the specific and potent effects of adenosine on Protein I phosphorylation provide additional evidence that suggests adenosine may play a pre-eminent neuromodulatory role in the regulation of synaptic activity in nerve terminals.

Taken together, the studies indicating the modulatory effects of adenosine and its analogues on neurotransmitter release, and its potent effects on neuronal firing, cAMP formation and synaptic protein phosphorylation, suggest that adenosine may play a major role in the regulation of neuronal sensitivity. In fact, adenosine may exert its effects by acting on extracellular receptors where it modulates the depolarization or stimulus-induced uptake of calcium which is associated with transmitter release from presynaptic terminals. In summary, it has been proposed (Phillis and Wu, 1981) that, following an action potential, ATP is released by CNS neurons as a co-transmitter, together with the transmitters with which it is stored, from a vesicular pool. Adenosine is then formed from released ATP by the action of ectonucleotidases and nucleosidases within the synaptic cleft; adenosine can then act presynaptically to limit further release of the transmitter as an autoregulatory mechanism. The net effect of these purinergic receptor mechanisms will be to enhance fidelity of neural information processing.

This phenomenon of neuromodulation, and particularly the control of neurotransmitter release from presynaptic terminals by specialized presynaptic receptor mechanisms (autoregulation of transmitter release), has received widespread acceptance for autonomic neuroeffector junctions and also in the CNS where similar phenomena have been observed. Those presynaptic receptors involved in the autoregulation of transmitter release are known as autoreceptors when they are activated by substances released from the same nerve terminal. Moreover, it has been repeatedly shown that adenosine inhibits the presynaptic release of a variety of neurotransmitters from peripheral and central synapses (Fredholm, 1988). Since adenosine does not polarize central nerve terminals (Stone, 1980), its action is unlikely to be a result of an increase in membrane permeability to specific ions. Rather, adenosine appears to act by a receptor-mediated reduction in the depolarization-evoked influx of Ca^{2+} into the nerve terminal and a resultant lack of intracellular Ca^{2+} availability for transmitter release. Adenosine's action therefore falls within the definition of that of a neuromodulator.

In summary, it is therefore possible that purine release may not be associated with a specific group of neurons, but it may be a general phenomenon of excitable tissues. Most tissues examined *in vivo* and *in vitro* have been shown to release purines when depolarized, including muscle tissue and isolated axonal segments. Moreover, in most of the CNS experiments, adenosine and adenosine analogues have consistently been shown to inhibit neuronal firing, EPSP, evoked potentials, population spikes, Ca^{2+} influx, and neurotransmitter release. These effects of adenosine and its analogues are mediated through extracellular adenosine receptors. As a result, for the CNS with its ubiquitous distribution of purines and its vast area of excitable membrane, the role of adenosine in the regulation of excitability provides a crucial focus for neurobiological research. Consequently, in view of the many actions of adenosine on neuronal firing, transmitter release, and receptor sensitivity, combined with the fact that adenosine levels in the extracellular space are apparently sufficient for the continuous purinergic control of cell excitability, the possibility exists that adenine derivatives as a group could form the basis of a more general synaptic control system in the CNS, and thus play a fundamental role in the intercellular regulation of bioreactivity.

In light of the proposed role for adenosine in the modulation of presynaptic calcium-dependent neurotransmitter release in mammals it was of interest to determine if this would occur in invertebrates (Barraco, unpublished). *Mytilus edulis* neural tissues were incubated with [^{3}H]monoamines and 50 mM KCl induced their release. In the absence of adenosine analogues the release caused by KCl was approximately 8% of the tissue content of radioactivity. Addition of 5′-*N*-ethylcarboxamidoadenosine (NECA) to the superfusing medium at the time of the K^+ application resulted in concentration-dependent inhibition of the evoked release of serotonin and dopamine, and at a higher dose of norephinephrine. Essentially complete inhibition of K^+-evoked release was achieved with nM concentrations of NECA. Superfusion with theophylline, an adenosine antagonist, produced a concentration-dependent reversal or prevention of the presynaptic inhibition of monoamine release by NECA. Theophylline alone at 10 μM to 100 μM had no observable effect on K^+-evoked release of the various monoamines. These preliminary observations may serve to indicate that adenosine modulation also occurs in 'simpler' animals.

Many signal systems found in invertebrates obviously have mammalian counterparts (Stefano, 1988). This indicates that these basic intercellular and intracellular signal systems must have originated early in the course of evolution. Stefano (1988) discusses how the evolutionary force may well be the stereoselective nature of the signal systems themselves that tend to maintain these major communication systems. Thus 'ancient', e.g. adenosine, communication systems would have tended to remain relatively intact in increasingly complex animal phyla.

References

Barker, J. L. (1978). Evidence for diverse cellular roles of peptides in neuronal function. In: *Peptides and Behavior: A Critical Analysis of Research Strategies. Neurosci. Res. Prog. Bull.*, **16**, 535–53.

Barchas, J. D., Akil, H., Elliot, G. R., Holman, R. B. and Watson, S. J. (1978). Behavioral neurochemistry: neuroregulators and behavioral states. *Science*, **200**, 964–73.

Barrett, E. F. and Barrett, J. N. (1976). Separation of two voltage-sensitive potassium currents, and demonstration of a tetrodotoxin-resistant calcium current in frog motoneurons. *J. Physiol.*, **255**, 737–74.

Bruns, R. F. (1988). Adenosine receptor binding assays. In: Cooper, D. M. F.

and Londos, C. (eds), *Receptor Biochemistry and Methodology*, vol. 11. Alan R. Liss, New York, pp. 43–62.

Busis, N. A. and Weight, F. F. (1976). Spike after-hyperpolarization of a sympathetic neurone is calcium sensitive and is potentiated by theophylline. *Nature*, **263**, 434–6.

Chan-Palay, V., Jonsson, G. and Palay, S. L. (1978). Serotonin and substance P co-exist in neurons of the rat's central nervous system. *Proc. Natl. Acad. Sci., USA*, **75**, 1582–6.

Daly, J. W. (1980). Mechanisms of regulation of neuron sensitivity. *Neurosci. Res. Prog. Bull.*, **18**, 333–45, 383–6.

Daly, J. W. (1985). Adenosine receptors. In: Cooper, D. M. F. and Seaman, K. B. (eds), *Advances in Cyclic Nucleotide and Protein Phosphorylation Research.* Raven Press, New York, pp. 29–46.

DeLorenzo, R. J. (1982). Calmodulin in neurotransmitter release and synaptic function. *Fed. Proc.*, **41**, 2265–72.

Dismukes, R. K. (1981). New concepts of molecular communication among neurons. *Behav. Brain Sci.*, **2**, 409–48.

Dolphin, A. C. and Greengard, P. (1981). Neurotransmitter- and neuromodulator-dependent alterations in phosphorylation of protein I in slices of rat facial nucleus. *J. Neurosci.*, **1**, 192–203.

Florey, E. (1960). Physiological evidence for naturally occurring inhibitory substances. In: Roberts, E. *et al.* (eds), *Inhibitions of the Nervous System and Gamma-aminobutyric Acid.* Pergamon Press, New York, pp. 21–43.

Fredholm, B. B. (1988). Presynaptic adenosine receptors. *ISI Atlas of Science (Pharmacol.)*, pp. 257–60.

Fredholm, B. B. and Dunwiddie, T. V. (1988). How does adenosine inhibit transmitter release? *Trends Pharmacol. Sci.*, **9**, 130–134.

Furshpan, E. J., Potter, D. D. and Landis, S. C. (1982). On the transmitter repertoire of sympathetic neurons in culture. *Harvey Lect.* **76**, 149–54.

Greengard, P. (1978). Phosphorylated proteins as physiological effectors. *Science*, **199**, 146–52.

Hökfelt, T., Ljungdahl, H., Steinbusch, H., Herhoptad, A., Nilsson, G., Brodin, E., Pernow, B. and Goldstein, M. (1978). Immunohistochemical evidence for substance-P like immunoreactivity in some 5-hydroxytryptamine-containing neurons in the central nervous system. *Neuroscience*, **3**, 517–38.

Kasakov, L., Ellis, J., Kirkpatrick, K., Milner, P. and Burnstock, G. (1988). Direct evidence for concomitant release of noradrenaline, ATP and neuropeptide Y from sympathetic nerve supplying the guinea-pig vas deferens. *J. Auton. Nerv. Syst.*, **22**, 75–82.

Milhorn, D. E. and Hökfelt, T. (1988). Chemical messengers and their coexistence in individual neurons. *NIPS*, **3**, 1–5.

Osborne, N. N. (1981). Communication between neurons: current concepts. *Neurochem. Int.*, **3**, 3–20.

Phillis, J. W. (1970). *The Pharmacology of Synapses*, Pergamon Press, New York.

Phillis, J. W. (1978). The actions of drugs on cells *in vitro*, or are isolated tissues a substitute for microiontophoresis. In: Ryall, R. W. and Kelly, J. S. (eds), *Iontophoresis and Transmitter Mechanisms in the Mammalian Central Nervous System.* Elsevier/North Holland, Amsterdam, pp. 169–78.

Phillis, J. W. and Barraco, R. A. (1985). Adenosine, adenylate cyclase, and transmitter release. In: Cooper, D. M. F. and Seamon, K. B. (eds), *Advances in Cyclic Nucleotide and Protein Phosphorylation Research.* Raven Press, New York, pp. 243–57.

Phillis, J. W. and Wu, P. H. (1981). The role of adenosine and its nucleotides in central synaptic transmission. *Prog. Neurobiol.*, **16**, 187–239.

Ribeiro, J. A. and Sebastiao, A. M. (1986). Adenosine receptors and calcium: Basis for proposing a third (A_3) adenosine receptor. *Prog. Neurobiol.*, **26**, 179–209.

Rubio, R., Bencherif, M. and Berne, R. M. (1988). Release of purines from postsynaptic structures of amphibian ganglia. *J. Neurochem.*, **51**, 1717–23.

Schultzberg, M., Hökfelt, T., Terenius, L., Elfvin, L. G., Lundberg, J. M., Brandt, J., Elde, R. P. and Goldstein, M. (1979). Enkephalin-immunoreactive nerve fibers and cell bodies in sympathetic ganglia of the guinea pig and rat. *Neuroscience*, **4**, 249–70.

Silinsky, E. M. (1975). On the association between transmitter secretion and the release of adenine nucleotides from mammalian motor nerve terminals. *J. Physiol. (Lond.)*, **247**, 145–62.

Stefano, G. B., (1982). Comparative aspects of opioid–dopamine interaction. *Cell. Mol. Neurobiol.*, **2**, 167–78.

Stefano, G. B. (1988). The evolvement of signal systems: conformational matching a determining force stabilizing families of signal molecules. *Comp. Biochem. Physiol.*, **90C**, 287–94.

Stone, T. W. (1980). Adenosine and related compounds do not affect nerve terminal excitability in rat CNS. *Brain Res.*, **182**, 198–200.

Ueda, T. and Greengard, P. (1977). Adenosine 3′-5′-monophosphate-regulated phosphoprotein system of neuronal membranes. I. Solubilization, purification, and some properties of an endogenous phosphoprotein. *J. Biol. Chem.*, **252**, 5155–63.

Vincent, S. R., Johansson, D., Hökfelt, T., Meyerson, B., Sachs, C., Elde, R. P., Terenius, L. and Kinnel, J. (1982). Neuropeptide coexistence in human cortical neurons. *Nature*, **298**, 65–7.

Watson, S. J., Akil, H., Fischli, W., Goldstein, A., Zimmerman, F., Nilaver, G. and Wimersma-Greidanus, T. B. van (1982). Dynorphin and vasopressin: common localization in magnocellular neurons. *Science*, **216**, 85–7.

Williams, M. (1989). Adenosine: the prototypic neuromodulator. *Neurochem. Int.*, **19**, 249–64.

Worley, P. F., Baraban, J. M. and Snyder, S. H. (1987). Beyond receptors: multiple second-messenger systems in the brain. *Ann. Neurol.*, **21**, 217–29.

Wu, P. H., Phillis, J. W. and Thierry, D. L. (1982). Adenosine receptor agonists inhibit K^+-evoked Ca^{2+} uptake by rat brain cortical synaptosomes. *J. Neurochem.*, **39**, 700–8.

21 *David O. Carpenter*

A common mechanism of excitation of area postrema neurons by several neuropeptides, hormones and monoamines

21.1. **Introduction**

The neurons that function as the chemosensitive trigger for the emetic reflex are located in the area postrema (Wang and Borison, 1952), one of the circumventricular organs of the brain, located in the floor of the fourth ventricle and outside the blood–brain barrier (Wislocki and Putnam, 1920). These neurons are known to respond to circulating toxins and neuroactive agents, then to project to deeper brainstem regions which activate the stereotypic autonomic and motor mechanisms which constitute the full reflex (Wang, 1980; Carpenter, 1989).

The receptors for neuroactive substances on these small neurons are of interest for several reasons. Firstly, the emetic reflex can be elicited by a great variety of substances, and there is an interesting question as to how such a system is organized. For example, do all neurons have receptors for all emetic substances, or are there subpopulations of neurons with a more restricted number of receptors? In addition, since the reflex is stereotyped and the sequence of events following activation of the chemosensitive trigger zone is apparently independent of which substance evokes the reflex, there is the interesting question of whether or not the receptors for the different neuroactive substances are distinct, and if they are distinct whether there is a common mechanism of excitation which allows the ultimate response to be independent of the specific excitatory substance.

Our studies on this system have been primarily performed in the dog, the animal which is most similar to man in terms of threshold sensitivity and pharmacological properties of the emetic reflex (Carpenter *et al.*, 1983, 1984a, 1988). These studies have been done by obtaining extracellular recordings from area postrema neurons in anaesthetized dogs, as well as investigation of emetic agents and their pharmacology by intravenous application of neuroactive substances and antagonists in conscious dogs. Our results suggest that neurons of the area postrema have a great number of specific receptors, and that only a single type of neuron exists. However, it appears that receptors for a great variety of neurotransmitters, neuropeptides and hormones are all coupled to a common, cyclic AMP-dependent second messenger system.

21.1.1. *Neuropeptide excitation*

The methods used have been reported for both the electrophysiological (Carpenter *et al.*, 1983, 1988) and behavioral studies (Carpenter *et al.*, 1984a, 1988). In brief, neurons were recorded through the centre barrel of a seven-barrelled ionophoretic array, and putative neuroactive substances were applied by ionophoresis from the other barrels. In addition to raw records of the extracellular action potentials, which are small and difficult to separate from the noise level, data were recorded as pulses triggered by the action potentials. In all cases the neurons were silent unless activated via receptors; thus they were located by ionophoretically applying glutamate, an excitatory amino acid transmitter which excites almost all mammalian central neurons.

Figure 21.1 shows a raw data recording of the response of an area postrema neuron to gastrin, a 17 amino acid gastrointestinal peptide which is known to be asymmetrically distributed in brain (Rehfeld *et al.*, 1979) and which is thought to be a functional neurotransmitter. When administered intravenously in conscious dogs, gastrin will induce emesis within 1–3 min (Carpenter *et al.*, 1984b). This neuron showed two quite different types of response to gastrin and glutamate. The response to glutamate is very similar to that seen in other central neurons, being a brief, high-frequency excitation. In contrast the response to gastrin has a long latency, a low frequency and a long duration. The considerable differences in the responses suggest that the mechanisms responsible for generation of these excitations are different.

Figure 21.2 shows responses to glutamate, leucine enkephalin and apomorphine, a classic emetic agent which is an agonist at dopamine receptors. In these records the extracellular action potential was used to trigger an electronic pulse, giving a record which is easier to illustrate given the small amplitude of the discharges. The response to glutamate is similar to that seen in Fig. 21.1; indeed this was characteristic of all neurons investigated. The responses to leucine enkephalin and apomorphine had characteristics similar to those recorded for gastrin in Fig. 21.1. Table 21.1 lists responses obtained from 308 neurons studied in anaesthetized dogs, and the percentage of neurons responsive to particular substances. A surprisingly large number of substances (but not all) were excitatory on these small neurons. The active substances include conventional small molecule neurotransmitters, neuropeptides and hormones, as well as several prostaglandins. The response to glutamate in every neuron studied was similar to that shown in Figs 21.1 and 21.2, but responses to all other substances that were excitatory were similar in character to those

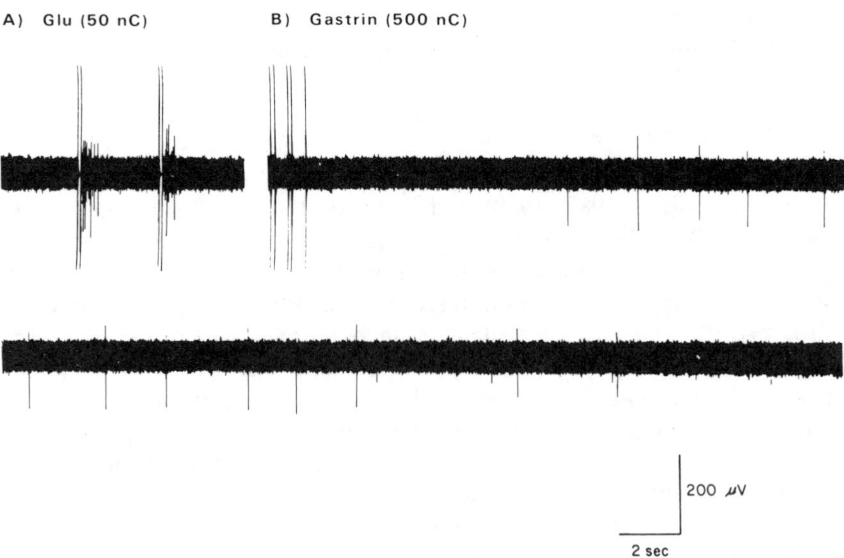

Fig. 21.1. Responses of an area postrema neuron to glutamate (Glu) and gastrin. The neuron was at 543 μm below the surface and did not respond to either VIP or angiotensin II. The large deflections are ionophoretic artifacts. Glutamate was applied as a single pulse of 50 nanocoulombs (nC). Gastrin was applied as five pulses, each 100 nC, of which only the last two are shown. (Reproduced with permission from Carpenter *et al.*, 1983.)

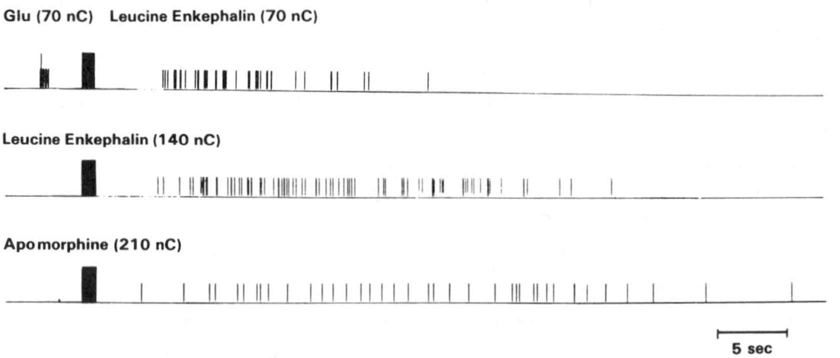

Fig. 21.2. Raster display of responses of an area postrema neuron 502 μm below the surface to ionophoretic application of glutamate (Glu), leucine enkephalin, and apomorphine. The time of ionophoretic application is indicated by the large upward deflection, whereas the smaller deflections are electronic pulses triggered by single enkephalin. There was a 7-min interval between the two leucine enkephalin applications to avoid receptor desensitization. (Reproduced with permission from Carpenter *et al.*, 1988).

shown for gastrin, leucine enkephalin and apomorphine in Figs 21.1 and 21.2.

As indicated in the far right column of Table 21.1, most of the substances which had excitatory responses have been found to be emetic. Those few which have not may be presumed to be emetic but have not been tested. Thus, in so far as it has been possible to test, there is a direct correlation between the ability of a substance to excite area postrema neurons and its being emetic.

In order to evaluate whether or not the receptors for the different substances are distinct, several experiments were performed in which antagonist drugs were tested for their ability to block emesis induced by intravenous administration of substances that excite area postrema neurons. Table 21.2 shows results of experiments using apomorphine, leucine enkephalin and angiotensin II. Three antagonists were used: naloxone, an opioid antagonist; domperidone, a dopamine D2 receptor antagonist which is known not to cross the blood–brain barrier (Laduron and Leysen, 1979); and chlorpromazine, also a dopamine receptor antagonist but one which can cross the blood–brain barrier and also has actions on at least some other types of receptor as well (Changeux *et al.*, 1986). Domperidone blocked the apomorphine response, but not the others. This is consistent with apomorphine being excitatory at a dopamine D2 receptor on the area postrema neurons as expected (Stefanini and Clement-Cormier, 1981). Naloxone blocked the leucine enkephalin response but not the others, consistent with the leucine enkephalin receptor being distinct from the others. In contrast, chlorpromazine blocked responses to all of the substances tested. The differences between domperidone and chlorpromazine suggest that the general action of chlorpromazine reflects actions at a chlorpromazine-sensitive site on the brain side of the blood–brain barrier, while domperidone and chlorpromazine block dopamine D2 receptors in the area postrema as well.

In another study we used receptor desensitization to evaluate whether or not receptors for emetic agents were distinct. A loss of response on a substance upon prolonged or repeated application within a brief period is characteristic of most active substances, and is a mechanism that is probably of protective value to the organism. Since desensitization is a property of the receptor, and not of the associated ionic channel, the specificity of desensitization is another way in which to evaluate whether or not receptors are specific. Table 21.3 shows results with leucine enkephalin, angiotensin II and apomorphine. The protocol used was to use a dose of each substance that consistently gave emesis, and test the same substance again 5 min later. For both leucine enkephalin and angiotensin II a second administration after 5 min failed to elicit a second emetic episode. However, apomorphine did not show desensitization. This may indicate a difference of the receptor, or may only reflect a dose which is

Table 21.1. Transmitter actions on area postrema neurons

Substance	*No. of units*	*Percentage excitation*	*Reference for emesis due to substance*
Glutamate	308	99	Levey *et al.*, 1949
Acetylcholine	15	38	Borison, 1959
Nicotine	6	0	
Pilocarpine	13	61	Borison, 1959
Serotonin	32	66	Douglas, 1975
Norepinephrine	15	40	Jenkins and Lahay, 1971
Histamine	45	64	Bhargava and Dixit, 1968
Epinephrine	22	32	Peng, 1963
Apomorphine	28	71	Hatcher and Weiss, 1923
Dopamine	11	73	Innes and Nickerson, 1975
Insulin	90	54	Carpenter and Briggs, 1986
Zinc	14	0	
Glucose	16	0	
Angiotensin II	73	47	Carpenter *et al.*, 1984a, b
Neurotensin	20	25	Ibid
TRH	64	66	Ibid
VIP	39	46	Ibid
Gastrin	11	36	Carpenter *et al.*, 1984a, b
Substance P	23	48	Carpenter *et al.*, 1984a, b
Vasopressin	10	50	Carpenter *et al.*, 1984a, b
Leucine	40	53	
Enkephalin	10	0	Carpenter *et al.*, 1984
Somatostatin	19	26	
CCK	2	0	
LHRH	4	50	Harding and McDonald, 1989
Neuropeptide Y	8	50	
Calcitonin	15	40	
Prostaglandin A_1	12	0	
Prostaglandin A_2	17	24	
Prostaglandin B_1	21	42	
Prostaglandin B_2	12	0	
Prostaglandin D_1	14	0	Kaul *et al.*, 1978
Prostaglandin D_2	14	47	
Prostaglandin E_1	23	26	
Prostaglandin $F_{1\alpha}$	16	50	
Prostaglandin $F_{1\alpha}$			

TRH, thyroid-releasing hormone; CCK, cholecystokinin; LHRH, leutinizing hormone-releasing hormone; VIP, vasoactive intestinal polypeptide.

Table 21.2. Effects of various antagonists on apomorphine and peptide emesis

Substance	*Angiotensin II (0.25 mg/kg)*	*Leu-enkephalin (0.15 mg/kg)*	*Apomorphine (0.015 mg/kg)*
Domperidone, 1.0 mg/kg (i.m.)	16/16	16/16	0/15
Chlorpromazine, 1.5 mg/kg (i.m.)	0/16	1/16	0/12
Naloxone hydrochloride, 0.2 mg/dog (i.m.)	12/12	0/12	12/12

All antagonists were given 30 min prior to test. Saralasin was given 30 s prior to test. Numbers reflect the number of emetic responses over the number of animals tested.

not optimal for demonstration of desensitization. However, for the two substances which did show desensitization, application of both of the other substances 5 min after the first did result in emesis. This experiment provides further proof that the receptors for at least these three substances are distinct. On the basis of this study, and the effects of antagonists, we conclude that there are distinct receptors for most or all of the substances which excite area postrema neurons and cause emesis.

The next question is related to the mechanism of excitation of area postrema neurons. The fact that so many substances excite these small neurons in an unusual manner (long latency, slow discharge frequency and long duration) suggests the possibility that, even though the receptors appear to be distinct, there is some commonality in the mechanism by which the response is elicited. The most obvious possibility is that there is a common second messenger involved, such as cyclic nucleotides. Figure 21.3 shows evidence consistent with the possibility that cyclic AMP (cAMP) is a common second messenger mediating these responses. 8-Bromo-cAMP, an analogue of cAMP which crosses cell membranes with relative ease, causes an excitation which looks very similar to that elicited by the other excitatory substances. In addition, forskolin, an activator of adenylate cyclase (Daly, 1975), showed a similar excitation. Table 21.4 shows results of ionophoretic application of these substances.

In several regards the conclusion that cAMP is a second messenger for so many neuroactive substances is surprising. While cAMP increases have been reported secondary to action of a variety of these substances, the apomorphine and dopamine receptor in the area postrema has been characterized as a D2 receptor (Stefanini and Clement-Cormier, 1981), which would not be expected to be coupled to adenylate cyclase (Kebabian and Calne, 1979). In addition insulin receptors at other sites are associated with an inhibition of adenylate cyclase (Illiano and Cuatrecasos, 1972) and an activation of phosphodiesterase (Fatemi, 1985).

In order to further test the hypothesis that all of the substances which give the slow excitation act through cAMP a series of behavioural studies

Table 21.3. Effects of multiple injections of emetic substances

		5 min.		30 min.		
Substance	Control	Angio	Leu-enkep	Apo	Angio	Leu-enkep
Angiotensin II, 0.25 mg/kg	11/11	0/4	4/4	4/4	4/4	
Leu-enkephalin, 0.075 mg/kg	10/10	4/4	0/4	4/4		4/4
Apomorphine hydrochloride, 0.015 mg/kg	5/5			4/4		

The ratios show the number of dogs showing emesis over the total number of tests. At 5 and 30 min. the initial test was made with the substance listed at the left at zero time, followed by the indicated substance at 5 or 30 min. Angio, angiotensin II; Leu-enkep, leucine-enkephalin; Apo, apomorphine, all at the same concentration listed at the left.

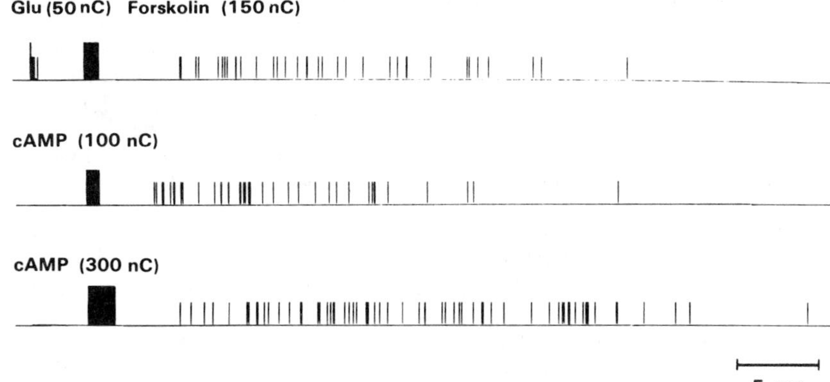

Fig. 21.3. Raster display of responses of an area postrema neuron 620 μm below the surface to glutamate (Glu), forskolin, and cAMP. The tall and/or thick black areas indicate time of ionophoresis. (Reproduced with permission from Carpenter *et al.*, 1988.)

Table 21.4. Effects of cyclic nucleotides and related substances

Substance	*No. of units*	*Percentage excitation*
8-Bromo-cAMP	48	56
cGMP	9	0
Forskolin	40	40
Theophylline	10	0

were performed in which animals were tested for emetic threshold for selected substances, and then pretreated with phosphodiesterase inhibitors and retested. If these agents act through synthesis of cAMP, systemic administration of phosphodiesterase inhibitors should result in a reduced breakdown of cAMP, and one would expect that the threshold concentration necessary for eliciting the reflex would be less than in the control. We used theophylline, IBMX and the non-methyl xanthine RP-1724. Table 21.5 shows results with theophylline on responses to insulin, angiotensin II and apomorphine. All three phosphodiesterase inhibitors caused a lowering of threshold to all substances tested (Carpenter *et al.*, 1988), consistent with the hypothesis that cAMP is a common second messenger.

Table 21.5. Effects of theophylline on emesis threshold

		Controls		*Pretreat with theophylline (25 mg/kg i.p.)*	
Substance	*Concentration*	*No. of trials*	*Percentage emesis*	*No. of trials*	*Percentage emesis*
Insulin (i.v.)	5 IU/kg	12	0	12	42
	10 IU/kg	24	46	12	66
	15 IU/kg	12	67	12	83
Angiotensin II (i.v.)	0.10 mg/kg	8	0	8	50
	0.15 mg/kg	8	0	8	75
	0.20 mg/kg	8	100	8	100
Apomorphine	0.0025 mg/kg	4	0	4	0
	0.0050 mg/kg	4	25	4	50
	0.0075 mg/kg	4	50	4	75
	0.0100 mg/kg	4	100	4	100

Effects were studied in 4 days, each before and after theophylline.

21.2. **Conclusions**

The vomiting reflex is an important one both in animals and man, having positive and negative effects. It has value to man by providing a mechanism for expulsion of harmful gastric contents, while at the same time if activated excessively it can cause severe discomfort and even death from dehydration and malnutrition. One of the problems in studying this important function is the variation in sensitivity to the reflex in different species.

While not a perfect model of human emesis, the dog is clearly most similar to man in this regard (Borison *et al.*, 1981). These studies in dog suggest that the small neurons of the area postrema have a great variety of specific receptors, but that all activate the response of the neurons by coupling to a common second messenger system. Such a system is not new, since it has been known since 1978 to occur in other cells (Tolkovsky and Levitzki, 1978). In fact such organization makes great sense for the organism, since one would want the output to be the same reflex but with many different and distinct triggers.

The fact that a great variety of small neurotransmitter substances, a number of neuropeptides and hormones and several prostaglandins all activate this pathway is an understanding of the nervous system. It seems likely that there are not clear distinctions in function or even in mechanisms of excitation among these different kinds of substances.

References

Bhargava, K. P. and Dixit, K. S. (1968). Role of the chemoreceptor trigger zone in histamine-induced emesis. *Br. J. Pharmacol.*, **34**, 508–13.

Borison, H. L. (1959). Effect of ablation of medullary emetic chemoreceptor trigger zone on vomiting responses to cerebral intraventricular injection of adrenaline, apomorphine and pilocarpine in the cat. *J. Physiol. Lond.*, **147**, 172–7.

Borison, H. L., Borison, R. and McCarthy, L. E. (1981). Phylogenic and neurologic aspects of the vomiting process. *J. Clin. Pharmacol.*, **21**, 235–95.

Carpenter, D. O. (1989). Central nervous system mechanisms in deglutition and emesis. In: Wood, J. D. (ed.), *Handbook of Physiology*, vol. IV, pp. 685–714.

Carpenter, D. O. and Briggs, D. B. (1986). Insulin excites neurons of the area postrema and causes emesis. *Neurosci. Lett.*, **68**, 85–9.

Carpenter, D. O., Briggs, D. B. and Strominger, N. (1983). Responses of neurons of canine area postrema to neurotransmitters and peptides. *Cell. Mol. Neurobiol.*, **3**, 113–26.

Carpenter, D. O., Briggs, D. B. and Strominger, N. (1984a). Peptide-induced emesis in dogs. *Behav. Brain Res.*, **11**, 277–81

Carpenter, D. O., Briggs, D. B. and Strominger, N. (1984b). Behavioral and electrophysiological studies of peptide-induced emesis in dogs. *Fed. Proc.*, **43**, 16–18.

Carpenter, D. O., Briggs, D. B., Knox, A. P. and Strominger, N. (1988). Excitation of area postrema neurons by transmitters, peptides, and cyclic nucleotides. *J. Neurophysiol.*, **59**, 358–369.

Changeux, J., Pinset, C. and Ribera, A. (1986). Effects of chlorpromazine and phencyclidine on mouse C2 acetylcholine receptor kinetics. *J. Physiol.*, **378**, 497–513.

Daly, J. W. (1975). Cyclic adenosine 3′5′-monophosphate role in the physiology and pharmacology of the central nervous system. *Biochem. Pharmacol.*, **24**, 159–64.

Douglas, W. W. (1975). Histamines and antihistamines; 5-hydroxytryptamine and antagonists. In: Goodman, L. S. and Gilman, A. (eds), *The Pharmacological Basis of Therapeutics*, 5th edn. Macmillan, New York, pp. 590–629.

Fatemi, S. H. (1985). Insulin-dependent cyclic AMP turnover in isolated rat adipocytes. *Cell. Molec. Biol.*, **31**, 153–61.

Harding, R. K. and McDonald, T. J. (1989). Identification and characterization of the emetic effects of peptide-YY. *Peptides*, **10**, 21–4.

Hatcher, R. A. and Weiss, S. (1923). Studies on vomiting, *J. Pharmacol.*, **22**, 139–93.

Illiano, G. and Cuatrecasos, P. (1972). Modulation of adenylate cyclase activity in liver and fat cell membranes by insulin. *Science, Wash., DC*, **175**, 906–8.

Innes, I. R. and Nickerson, M. (1975). Norepinephrine, epinephrine and the sympathomimetic amines. In: Goodman, L. S. and Gilman, A. (eds), *The Pharmacological Basis of Therapeutics*, 5th edn. Macmillan, New York, pp. 477–513.

Jenkins, L. C. and Lahay, D. (1971). Central mechanisms of vomiting related to catecholamine response: anesthetic implications. *Can. Anesth. Soc. J.*, **18**, 434–41.

Kaul, A. F., Federschneider, J. M. and Stubblefield, M. D. (1978). A controlled trial of antiemetics in abortion of $PGF_{2\alpha}$ and Laminaria. *J. Reprod. Med.*, **20**, 213–18.

Kebabian, J. W. and Calne, D. B. (1979). Multiple receptors for dopamine.

Nature, Lond., **277**, 93–6.

Laduron, P. M. and Leysen, J. E. (1979). Domperidone, a specific *in vitro* dopamine antagonist, devoid of *in vivo* central dopaminergic activity. *Biochem. Pharmacol.*, **28**, 2161–5.

Levey, S., Harroun, J. E. and Smyth, C. J. (1949). Serum glutamic acid levels and the occurrence of nausea and vomiting after the intravenous administration of amino acid mixtures. *J. Lab. Clin. Med.*, **34**, 1238–48.

Peng, M. T. (1963). Locus of emetic action of epinephrine and dopa in dogs. *J. Pharmacol. Exp. Ther.*, **139**, 345–9.

Rehfeld, J. F., Goltermann, N., Larsson, L.-I., Emson, P. M. and Lee, C. M. (1979). Gastrin and cholecystokinin in central and peripheral neurons. *Fed. Proc.*, **38**, 2325–9.

Stefanini, E. and Clement-Cormier, Y. (1981). Detection of dopamine receptors in the area postrema. *Eur. J. Pharmacol.*, **74**, 257–60.

Tolkovsky, A. M. and Levitzki, A. (1978). Coupling of a single adenylate cyclase to two receptors: adenosine and catecholamine. *Biochemistry*, **17**, 3811–17.

Wang, S. C. (1980). *Physiology and Pharmacology of the Brain Stem*. Futura, Mount Kisco, New York.

Wang, S. C. and Borison, H. L. (1952). A new concept of organization of the central emetic mechanism: recent studies on the sites of action of apomorphine, copper sulfate and cardiac glycosides. *Gastroenterology*, **22**, 1–12.

Wislocki, G. B. and Putnam, T. J. (1920). Note on the anatomy of the area postrema. *Anat. Rec.*, **19**, 281–7.

Regulation by neuropeptides of amino acid transmitter release in mammalian brain and retina

22.1. Introduction

γ-Aminobutyric acid (GABA), the principal inhibitory transmitter of the vertebrate CNS, is present in most CNS regions, primarily in interneurons but also in certain projection pathways. In addition, the morphological and presumed biochemical characteristics of GABAergic interneurons vary appreciably. Neurons may release GABA not only from axon terminals, but in some instance from dendrites as well. Also, certain neuropeptides, e.g. somatostatin and enkephalin, may be co-localized with GABA in subpopulations of GABAergic neurons. Glutamic acid (Glu) and probably also aspartic acid (Asp) are the principal excitatory transmitters of the vertebrate CNS. Our studies of transmitter regulation of GABA synthesis and release and of Glu release have been carried out primarily with synaptosome or synaptoneurosome preparations of calf and rat retina and rat olfactory bulb and striatum; additional studies have utilized rat frontal cortex, hippocampus, substantia nigra, hypothalamic regions and preoptic area (Makman *et al.*, 1984; Cubells *et al.*, 1986b; Fleischmann *et al.*, 1990, manuscripts in preparation). In striatum there are GABA interneurons, GABA neurons projecting to substantia nigra, as well as Glu-containing axon terminals of neurons projecting from cortex (Fonnum, 1988). The strionigral output neurons may themselves be influenced within the striatum by GABAergic neurons containing enkephalin as well as GAD immunoreactivity (Aronin *et al.*, 1986). In mammalian retina GABA is primarily or exclusively (depending on species) in amacrine cells which make dendrodendritic contacts with other neurons (Voaden, 1988; Cubells *et al.*, 1988); Glu and Asp are transmitters of photoreceptors and are also present in certain amacrines (Voaden, 1988). In olfactory bulb a major component of GABA is released at dendrodendritic synapses (Shepherd, 1988; Jaffe and Cuello, 1980; Quinn and Cagan, 1980, 1982). Much less is known about the regulation of release of excitatory amino acids than of GABA. The functional role of Gluergic neurons, and the relationship of Gluergic neurons to GABAergic and other neurons is of particular interest in view of the putative role of Glu in processes such as memory, initiation of seizure activity and neuronal damage due to anoxia and other causes.

22.2. Characteristics of GABAergic and Gluergic neurons

The studies described here utilize primarily an *in vitro* synaptosomal preparation for study of receptor- and second messenger-mediated regulation of release of [^{3}H]GABA that has been newly synthesized from [^{3}H]Glu, as well as the release of [^{3}H]Glu itself. These studies have enabled us to delineate certain regulatory properties that may be shared by GABAergic and Gluergic neurons in several different regions of the CNS. One characteristic of GABAergic neurons, probably also shared by Gluergic neurons, is that regulation of neuronal activity by receptor-signal transduction occurs primarily, if not exclusively, at the level of transmitter release. In marked contrast, regulation of catecholamine and serotonin systems is readily evident both at the level of transmitter synthesis and transmitter release (Garber and Makman, 1987).

A major consideration in our studies of release of the amino acids has been the possibility that in addition to vesicular release, functionally significant non-vesicular, possibly Ca^{2+}-independent, release also occurs. An important difference between amine release and amino acid release from slices or synaptosomes is the extent of dependence on extracellular Ca^{2+}. Numerous studies have indicated that there is a less complete dependence upon extracellular Ca^{2+} for amino acid than for amine or for ACh release (Levi, 1984; Tapia, 1983; Nicholls, 1989). For GABA as well as for other amino acid transmitters there appears to exist a non-vesicular as well as a vesicular pool (Tapia, 1983; Wood *et al.*, 1988; Dagani and Erecinskia, 1987; Miller, 1988). While depolarization may cause GABA and Glu release independent of extracellular Ca^{2+}, the actual mechanism for this release is not known; moreover, evidence fails to clearly support either a carrier (Na^+-independent) release or release related to mobilization of intracellular Ca^{2+} (Tapia, 1983). The Na^+- and Cl^--coupled GABA transporter has been purified from rat brain to homogeneity, reconstituted functionally into liposomes and many of its properties investigated (Radian *et al.*, 1986; Kanner *et al.*, 1987). A novel Na^+-dependent exchange mechanism has been proposed to account for Glu-evoked release of GABA from cultured avian retinal cells (de Mello *et al.*, 1988). We propose that non-vesicular release of GABA may be an important mechanism for dendritic release. Also, the studies described here support the possibility that this type of release may be transmitter-regulated and in some, but not all, instances (Schwartz, 1982; de Mello *et al.*, 1988) require mobilization of intracellular Ca^{2+}.

22.3. Regulation of GABA release by neuropeptides and by dopamine

Neurons and synaptosomes may be distinguished not only on the basis of co-transmitter but also by the particular set of receptors and signal transduction systems regulating transmitter synthesis and release. GABAergic

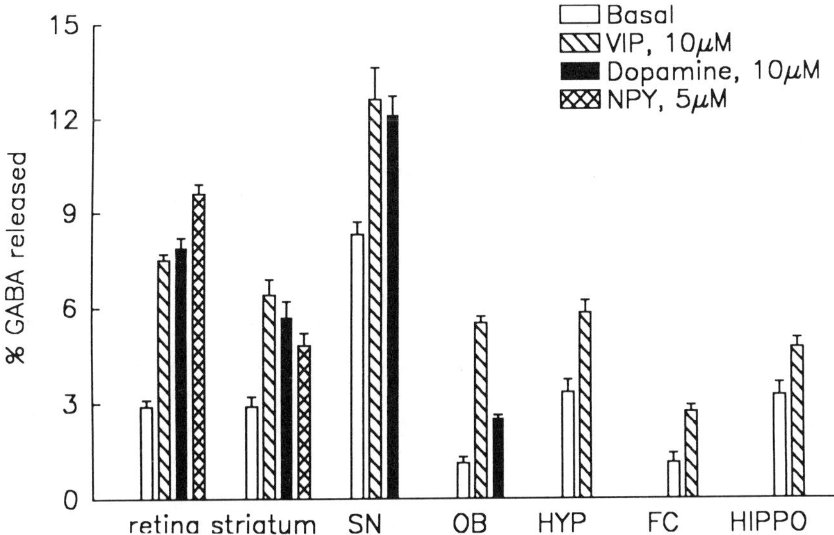

Fig. 22.1. Influence of VIP, NPY and dopamine (DA) on release of newly synthesized GABA from rat CNS synaptosomes. SN: substantia nigra; OB: olfactory bulb; HYP: medial hypothalamus; FC: frontal cortex; HIPPO: hippocampus. Tissues were homogenized directly in Krebs-bicarbonate buffer with glucose (KRB) for preparation of 'synaptoneurosomes' (Garber and Makman, 1987). Synaptosomes were incubated at 37°C for 30 min with 1 μM Glu plus 2–10 μCi/ml 2,3-[^{3}H]GLU. The percentage of intrasynaptosomal labelled glutamate converted to GABA under these conditions generally ranged from 5% to 12% but exceeded 20% in synaptosomes derived from carefully dissected SN. The labelled synaptosomes were then washed by repeated centrifugation and resuspended in KRB. The final pellet was divided into 200 μl aliquots and reincubated at 30°C for 10 min, with or without various agents as indicated. Radioactivity in the medium of samples kept on ice for the same time period was also determined for subsequent subtraction from values for release at 30°C. Following incubation, samples were chilled in an ice bath and rapidly centrifuged to separate synaptosomes from medium. Boiled and pH-adjusted synaptosomal extracts and medium fractions were then applied to columns of Bio-Rad AG50WX8 (200–400 mesh) resin (bed volume 0.8 ml) previously equilibrated with 0.1 M citrate buffer, pH 4.9. After washing columns with 0.1 M citrate buffer to elute [^{3}H]glutamate and deaminated metabolites, [^{3}H]GABA was subsequently eluted with 0.1 M citrate at pH 5.5. Values represent percentage of newly synthesized synaptosomal [^{3}H]GABA released into the medium and are means ± SEM (n = 5–15); VIP, NPY and DA produced significant stimulation in all regions studied.

nerve terminals contain not only presynaptic receptors for GABA itself, but for other receptors as well. Our studies indicate that in several CNS regions two distinct families of neuropeptides, that which includes vasoactive intestinal peptide (VIP) and that which includes neuropeptide Y (NPY) (Polok and Bloom, 1984), produce receptor-mediated stimulation of the release of GABA and of Glu. Dopamine is also stimulatory but exhibits a somewhat different regional pattern of stimulation. In striatum, dopamine D_1 receptors may be located primariliy on GABAergic neurons, whereas D_2 receptors may be primarily on cholinergic neurons. Both GABAergic striatal interneurons and GABAergic strionigral neurons appear to possess D_1 receptors, based on our own studies, described below, as well as those of others (Nester and Greengard, 1989; Hemmings *et al.*, 1987).

These studies began as part of an attempt to localize further and characterize VIP receptors, VIP-stimulated adenylate cyclase and VIP action. VIP and structurally related peptides were found to stimulate the release of newly synthesized [^{3}H]GABA from rat and calf retinal synaptosomes, as well as from synaptosomes derived from rat striatum, olfactory bulb, and the several other CNS regions investigated (Fig. 22.1 and 22.2). As little as 10 nM VIP caused significant stimulation of GABA release from retinal synaptosomes (Fig. 22.2). Stimulation was most marked in retina, striatum and olfactory bulb. Relative potencies in bovine retina (Fig. 22.2) and rat striatum were VIP ≫ PHI > secretin ≫

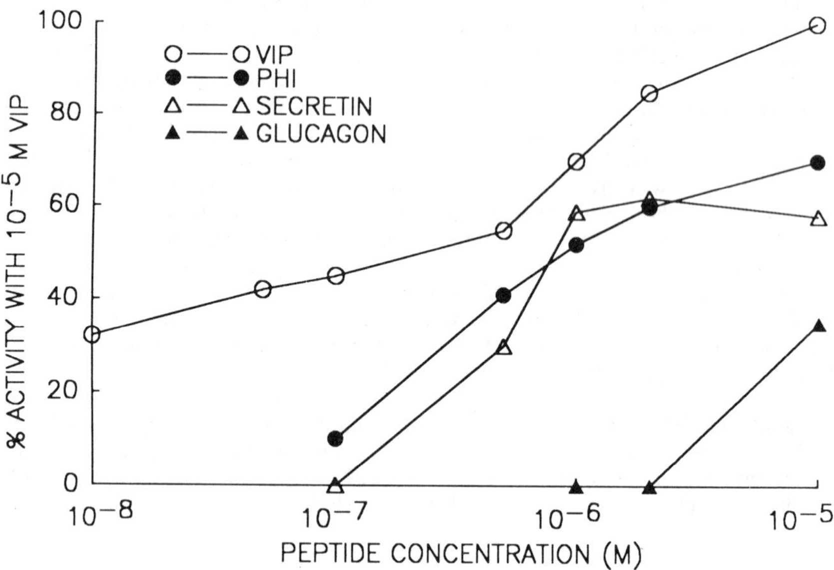

Fig. 22.2. Stimulation of release of newly synthesized [^{3}H]GABA from calf retinal synaptosomes by VIP and related neuropeptides: concentration dependence. Conditions were as described in Fig. 22.1. Values represent the percentage of the stimulation over basal activity produced by 10 μM VIP (= 100%). Stimulation by 10 μM VIP averaged 7.6-fold. Values are means of four to seven separate experiments.

glucagon; secretin and PHI were slightly more potent than VIP in rat olfactory bulb (Smith *et al.*, 1985). In contrast, VIP did not influence the synthesis or release of dopamine in retina and striatum, or the synthesis or release of serotonin in striatum (data not shown). PHI is a peptide with strong homology to VIP and contained together with VIP in the prepro-VIP molecule. Cortex in particular, and also other CNS regions (including retina), contain VIP and PHI (Brecha and Karten, 1983). Striatum, retina, cortex and hippocampus are all rich in VIP receptors and in VIP-stimulated adenylate cyclase activity (Deschodt-Lanckman *et al.*, 1977; Longshore and Makman, 1981).

Neuropeptide Y (NPY) and the structurally related avian pancreatic polypeptide (APP) also markedly stimulated release of GABA from striatal and retinal synaptosomes (Fig. 22.). FMRFamide, with sequence related to the terminal sequence of NPY, was much less effective. NPY itself is known to be present in relatively high concentration in striatum, and also to be present in retina (Polok and Bloom, 1984; Bruun *et al.*, 1984). A number of other peptides, including somatostatin, substance P, neurotensin and enkephalin, did not influence GABA release in retina (Fig. 22.3), striatum, olfactory bulb or cortex. The effects of VIP and

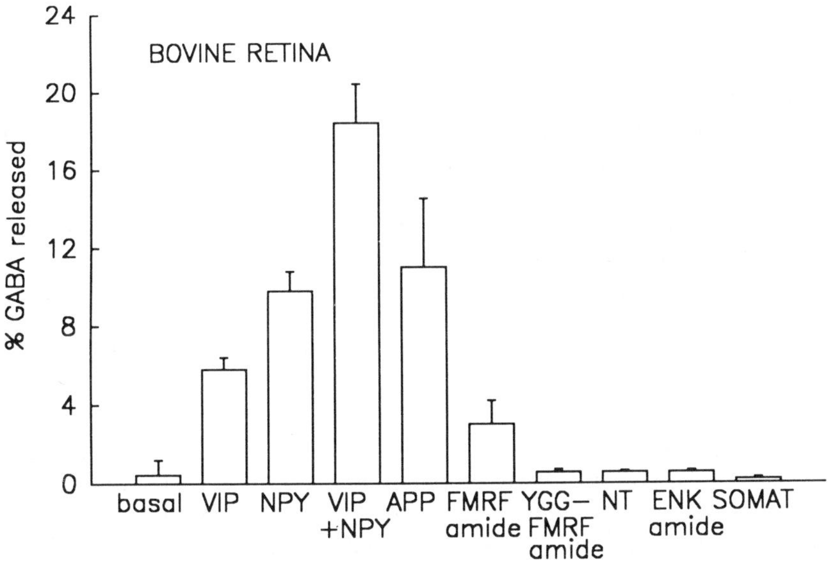

Fig. 22.3. Influence of NPY and other neuropeptides on release of newly synthesized [^{3}H]GABA from bovine retinal synaptosomes. APP: avian pancreatic peptide; NT : neurotensin; ENKamide: met-enkephalinamide; SOMAT: somatostatin. Each peptide was present at 5 μM. Other conditions were as described in Fig. 22.1. Values are means ± SEM (n = 4–7).

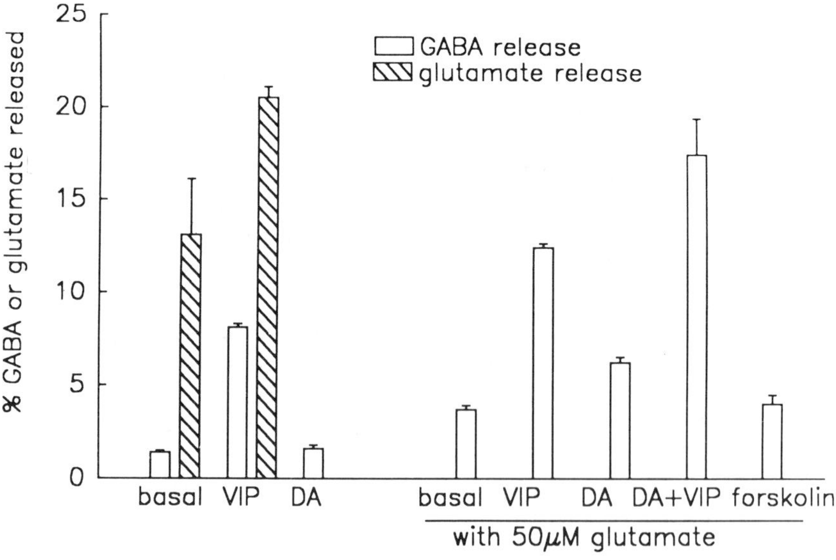

Fig. 22.4. Regulation of release of newly synthesized GABA and of glutamate (Glu) from bovine retinal synaptosomes. Synaptosomes were incubated either without or with 50 μM Glu, and with 10 μM dopamine (DA) or 10 μM forskolin as indicated during the release incubation. Other conditions were as in Fig. 22.1. Values are means ± SEM (n = 7–12).

NPY at 5 or 10 μM were additive (Fig. 22.3), suggesting that these two peptides act at distinct receptor--second messenger systems and/or on separate subpopulations of synaptosomes.

Also, dopamine was effective in rat retina but had no effect in calf retina unless Glu was also present at a threshold stimulatory concentration (VIP and/or NPY did not require the presence of Glu in order to stimulate release in calf retina). The effects of dopamine and VIP on retinal GABA release were found to be additive (Fig. 22.4), suggesting separate mechanisms of action and/or effects on separate subpopulations of synaptosomes.

Basal release during the standard 10 min incubation period was generally less than 3% of the newly synthesized synaptosomal GABA (except for substantia nigra), whereas release in the presence of stimulatory transmitters or neuromodulators at maximally effective concentrations was often more than 10% of synaptosomal GABA. Generally from 20% to 40% of the newly synthesized GABA was released during the standard 10 min incubation with veratridine, and KCl-induced release of GABA was generally less than that caused by veratridine (see Fig. 22.8 for representative data for calf retinal synaptosomes).

Dopamine also stimulated release of newly synthesized GABA from synaptosomes prepared from several regions, although the magnitude of the effect of dopamine did not parallel that of VIP (Figs 22.1 and 22.4).

22.4. **Stimulation of Glu release by VIP**

VIP stimulated release of Glu as well as GABA from synaptosomes prepared from retina (Fig. 22.4) and from olfactory bulb. Also, in other studies involving retinal synaptosomes doubly labelled with [^{14}C]Glu and [^{3}H]GABA, VIP caused comparable release of Glu and GABA from calf retinal synaptosomes (data not shown).

22.5. **Stimulation of somatostatin release by VIP, NPY and dopamine**

In addition to their effects on GABA and Glu release, VIP and NPY (Fig. 22.5) but not glucagon at 10 μM stimulated the release of somatostatin-like immunoreactivity (SL-IR) from striatal and retinal synaptosomes. Dopamine stimulated SL-IR release from striatal synaptosomes, but not from calf retinal synaptosomes (Fig. 22.5). It may be noted that dopamine was tested in this retinal preparation only under conditions in which GABA release also was not stimulated by dopamine. VIP was also

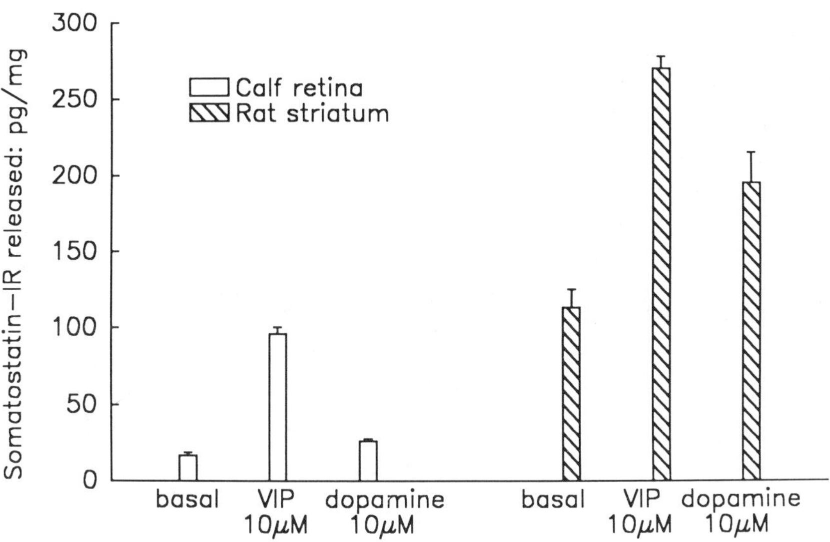

Fig. 22.5. Influence of VIP, NPY and dopamine on release of somatostatin-like immunoreactivity (SL-IR) from calf retinal and rat striatal synaptosomes. Washed (unlabelled) synaptosomes were incubated at 30°C for 10 min in the presence of VIP, NPY or dopamine as indicated. SL-IR in the medium and synaptosomal pellets was measured by radioimmunoassay as previously described (Thal *et al.*, 1983). Values for pmol SL-IR/mg protein are means ± SEM (n = 4).

effective with synaptosomes from olfactory bulb (Smith *et al.*, 1985) and frontal cortex (NPY not tested), and we have previously shown that dopamine stimulates SL-IR release from slices of frontal cortex (Thal *et al.*, 1986). It may be that GABA, somatostatin and possibly also glutamate are co-released, at least in part, from the same synaptosomes in response to these stimulatory agents. This question is not yet resolved (see discussion below).

22.6. Uptake of GABA, uptake of Glu and synthesis of GABA

VIP and NPY did not influence uptake of Glu (Fig. 22.6), synthesis of GABA from Glu (not shown) or uptake of GABA (Fig. 22.6). Thus, the neuropeptides act in a selective fashion in that they influence release but not uptake. This is an important point, since mechanisms for amino acid release might possibly involve the uptake carrier.

GABA uptake or reuptake has been further investigated with the use of selective inhibitors of GABA uptake: 1-(4,4-diphenyl-3-butenyl)-3-

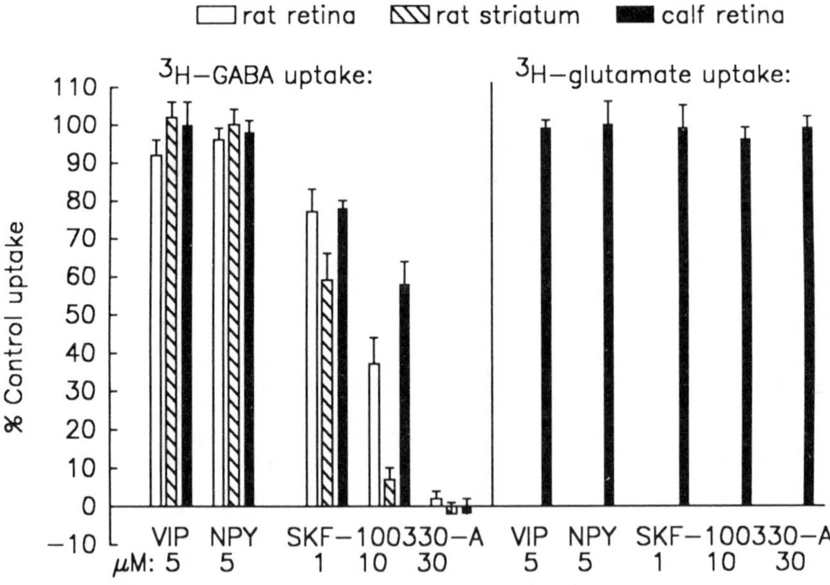

Fig. 22.6. Synaptosomal uptake of [^{3}H]GABA and [^{14}C]glutamate (Glu): influence of VIP, NPY and SKF100330-A. Synaptosomes were incubated with [^{3}H]GABA and [^{14}C]Glu (final concentrations of each 1 μM) for 30 min at 37°C with agents as indicated; aliquots of synaptosomes were also kept on ice for later subtraction of nonspecific binding or uptake. Incubations were terminated by chilling the samples and collecting the doubly labelled synaptosomes on glass-fibre filters. Values are means ± SEM (n = 3).

piperidine-carboxylic acid (SKF 89976-A) and 4-(3-(1,2,5,6-tetrahydro)-carboxypyridino-1,1-diphenyl-1-butene) HCl (SKF 100330-A) (Yunger *et al.*, 1983). These compounds prevent uptake of labeled GABA but not Glu into retinal and striatal synaptosomes (data for SKF 100330-A shown in Fig. 22.6). Experiments were also carried out using retinal synaptosomes which had been labelled with [^{3}H] Glu either in the usual manner or with 30 μM SKF 100330-A, a concentration that completely blocked GABA uptake, also present (Fig. 22.7). Normally during the 30 min incubation with labelled Glu, about 30% of the labelled GABA formed is released into the medium and therefore removed when the synaptosomes are washed prior to the final incubation for study of release. Incubation of synaptosomes with SKF 100330-A during this period of GABA synthesis increased the fraction of GABA found in the medium to about 55%, presumably by preventing GABA reuptake. When SKF 100330-A was present during the initial incubation for GABA synthesis, the percentage of GABA remaining in the synaptosomes that is then released under basal conditions was found to be appreciably reduced (Fig. 22.7). However, the stimulatory effects of NPY and VIP on release of GABA from retinal synaptosomes were still readily manifest in the synaptosomes

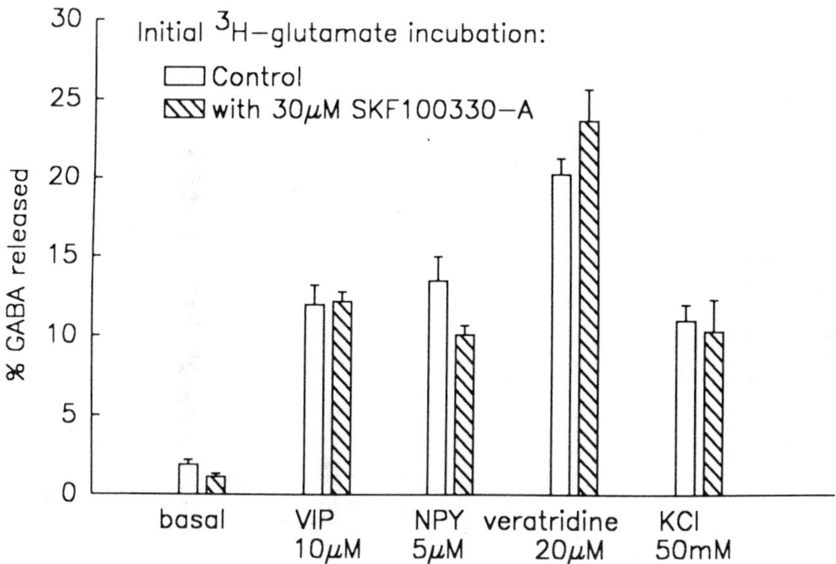

Fig. 22.7. Influence of SKF 100330-A, a selective inhibitor of GABA uptake, on responsiveness of bovine retinal synaptosomes to VIP and NPY. Synaptosomes were incubated without or with SKF 100330-A during the standard 30 min at 37°C period of labelling with [^{3}H]Glu. Synaptosomes were then washed and reincubated without SKF 100330-A for study of release. Other conditions were as in Fig. 22.1. Values are means ± SEM (n = 3).

pretreated with SKF 100330-A. Since the synthesis of [^{3}H]GABA in the presence of SKF 100330-A precludes release and synaptosomal reuptake during the period of synthesis, the effects of VIP and NPY on release then most assuredly must involve release from GABAergic nerve endings. Furthermore, on the basis of these results it seems likely that if the GABA carrier were involved in the effects of VIP and NPY, it would be involved only indirectly or passively. Nevertheless, these studies do not entirely rule out participation of the GABA carrier in the effects of VIP and NPY.

22.7. **Role of Na^+ channels and external Ca^{2+}**

Blockade of Na^+ conductance by tetrodotoxin nearly abolished the potent effect of veratridine on retinal GABA release; however tetrodotoxin had no influence on the effects of VIP or NPY (Fig. 22.8). Furthermore *external* Ca^{2+} did not appear to be required for stimulation of GABA release by VIP or NPY, since stimulation was fully manifest when Ca^{2+} in the release medium was replaced by Co^{2+} (Fig. 22.8). In contrast, replacement of Ca^{2+} by Co^{2+} partly blocked the stimulatory effect of 50 mM KCl.

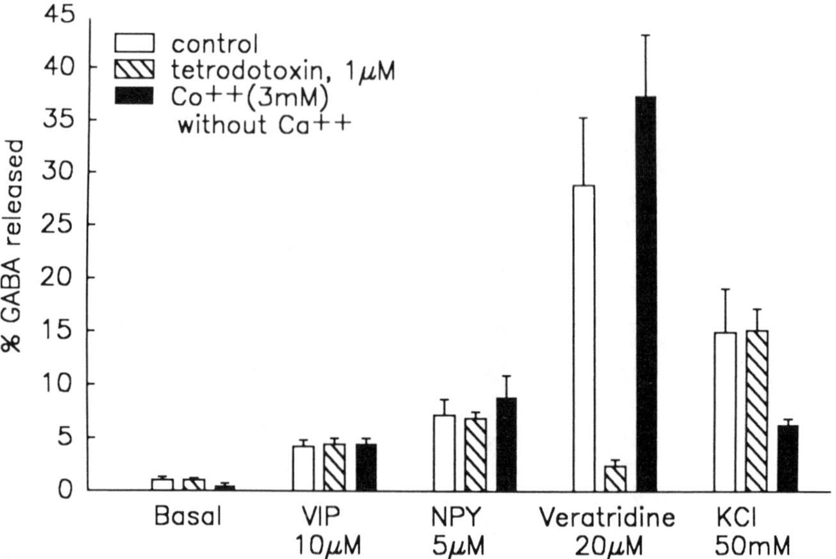

Fig. 22.8. Influence of tetrodotoxin and of replacement of Ca^{2+} with Co^{2+} on the responsiveness of bovine retinal synaptosomes. Conditions were as described in Fig. 22.1. Release of newly synthesized [^{3}H]GABA was carried out in the presence of agents as indicated. Values are means ± SEM (n = 4).

22.8. **Role of cyclic AMP; influence of other receptor agonists on the response to VIP**

VIP is a potent stimulator of adenylate cyclase activity, as mentioned above. VIP and also forskolin stimulate cyclic AMP (cAMP) formation in retinal and striatal synaptosomes (data not shown). However, cAMP did not appear to mediate the effect of VIP on retinal GABA release, since cAMP analogues and forskolin were without effect (Fig. 22.9). Preliminary experiments indicate that this finding also applies to striatum. In contrast, both forskolin and cAMP analogues are capable of activating tyrosine hydroxylase activity of striatal and retinal synaptosomes (Katz *et al.*, 1982; Garber and Makman, 1987). Several agents that might have the capability of antagonizing the effect of VIP on GABA release via retinal receptors coupled to and inhibitory to adenylate cyclase, or coupled to other second-messenger systems, were also tested and found to be inactive. Included were somatostatin (data not shown), dopamine (at D_2 receptors), carbachol (at muscarinic receptors) and $N^2$6(2-phenylisopropyl)adenosine (at A_1 receptors) (Fig. 22.9).

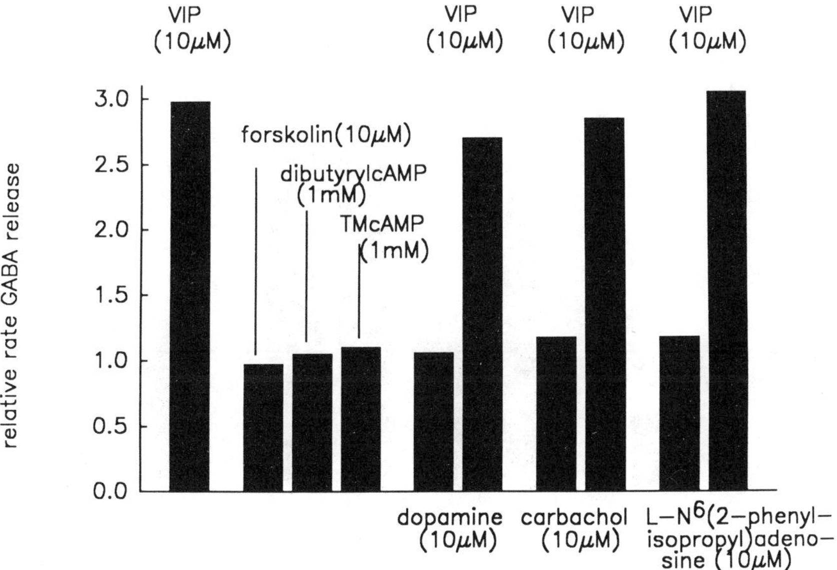

Fig. 22.9. Influence of cAMP analogues, forskolin and selected receptor agonists on the responsiveness of calf retinal synaptosomes. Conditions were as described in Fig. 22.1. Release of newly synthesized [^{3}H]GABA was carried out in the presence of agents as indicated. Values are expressed as relative rate of GABA released (ratio of experimental to basal release) and are means ± SEM (n = 5).

22.9. Role of other second-messenger systems and of protein phosphorylation

More recently we have obtained evidence to suggest that the effect of VIP on GABA release in retina and striatum is mediated by the phosphatidyl inositol (PI) system and protein kinase C (manuscript in preparation). Thus, the effect of VIP in GABA release is blocked by an inhibitor of protein kinase C (1-(5-isoquinolinyl-sulphonyl)-2-methylpiperazine (H7)) (Hikada *et al.*, 1984). In contrast, inhibitors of calmodulin-dependent processes (N-(6-aminohexyl)-5-chloro-1-naphthalene sulphonamide (W7) and calmidazolium) (Gietzen *et al.*, 1981; Mills *et al.*, 1985; Schatzman *et al.*, 1983) failed to block the VIP effect (Fig. 22.10). Furthermore, GABA release is stimulated by the phorbol ester, 12-*o*-tetradecanoylphorbol-13-acetate (TPA), and that stimulation is also blocked by H7 (Fig. 22.10). TPA is known to activate protein kinase C. Although H7 may also inhibit cyclic AMP-dependent protein kinase (protein kinase A) at the 10 μM concentration used, the other studies referred to above argue against cAMP mediation of GABA release. The lack of influence of W7 and calmidazolium on the VIP effect indicates

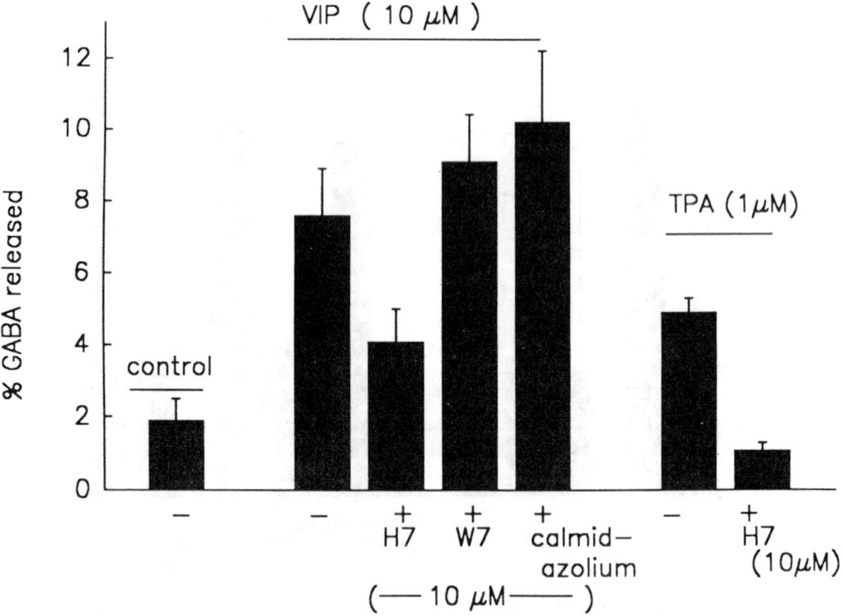

Fig. 22.10. Influence of protein kinase inhibitors and of 12-*o*-tetradecanoylphorbol-13-acetate (TPA) on the responsiveness of calf retinal synaptosomes. Conditions were as described in Fig. 22.1. Release of newly synthesized [^{3}H]GABA was carried out in the presence of agents as indicated. Values are means ± SEM (n = 3).

that calmodulin-activated processes are also not involved. It may be noted that we find other synaptosomal processes, such as serotonin and dopamine synthesis and release and also intrasynaptosomal [Ca^{2+}] (Makman *et al.*, 1989), to be markedly influenced by W7 and calmidazolium.

22.10. **Conclusions: relationships of the present findings to other transmitter systems and future direction**

VIP appears to influence release of GABA from nerve endings in many regions of the CNS, most likely by a process that involves the PI system and activation of protein kinase C. VIP has been found to increase inositol phospholipid breakdown in the rat superior cervical ganglion (Audigier *et al.*, 1986), although such an effect of VIP has not yet been demonstrated for brain or retina. The related protein, glucagon, has been found to activate both the adenylate cyclase and the PI systems in hepatocytes (Wakelam *et al.*, 1986). The same nerve endings that release GABA in response to VIP in all probability contain a VIP-stimulated adenylate cyclase. It remains possible that VIP action *in vitro* does not involve cAMP, because cAMP levels are already sufficiently high in the synaptosomes, whereas *in vivo* both cAMP and the PI system may be involved and facilitate one another in the process of GABA release. Use of selective inhibitors of cAMP-dependent protein kinase in the *in vitro* studies may help resolve this issue.

The role of Ca^{2+} in the action of VIP also requires further analysis. While extracellular Ca^{2+} is apparently not involved, mobilization of intracellular Ca^{2+} following activation of the PI system may possibly be a necessary component of the VIP effect. VIP does not appear to act, however, by causing Ca^{2+} entry via depolarization.

It seems likely that at least some nerve endings release both Glu and GABA. Whether or not VIP stimulates release of other amino acids in addition to Glu and GABA is not known. HPLC techniques may be useful in this regard to separate and analyse endogenous amino acids released from synaptosomes in response to VIP and other agents. In previous studies gabaculine and other irreversible inhibitors of GABA transaminase, administered subcutaneously *in vivo* to rats, were found to potently and almost completely inactivate GABA transaminase and to markedly increase GABA concentration in the retina (Cubells *et al.*, 1986a, 1987, 1988). We now find that gabaculine treatment *in vivo* results in nearly 10-fold increase in the endogenous GABA content of retinal synaptosomes (unpublished studies), making much more feasible the investigation of mechanisms for release of endogenous GABA from these synaptosomes. Co-release of Glu together with GABA has been reported to occur from purified cortical GABAergic synaptosomes (Docherty *et al.*, 1987). Those investigators had separated GABAergic from other synaptosomes using an antibody to glutamic acid decarboxylase that linked GABAergic synaptosomes to magnetic microspheres coupled to protein A. We are currently exploring a similar approach for purification of GABAergic and Gluergic synaptosomes, in order to examine further the influence of neuropeptides and dopamine on release of amino acids.

The widespread occurrence of Glu and the ready formation of GABA in certain neurons, together with the sensitivity of most neurons in many species to one or both of these acids, provide a very basic system for neuronal connectivity. While the present studies are concerned with the mammalian CNS and with VIP and NPY, it seems likely that similar mechanisms, involving different neuropeptides (possibly FMRFamide), but possibly also dopamine, are involved in regulation of amino acid transmitters in other vertebrates and in invertebrates.

References

Aronin, N., Chase, K. and Difiglia, M. (1986). Glutamic acid decarboxylase and enkephalin immunoreactive axon terminals in the rat neostriatum synapse with strionigral neurons. Brain Res., **365**, 151–8.

Audigier, S., Barberis, C. and Jard, S. (1986). Vasoactive intestinal polypeptide increases inositol phospholipid breakdown in the rat superior cervical ganglion. *Brain Res.*, **376**, 363–7.

Brecha, N. C. and Karten, H. J. (1983). Identification and localization of neuropeptides in the vertebrate retina. In: Krieger, D. T., Brownstein, M. J. and Martin, J. B. (eds), *Brain Peptides*. John Wiley, New York, pp. 437–62.

Bruun, A., Ehinger, B., Sundler, F., Tornqvist, K. and Uddman, R. (1984). Neuropeptide Y immunoreactive neurons in the guinea pig uvea and retina. *Invest. Ophthalmol. Vis. Sci.*, **25**, 1113–23.

Cubells, J. F., Blanchard, J. S., Smith, D. M., and Makman, M. H. (1986a). *In vivo* action of enzyme-activated irreversible inhibitors of glutamic acid decarboxylase and γ-aminobutyric acid transaminase in retina *vs.* brain. *J. Pharmacol. Exp. Ther.*, **238**, 508–14.

Cubells, J. F., Smith, D. M., Horowitz, S. G. and Makman, M. H. (1986b). Neuropeptide-Y stimulates the release of GABA and somatostatin-like immunoreactivity from bovine retinal synaptosomes. *Soc. Neurosci. Abstr.*, **12**, 33.

Cubells, J. F., Blanchard, J. S. and Makman, M. H. (1987). The effects of *in vivo* inactivation of GABA-transaminase and glutamic acid decarboxylase on level of GABA in the rat retina. *Brain Res.* **419**, 208–15.

Cubells, J. F., Walkley, S. U. and Makman, M. H. (1988). The effects of gabaculine *in vivo* on the distribution of GABA-like immunoreactivity in the rat retina. *Brain Res.*, **458**, 82–90

Dagani, F. and Erecinska, M. (1987). Relationships among ATP synthesis, K^+ gradients and neurotransmitter amino acid levels in isolated rat brain synaptosomes. *J. Neurochem.*, **49**, 1229–40.

de Mello, M. C. F., Klein, W. L. and de Mello, F. G. (1988). L-Glutamate evoked release of GABA from cultured avian retina cells does not require glutamate receptor activation. *Brain Res.*, **443**, 166–72.

Deschodt-Lanckman, M., Robberecht, P. and Christophe, J. P. (1977). Characterization of VIP-sensitive adenylate cyclase in guinea pig brain. *FEBS Lett.*, **83**, 76–80.

Docherty, M., Bradford, H. F. and Wu, J.-Y. (1987). Co-release of glutamate and aspartate from cholinergic and GABAergic synaptosomes. *Nature*, **330**, 64–6

Fleischmann, A., Makman, M. H. and Etgen, A. M. (1990). Ovarian steroids increase veratridine-induced release of amino acid neurotransmitters in preoptic area synaptosomes. *Brain Res.*, **507**, 161–3.

Fonnum, F. (1988), Transmitter glutamate in mammalian hippocampus and striatum. In: Kuamme, E. (ed.), *Glutamine and Glutamate in Mammals*, vol. 1. CRC Press, Boca Raton, FL, pp. 51–69.

Garber, S. L. and Makman, M. H. (1987). Regulation of tryptophan hydroxylase activity by a cyclic AMP-dependent mechanism in rat striatum. *Molec. Brain Res.*, **3**, 10–20

Gietzen, K., Wüthrich, A. and Bader, H. (1981). R-24571: A new powerful inhibitor of red blood cell Ca^{++}-transport ATPase and of calmodulin-regulated functions. *Biochem. Biophys. Res. Commun.*, **101**, 418–25.

Hemmings, H. C., Jr, Nestler, E. J., Waalaas, I., Ouimet, C. C. and Greengard, P. (1987). Protein phosphorylation and neuronal function: DARPP-32, an illustrative example. In Edelman, G. M., Gall, W. E. and Cowan, W. M. (eds), *Synaptic Function*. Wiley, New York, pp. 213–40.

Hikada, H., Inagaki, M., Kawamoto, S. and Sasaki, Y. (1984). Isoquinoline sulfonamides, novel and potent inhibitors of cyclic nucleotide dependent protein kinase and protein kinase C. *Biochemistry*, **23**, 5036–41.

Jaffe, E. H. and Cuello, A. C. (1980). Release of γ-aminobutyrate from the external plexiform layer of the olfactory bulb: possible dendritic involvement. *Neuroscience*, **5**, 1859–69.

Kanner, B. I., Schuldiner, S. and Rudnick, G. (1987). Mechanism of transport and storage of neurotransmitters. *CRC Crit. Rev. Biochem.*, **22**, 1–38.

Katz, I. R., Smith, D. and Makman, M. H. (1982). Forskolin stimulates the conversion of tyrosine to dopamine in catecholaminergic neural tissue. *Brain Res.*, **264**, 173–7.

Levi, G. (1984). Release of putative transmitter amino acids. In: Lajtha, A. (ed.), *Handbook of Neurochemistry*, vol. 6, 2nd edn. Plenum Press, New York, pp. 463–509.

Longshore, M, and Makman, M. H. (1981). Stimulation of retinal adenylate cyclase by vasoactive intestinal peptide (VIP). *Eur. J. Pharmacol.*, **70**, 237–40.

Makman, M. H., Cubells, J. F., Smith, D. and Dvorkin, B. (1984). Vasoactive intestinal peptide stimulates release of GABA and glutamate from mammalian retinal synaptosomes. *Soc. Neurosci. Abstr.*, **10**, 837.

Makman, M. H., Garber, S. L. and Dvorkin, B. (1989). Inhibitors of calmodulin and protein kinase C influence intrasynaptosomal calcium level and transmitter release. *Soc. Neurosci. Abstr.* **15**, 1003.

Miller, R. F. (1988). Are single retinal neurons both excitatory and inhibitory? *Nature*, **366**, 517–18.

Mills, J. S., Bailey, B. L. and Johnson, J. D. (1985). Cooperativity among calmodulin's drug binding sites. *Biochemistry*, **24**, 4897–902.

Nester, E. J. and Greengard, P. (1989). Protein phosphorylation and the regulation of neuronal function. In: Siegel, G. J. *et al.* (eds), *Basic Neurochemistry: Molecular, Cellular and Medical Aspects*, 4th edn. Raven Press, New York, pp. 373–98

Nicholls, D. G. (1989). Release of glutamate, aspartate and γ-aminobutyric acid from isolated nerve terminals. *J. Neurochem.*, **52**, 331–41.

Polok, J. M. and Bloom, S. R. (1984). Regulatory peptides – the distribution of two newly discovered peptides: PHI and NPY. *Regulat. Peptides*, **5** (Suppl. 1), 79–89.

Quinn, M. R. and Cagan, R. H. (1980). Subcellular distribution of glutamate decarboxylase in rat olfactory bulb: high content in dendrodendritic synaptosomes. *J. Neurochem.*, **35**, 583–90.

Quinn, M. R. and Cagan, R. H. (1982). High specific binding of [^{3}H]GABA and [^{3}H]muscimol to membranes from dendrodendritic synaptosomes of the rat olfactory bulb. *J. Neurochem.*, **39**, 1381–6.

Radian, R., Bendahan, A. and Kanner, B. I. (1986). Purification and identification of the functional sodium- and chloride-coupled γ-aminobutyric acid transport glycoprotein from rat brain. *J. Biol. Chem.*, **261**, 15437–41.

Schatzman, R. C., Raynor, R. L. and Kuo, J. F. (1983). N-(6-Aminohexyl)-5-chloro-1-naphthalenesulfonamide (W-7), a calmodulin antagonist, also inhibits phospholipid-sensitive calcium-dependent protein kinase. *Biochim. Biophys. Acta*, **755**, 144–7.

Schwartz, E. A. (1982). Calcium-independent release of GABA from isolated horizontal cells of the toad retina. *J. Physiol.*, **323**, 211–27.

Shepherd, G. M. (1988) *Neurobiology*, 2nd edn. Oxford University Press, New York.

Smith, D. M., Horowitz, S. G., Cubells, J. F. and Makman, M. H. (1985). Regulation of GABA and somatostatin release in retina and olfactory bulb by vasoactive intestinal peptide (VIP) and dopamine. *Abstr. Soc. Neurosci.*, **11**, 242.

Tapia, R. (1983) γ-Aminobutyric acid. Metabolism and biochemistry of synaptic transmission. In: Lajtha, A. (ed.), *Handbook of Neurochemistry*, vol. 3, 2nd edn. Plenum Press, New York, pp. 423–466.

Thal, L. J., Sharpless, N. S., Hirschhorn, I. D., Horowitz, S. G. and Makman, M. H. (1983). Striatal met-enkephalin concentration increases following nigrostriatal denervation. *Biochem. Pharmacol.*, **32**, 3297–301.

Thal, L. J., Laing, K., Horowitz, S. G. and Makman, M. H. (1986). Dopamine stimulates rat cortical somatostatin release. *Brain Res.*, **372**, 205–9.

Voaden, M. J. (1988). Glutamine and its neuroactive derivatives in the retina. In: Kvamme, E. (ed.), *Glutamine and glutamate in Mammals*, vol. II. CRC Press, Boca Raton, FL, pp. 71–88.

Wakelam, M. J. O., Murphy, G. J., Hruby, V. J. and Houslay, M. D. (1986). Activation of two signal-transduction systems in hepatocytes by glucagon. *Nature*, **323**, 68–71.

Wood, J. D., Kurylo, E. and Lane, R. (1988). γ-Aminobutyric acid release from synaptosomes prepared from rats treated with isonicotinic acid hydrazide and gabaculine. *J. Neurochem.*, **50**, 1839–43.

Yunger, L. M., Lafferty, J. J., Rush, J. A. and Lester, B. R. (1983). Anticonvulsant actions of a novel series of GABA uptake inhibitors. *Neurosci. Abstr.*, **9**, 1040.

23 *Silvia De Biasi and Laura Vitellaro–Zuccarello*

Comparative aspects of the submicroscopic immunocytochemical localization of neuropeptides

23.1. **Introduction**

The increasing number of neuropeptides identified in both vertebrate and invertebrate nervous tissues poses the problem of correlating the biochemical and functional characteristics of different neuronal populations with their morphological features. In this regard the recent developments of immunocytochemistry offer a powerful tool to correlate ultrastructure and expression of different neurotransmitters. In particular, the colloidal gold immunostaining method (Varndell and Polak, 1984) is especially useful as it enables the identification and submicroscopic localization of immunoreactive substances without masking the morphological details of the labelled elements. Moreover, the method is particularly suitable for study of neurochemical interactions at different types of synapses as it permits multiple immunolabelling of the same preparation and can be combined with other neuroanatomical procedures, such as tract tracing (De Biasi and Rustioni, 1988).

In our laboratory we have previously applied the immunogold staining method to the study of the synaptic circuitry of aminoacidergic systems in different areas of the rat somatosensory system (De Biasi *et al.*, 1986a; De Biasi and Rustioni, 1988; Spreafico *et al.*, 1988) and to the investigation of the subcellular localization of serotonin and GABA in the pedal ganglion of a bivalve mollusc, *Mytilus galloprovincialis* (Vitellaro-Zuccarello *et al.*, 1988; Vitellaro-Zuccarello and De Biasi, 1988a). We here report the application of the immunogold staining method to investigate the submicroscopic localization of two neuropeptides, substance P (SP) and cholecystokinin (CCK), in the superficial laminae of the rat spinal cord and in the pedal ganglion of *Mytilus*. The procedures employed are described in detail in the previous publications.

For the immunolocalization of SP, two different antisera were used: a polyclonal antiserum raised in rabbit (gift from Dr P. Petrusz) and a rat monoclonal antibody (from Sera Lab) directed against the carboxy-terminal portion of vertebrate SP, common to all members of the tachykinin family. For the immunolocalization of CCK we have used an antiserum (from Immuno Nuclear Corp.) directed against the unsulphated carboxy-terminal octapeptide common to CCK and gastrin, which represents the active part of the molecule in mammals.

23.2. **Immunolocalization of peptides in the rat dorsal horn**

23.2.1. *SP-like immunoreactivity (SP-LI)*

In the superficial laminae (I and II) of the rat spinal cord, with both anti-SP sera, labelling is exclusively found over dense core vesicles (DCV) with a diameter of 80–120 nm ($\bar{x}$ = 100) (Figs 23.1 and 23.2). Labelled DCV are present in a few neuronal cell bodies and small unmyelinated or thinly myelinated axons, and in many axonal boutons. Neurons typically contain very few labelled DCV, whereas boutons can contain variable numbers (1–20) of DCV, together with small clear vesicles that are not labelled (Figs 23.1 and 23.2). In the same bouton not all the DCV are labelled and those labelled are tagged by a different number of gold particles. Labelled boutons with synaptic specialization are of two morphologically different types: (a) small and dome-shaped, preferentially contacting a single dendrite (Fig. 23.1); (b) large and scalloped, generally contacting several profiles. Many labelled boutons without synaptic specializations in the plane of the section are also present (Fig. 23.2).

These results are in agreement with those of previous studies on SP-LI in the spinal dorsal horn, performed with pre-embedding methods (Di Figlia *et al.*, 1982). However, the particulate nature of the label and the good ultrastructural preservation offered by the post-embedding immunogold staining method allow better morphological classification of the labelled terminals and therefore their correlation with terminals whose origin has been inferred in previous studies with other approaches. In the superficial laminae of the spinal cord several lines of evidence (Maxwell and Réthelyi, 1987) indicate that the large, scalloped terminals are primary afferents, i.e. they originate from dorsal root ganglia (DRG). Three types of large scalloped terminals have been described in the rat dorsal horn on the basis of their ultrastructural characteristics (Ribeiro-da-Silva and Coimbra, 1982). These morphologically distinct types of terminals show different degeneration patterns after rhizotomy (Coimbra *et al.*, 1984) or after treatment with capsaicin, that selectively destroys small unmyelinated fibres (Ribeiro-da-Silva and Coimbra, 1984), and therefore are thought to convey different sensory modalities. Immunogold staining demonstrates that SP-LI is present only in terminals that are thought to originate from unmyelinated C fibres and therefore mostly convey nociceptive inputs (Lynn and Hunt, 1984).

Moreover, a direct visualization of SP-LI in identified primary afferent terminals is obtained by combining immunogold staining with tract tracing methods such as injection of horseradish peroxidase (HRP) in the DRG,

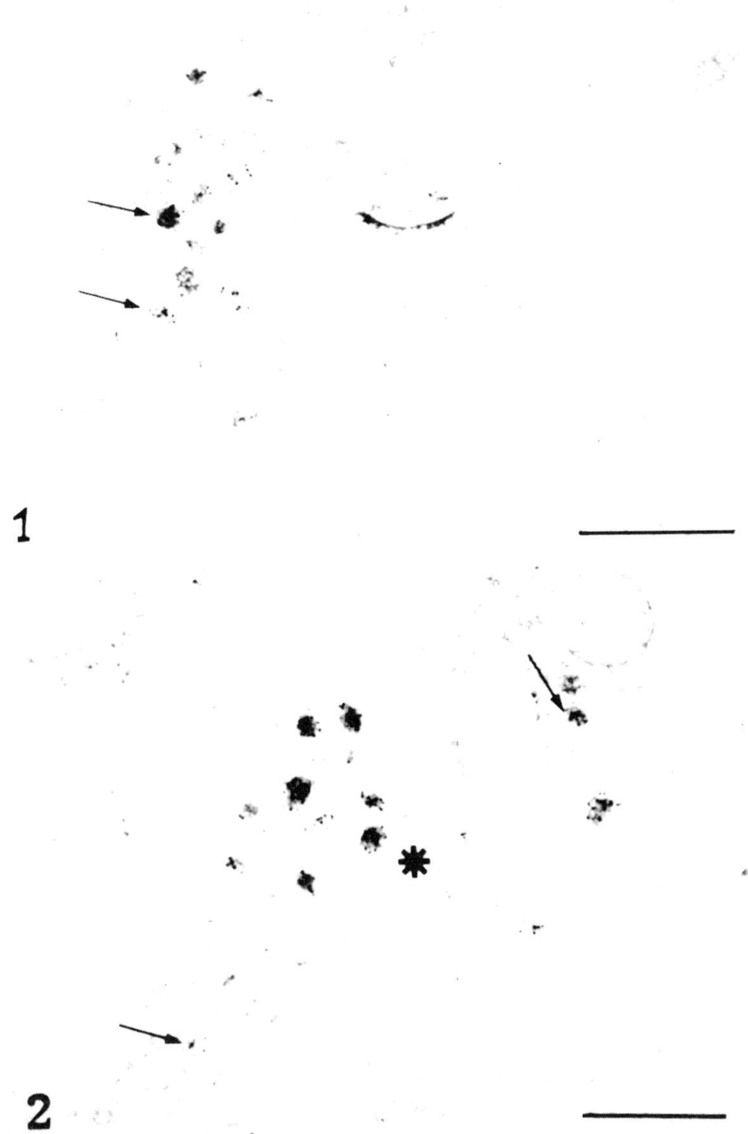

Fig. 23.1. SP-LI in rat dorsal horn. A bouton containing labelled (arrows) and unlabelled dense core vesicles, together with unlabelled clear vesicles, makes an asymmetric synaptic contact with a small dendrite. Scale bar = 0.5 μm.

Fig. 23.2. SP-LI in rat dorsal horn. SP-labelling is present over dense core vesicles contained in a large bouton (asterisk) that does not show synaptic specializations in the plane of the section, and in adjacent profiles (arrows). Scale bar = 0.5 μm.

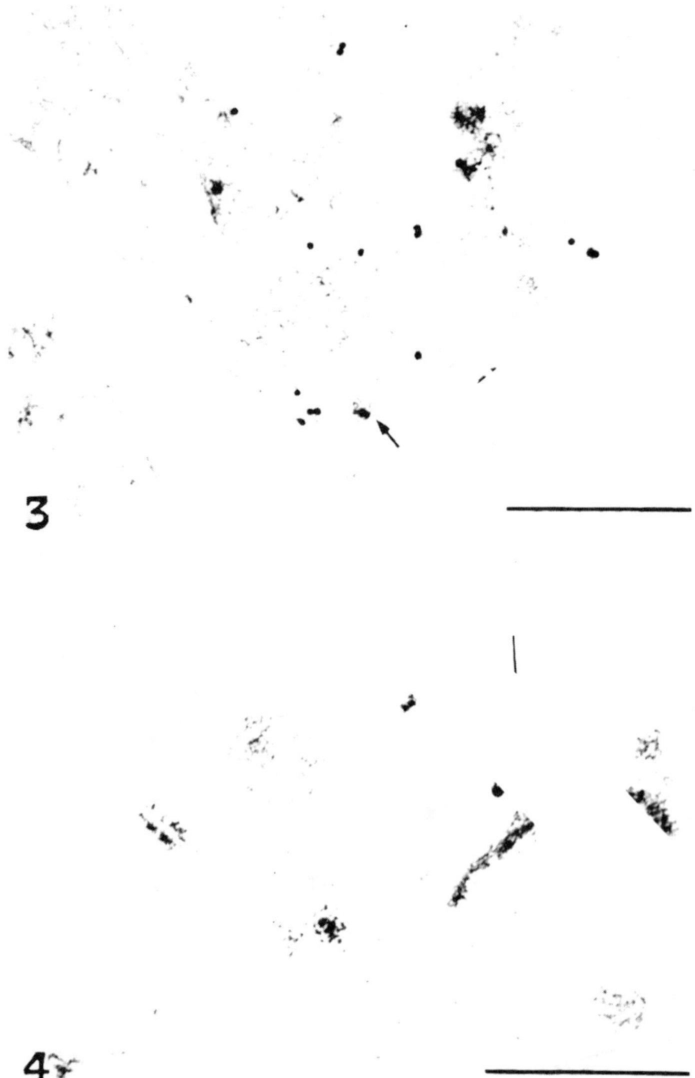

Fig. 23.3. Multiple labelling in rat spinal cord. A large, scalloped bouton in lamina II is identified as a primary afferent for the presence of HRP reaction product (arrowhead). The bouton shows SP-LI over a dense core vesicle (arrow, 10 nm gold particles) and is also immunoreactive for glutamate (20 nm gold particles). The triple labelling can be followed in adjacent serial sections. Scale bar = 0.5 μm.

Fig. 23.4. Multiple labelling in rat spinal cord. A small bouton in lamina I shows co-localization of CCK-LI (10 nm gold particles) and SP-LI (20 nm gold particles) in the same dense core vesicle (arrow). Single labelled vesicles are also present (arrowheads). Scale bar = 0.5 μm.

or dorsal rhizotomy that causes degeneration of the afferent fibres. When the immunoreaction is carried out on thin sections of spinal cord obtained from rats that underwent one of these two experimental procedures, SP-LI is found in terminals that are identified as primary afferents because they contain HRP reaction product (Fig. 23.3) or show signs of degeneration.

As for the nature of the peptide detected by the antisera used, control experiments cannot completely exclude the possibility of cross-reactivity with other tachykinins, in particular neurokinin A and B. However, since neurokinin A – whose distribution in the central nervous system is similar to that of SP – is present in limited amounts in the rat spinal cord, and has excitatory actions on spinal neurons similar to those of SP (Otsuka and Yanagisawa, 1987), a possible cross-reaction does not represent an actual risk of misinterpretation of the results. Neurokinin B is not present in DRG (Warden and Young, 1988) and therefore it should not be present in their identified terminals. Moreover, in mammalian spinal cord the presence of SP has been confirmed by several different experimental approaches (Salt and Hill, 1983). In particular, the *in situ* hybridization technique has confirmed the distribution of SP positive cells in both DRG and spinal cord of rats (Warden and Young, 1988). The function of SP as a neurotransmitter exciting dorsal horn neurons has been indicated by several lines of evidence (Salt and Hill, 1983).

23.2.2. *CCK-like immunoreactivity (CCK-LI)*

Labelling with the anti-CCK serum is always associated with DCV present in boutons of variable size and shape, also containing unlabelled SCV. CCK-positive DCV vesicles are morphologically similar in shape and size to those labelled for SP in the rat dorsal horn. In fact when double labelling for the two neuropeptides is performed on sections of spinal cord, using gold particles of different size to reveal the two antigens, it is possible to demonstrate colocalization of SP-LI and CCK-LI not only in the same bouton, but also in the same vesicle (Fig. 23.4). In the same terminal, double labelled DCV coexist with unlabelled or single labelled DCV.

As for the nature of the visualized peptide, Hökfelt *et al.* (1988) performed an extensive immunohistochemical investigation on the rat CNS using several sequence specific anti-CCK sera, and concluded that at least some of the CCK-LI observed in the dorsal horn is due to true CCK. Both excitatory and inhibitory effects of CCK in the spinal cord have been demonstrated, with sensitivity found for both nociceptive and non-nociceptive dorsal horn neurons (Willetts *et al.*, 1985).

Coexistence of CCK and SP has been previously demonstrated by light microscopy in varicosities in rat dorsal horn (Tuchscherer *et al.*, 1987) and in a population of periaqueductal neurons projecting to the spinal cord

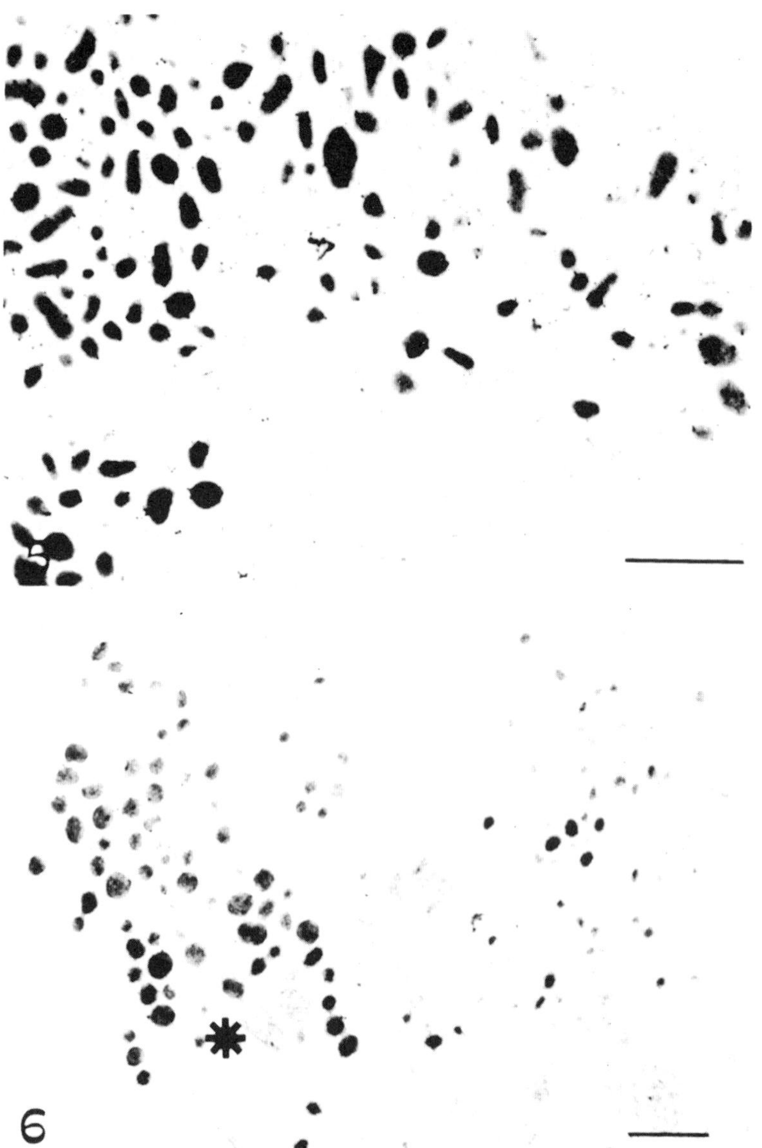

Fig. 23.5. Immunolabelling in *Mytilus* pedal ganglia. Detail of an SP-labelled neuron in the cortical region, containing many dense core vesicles tagged by 10 nm gold particles. No labelling is found over other cytoplasmic organelles. Scale bar = 0.5 μm.

Fig. 23.6. Immunolabelling in *Mytilus* pedal ganglia. A bouton containing CCK-labelled dense core vesicies (asterisk) is present in the neuropile among many other varicosities showing various types of unlabelled vesicles. Scale bar = 0.5 μm.

(Skirboll *et al.*, 1983). The functional significance of this colocalization is, however, far from clear.

23.3. **Immunolocalization of peptides in *Mytilus* pedal ganglia**

23.3.1. *SP-like immunoreactivity (SP-LI)*

Cell bodies and processes displaying SP-LI have the same ultrastructural characteristics, whatever the antiserum used. The distinctive feature of the labelled elements is the presence of DCV overlaid by gold particles (Fig. 23.5). SP-labelled DCV in *Mytilus* are slightly different from those detected in rat dorsal horn, in that they are larger (80–140 nm in diameter, $\bar{x}$ = 110), more electron-dense and often oval in shape. Labelled cell bodies typically contain many DCV tagged by a variable number of gold beads. Gold is found exclusively over the vesicles and no other cytoplasmic organelle is labelled (Fig. 23.5). The same type of labelled vesicles is also present in varicosities in the ganglionic neuropil. So far we have not found SP labelled boutons with typical synaptic specializations in the plane of the section, but these are known to be rare in invertebrates.

Although ultrastructural studies are very scanty, previous light microscopic investigations have shown that SP-LI is present in specific neurons in many invertebrates, including molluscs (Verhaert and De Loof, 1985; Boyd *et al.*, 1986; Vitellaro-Zuccarello and De Biasi, 1988b). Further studies are, however, needed to characterize the exact nature of the endogenous peptide. Most of the available evidence, obtained by radioimmunoassay and gel chromatography, indicates that SP-like peptides of invertebrates are distinct from mammalian SP (Mancillas and Brown, 1984; Osborne *et al.*, 1982). Moreover, true SP could not be demonstrated in *Mytilus* extracts (Kream *et al.*, 1986). Only Grimmelikhuijzen *et al.* (1981) reported that the SP-like material in *Hydra* did show a parallel displacement curve to authentic SP in a radioimmunoassay, and also elute with authentic SP using gel chromatography.

As for the functional role of this peptide, several pharmacological actions of SP on molluscan tissues have been reported. It produces excitation on the heart of *Helix* (Boyd *et al.*, 1984) and contracts the rectum of *Mercenaria* (Greenberg, 1983), and in *Helix*, depending on the physiological state, it alters the normal rhythmic bursting pattern by prolonging the bursting period (Salanki *et al.*, 1983). A sensory or visceral function of the detected SP-like peptide in *Mytilus* is suggested by the visualization of dense plexuses of positive fibres, with intervening unipolar neurons, in most of the foot tissues: under the surface epithelium, around blood vessels, among glandular cells. No SP-LI has been detected in foot muscles or in the anterior byssus retractor muscle, thus suggesting that the SP-like neuropeptide detected is not involved in motor control (Vitellaro-Zuccarello and De Biasi, 1987).

23.3.2. *CCK-like immunoreactivity (CCK-LI)*

With the electron microscope, labelling with the anti-CCK serum is specifically associated with a morphologically distinct type of cytoplasmic organelle, represented by very electron-dense vesicles, with round or elliptical shape and large size (120–180 nm, $\bar{x}$ = 150) (Fig. 23.6). Labelled DCV are present in cell bodies and in varicosities in the neuropile. Not all vesicles in the neuropile are labelled, thus indicating selectivity of the immunostaining (Fig. 23.6). DCV labelled with the anti-CCK serum are morphologically different from those labelled with the anti-SP serum. This is also confirmed in double-labelling experiments in which sections of pedal ganglion were sequentially incubated with anti-SP and anti-CCK sera: labelling for the two antigens is always confined into separate neurons and terminals.

As for the nature of the peptide detected by the anti-CCK serum, radioimmunoassay and gel chromatography from neural tissues of *Helix* (Osborne *et al.*, 1982) show that the CCK-LI material in molluscs is distinguishable from the main forms of mammalian CCK and gastrin. Using several antisera recognizing various epitopes of the peptides CCK and gastrin, Ono (1986) found that only the antisera that recognized the carboxy-terminal sequence common to CCK and gastrin reacted with *Aplysia* tissues. The similarities between the molluscan CCK-like neuropeptide and its mammalian counterparts in their carboxy-terminal regions are compatible with the idea that these regions are conserved because they include the minimum fragment needed for biological activity (Osborne *et al.*, 1982).

A physiological role of CCK in invertebrates has been suggested (Dhainaut-Courtois *et al.*, 1986). As for the molluscs, Ono (1986), with intracellular electrophysiological studies of identified *Aplysia* neurons, showed that the mammalian form of CCK and gastrin did not mimic the synaptic responses. However, Hammond *et al.* (1987) demonstrated that CCK modulates the Ca^{2+} current in identified neurons of *Helix* by means of protein kinase C and that it acts on identified snail neurons at concentrations similar to those at which it binds to receptors in both pancreatic and brain membranes. Finally, as in mammals (Morley, 1982), a role for CCK in feeding behaviour has been suggested by the demonstration that in *Navanax* (an opistobranch mollusc) CCK selectively inhibits buccal expansion motoneurons responsible for prey capture (Zimering *et al.*, 1988).

23.4. Conclusions

The immunocytochemical methods here applied provide evidence for a similar submicroscopic localization of the immunoreactivity for two different neuropeptides in nervous tissues of phylogenetically distant animals. Label is in fact always selectively localized over electron-dense vesicles, which therefore can be considered as the major storage organelles for these peptides, whereas small clear vesicles are never labelled. Our observations are consistent with data obtained with subcellular fractionation techniques in peripheral nerves, which showed peptides restricted to the fraction contanining DCV (Fried, 1982). Amino-acidic neurotransmitters such as GABA (De Biasi *et al.*, 1986a) and glutamate are not localized over DCV (Fig. 26.6; De Biasi and Rustioni, 1988). The absence of gold particles from cytoplasmic organelles such as rough endoplasmic reticulum and Golgi apparatus – that are likely to be involved in the biosynthesis of the detected SP- and CCK-like peptides – suggests that the antisera used may not recognize the precursor peptides.

Vesicular localization of the peptides is consistent with their functioning as extracellular messengers. Their release can occur from varicosities endowed with typical synaptic specializations and contacting various types of postsynaptic profiles, as shown in several regions of the vertebrate central nervous system. However, as nonsynaptic release of peptides has been proposed in both vertebrate and invertebrate nervous tissues (Zhu *et al.*, 1986) it is reasonable to assume that the detected SP- and CCK-like substances could also be released from the numerous labelled boutons lacking typical synaptic specializations observed both in *Mytilus* pedal ganglia and in rat spinal cord. The presence of SP- and CCK-LI over the same DCV indicates that peptides encoded by distinct genes can be co-stored into the same vesicle, and poses the problem of their simultaneous co-release and subsequent action on specific receptors.

Significative differences in the ultrastructure of peptide-immunoreactive neuronal cell bodies exist between the two nervous tissues examined: in *Mytilus* they contain many DCV, whereas these organelles are scarce in rat spinal cord. This may reflect differences in the rate of synthesis, transport and release of the peptides. In this regard it is worth noting that immunocytochemical visualization of peptides in vertebrate neurons generally requires a pretreatment of the animals with colchicine to block axoplasmic transport, whereas this procedure is not required for invertebrate neurons. Alternatively, the extreme abundance of labelled DCV in the perikarya of pedal ganglia may indicate the possibility of a paracrine release of peptides that can directly influence neighbouring neurons. Axosomatic contacts are rare – in fact extremely rare – in the cortical region of *Mytilus* pedal ganglia, and in invertebrates in general.

The differences in the morphology of labelled vesicles observed between *Mytilus* and rat may reflect different modalities of packaging (e.g. association with different proteins), or differences in the nature of the endogenous peptide. Further studies combining different methodological approaches are needed, especially in *Mytilus*, to identify the real nature of the detected immunoreactive peptides and to elucidate their possible physiological roles.

In conclusion, the visualization of neuropeptides and other neuroactive substances by means of immunogold staining may lead to a substantial increase of our understanding of the nervous system organization and connectivity. The technique in fact not only enables submicroscopic localization of neuropeptides with a degree of resolution not previously attainable, it also offers the possibility of multiple immunolabelling and of combination with tracing techniques. The simultaneous detection of multiple neurotransmitters in pre- and postsynaptic neuronal elements is particularly useful to investigate spatial interrelationships between different molecules, and therefore to draw conclusions about neuronal associations. Moreover, the combination of immunolabelling with classical tracing methods allows localization of chemically characterized neurons in an anatomically defined circuit. All these strategies, so far only scarcely exploited in invertebrates, may prove especially suited for providing data of comparative interest.

Acknowledgements

Part of the original work on the rat spinal cord described here was performed in collaboration with Dr A. Rustioni in the Department of Cell Biology and Anatomy at the University of North Carolina, Chapel Hill, NC, USA. The authors wish to thank Dr P. Petrusz (University of North Carolina, Chapel Hill, NC, USA) for kindly supplying polyclonal anti-SP serum. This work has been supported by grants (40% and 60%) from the Italian Ministry of Public Education and by grant 88.00685.04 from CNR (Rome)

References

Boyd, P. J., Osborne, N. N. and Walker, R. J. (1984). The pharmacological actions of 5-hydroxytryptamine, FMRF-amide and substance P and their possible occurrence in the heart of the snail *Helix aspersa*. *Neurochem. Int.*, **6**, 633–44.

Boyd, P. J., Osborne, N. N. and Walker, R. J. (1986). Localization of a substance P-like material in the central nervous system of the snail *Helix aspersa*. *Histochemistry*, **84**, 97–103.

Coimbra, A., Ribeiro-da-Silva, A. and Pignatelli, D. (1984). Effects of dorsal rhizotomy on the several types of primary afferent terminals in laminae I–III of

the rat spinal cord. *Anat. Embryol.*, **170**, 279–87.
De Biasi, S. and Rustioni, A. (1988). Glutamate and substance P coexist in primary afferent terminals in the superficial laminae of spinal cord. *Proc. Natl. Acad. Sci. USA*, **85**, 7820–4.
De Biasi, S., Frassoni, C. and Spreafico, R. (1986a). GABA immunoreactivity in the thalamic reticular nucleus of the rat. A light and electron microscopic study. *Brain Res.*, **399**, 143–7.
De Biasi, S., Frassoni, C. and Vitellaro-Zuccarello, L. (1986b). Glutamic acid decarboxylase (GAD)-like immunoreactivity in the pedal ganglia of *Mytilus galloprovincialis*. *Cell Tiss. Res.*, **244**, 591–3.
Dhainaut-Courtois, N., Tramu, G., Marcel, R., Malécha, J., Verger-Bocquet, M., Andriès, J. C., Masson, M., Selloum, L., Belemtougri, G. and Beauvillain, J. C. (1986). Cholecystokinin in the nervous systems of Invertebrates and Protochordates. Immunohistochemical localization of a cholecystokinin-8-like substance in Annelids and Insects. *Ann. N.Y. Acad. Sci.*, **448**, 167–87.
Di Figlia, M., Aronin, N. and Leeman, S. E. (1982). Light microscopic and ultrastructural localization of immunoreactive substance P in the dorsal horn of the monkey spinal cord. *Neuroscience*, **7**, 1127–39.
Fried, G. (1982). Neuropeptide storage in vesicles. In: Klein, R. L. *et al.* (eds), *Neurotransmitter Vesicles*. Academic Press, London, pp. 361–72.
Greenberg, M. J. (1983). The responsiveness of molluscan muscles to FMRF-amide, its analogs and other neuropeptides. In: Lever, J. and Boer, H. H. (eds), *Molluscan Neuroendocrinology*. North Holland, Amsterdam, pp. 190–5.
Grimmelikhuijzen, C. J. P., Balfen, A., Emson, P. C., Powell, D. and Sundler, F. (1981). Substance P-like immunoreactivity in the nervous system of *Hydra*. *Histochemistry*, **71**, 325–33.
Hammond, C., Paupardin-Tritsch, D., Nairn, A. C., Greengard, P. and Gerschenfeld, H. M. (1987). Cholecystokinin induces a decrease in Ca^{++} current in snail neurons that appears to be mediated by protein kinase C. *Nature*, **325**, 809–11.
Hökfelt, T., Herrera-Marschitz, M., Seroogy, K., Staines, M. A., Holets, V., Schalling, M., Ungerstedt, U., Post, C., Rehfeld J., Frey, P., Fischer, J., Dockray, G., Hamaoka, T., Walsh, J. H. and Goldstein, M. (1988). Immunohistochemical studies on cholecystokinin (CCK)-immunoreactive neurons in the rat using sequence specific antisera and with special reference to the caudate nucleus and primary sensory neurons. *J. Chem. Neuroanat.*, **1**, 11–52.
Kream, R. M., Leung, M. K. and Stefano, G. B. (1986). Is there authentic substance P in invertebrates? In: Stefano, G. B. (ed.), *CRC Handbook of Comparative Opioid and Related Neuropeptide Mechanisms*, vol. I. CRC Press, Boca Raton, FL, pp. 65–72.
Lynn, B. and Hunt, S. P. (1984). Afferent C-fibers: physiological and biochemical correlations. *Trends Neurosci*, **7**, 186–8.
Mancillas, J. R. and Brown M. R. (1984). Neuropeptide modulation of photosensitivity. I. Presence, distribution and characterization of a substance P-like peptide in the lateral eye of *Limulus*. *J. Neurosci.*, **4**, 832–46.
Maxwell, D. J. and Réthelyi, M. (1987). Ultrastructure and synaptic connections of cutaneous afferent fibers in the spinal cord. *Trends Neurosci.*, **10**, 117–23.
Morley, J. E. (1982). Minireview: the ascent of cholecystokinin (CCK) – from gut to brain. *Life Sci.*, **30**, 479–93.
Ono, J. K. (1986). Localization and identification of neurons with cholecystokinin and gastrin-like immunoreactivity in wholemounts of *Aplysia* ganglia. *Neuroscience*, **18**, 957–74.

Osborne, N. N., Cuello, A. C. and Dockray, G. J. (1982). Substance P and cholecystokinin-like peptides in *Helix* neurons and cholecystokinin and serotonin in a giant neuron. *Science*, **216**, 409–11.

Otsuka, M. and Yanagisawa, M. (1987). Does substance P act as a pain transmitter? *Trends Pharmacol. Sci.*, **8**, 506–10.

Ribeiro-da-Silva, A. and Coimbra, A. (1982). Two types of synaptic glomeruli and their distribution in laminae I–III of the rat spinal cord. *J. Comp. Neurol.*, **209**, 176–86.

Ribeiro-da-Silva, A. and Coimbra, A. (1984). Capsaicin causes selective damage to type I synaptic glomeruli in rat substantia gelatinosa. *Brain Res.*, **290**, 380–3.

Salanki, J., Verhovszky, A. and Stefano, G. (1983). Interaction of substance P and opiates in the CNS of *Helix pomatia. Comp. Biochem. Physiol.*, **75C**, 387–90.

Salt, T. E. and Hill, R. G. (1983). Neurotransmitter candidates of somatosensory primary afferent fibers. *Neuroscience*, **10**, 1083–103.

Skirboll, L., Hökfelt, T., Dockray, G., Rehfeld, J., Brownstein, M. and Cuello, A. C. (1983). Evidence for periaqueductal cholecystokinin-substance P neurons projecting to the spinal cord. *J. Neurosci.*, **3**, 1151–7.

Spreafico, R., De Biasi, S. and Rustioni, A. (1988). Ultrastructural investigation on glutamate immunoreactivity in the ventro postero lateral nucleus of the rat thalamus. *Neurosci. Lett.*, Suppl. 33, 185.

Tuchscherer, M. M., Knox, C. and Seybold, V. S. (1987). Substance P and cholecystokinin-like immunoreactive varicosities in somatosensory and autonomic regions of the rat spinal cord: a quantitative study of coexistence. *J. Neurosci.*, **7**, 3984–95.

Varndell, I. M. and Polak, J. M. (1984). Double immunostaining procedures: techniques and applications. In: Polak, J. M. and Varndell, I. M. (eds), *Immunolabeling for Electron Microscopy*. Elsevier, Amsterdam, pp. 155–77.

Verhaert, P. and De Loof, A. (1985). Substance P-like immunoreactivity in the central nervous system of the blattarian insect *Periplaneta americana L.* revealed by a monoclonal antibody. *Histochemistry*, **83**, 501–7.

Vitellaro-Zuccarello, L. and De Biasi, S. (1987). Distribution of SP- and 5HT-like immunoreactivity in peripheral tissues of *Mytilus galloprovincialis. Basic Appl. Histochem.*, **31** (Suppl), 188.

Vitellaro-Zuccarello, L. and De Biasi, S. (1988a). GABA-like immunoreactivity in the pedal ganglia of *Mytilus galloprovincialis*: light and electron microscopic study. *J. Comp. Neurol.*, **267**, 516–24.

Vitellaro-Zuccarello, L. and De Biasi, S. (1988b). Distribution of substance P-like immunoreactivity in the pedal ganglion of *Mytilus galloprovincialis. Basic Appl. Histochem.*, **32**, 109–13.

Vitellaro-Zuccarello, L., De Biasi, S. and Bairati, A. (1988). Subcellular localization of serotonin-immunoreactivity in the pedal ganglion of *Mytilus galloprovincialis* (Mollusca, Bivalvia). *J. Submicr. Cytol.*, **20**, 109–13.

Warden, M. K. and Young, W. S. III (1988). Distribution of cells containing mRNAs encoding substance P and neurokinin B in the rat central nervous system. *J. Comp. Neurol.*, **272**, 90–113.

Willetts, J., Urban, K., Murase, K. and Randic, M. (1985). Actions of cholecystokinin octapeptide on rat spinal dorsal horn neurons. *Ann. N.Y. Acad. Sci.*, **448**, 385–402.

Zimering, M. B., Madsen, A. J. Jr and Elde, R. P. (1988). CCK-8 inhibits feeding-specific neurons in *Navanax*, an opisthobranch mollusc. *Peptides*, **9**, 133–9.

Zhu, P. C., Thureson-Klein, A. and Klein, R. L. (1986), Exocytosis from large dense cored vesicles outside the active synaptic zones within the trigeminal subnucleus caudalis: a possible mechanism for neuropeptide release. *Neuroscience*, **19**, 43–54.

24 *Gábor L. Kovàcs*

The interaction of posterior pituitary neuropeptides with dopamine in experimental drug addiction: the relevance of rodent models

24.1. Introduction: the relevance of the rodent model

In order to study the pathophysiology of drug dependence it is necessary to evolve an appropriate animal model. Rodents (rats and mice) have proven to be useful subjects for the study of the mechanism of narcotic addiction, since they are easily made dependent on narcotics by a variety of methods. One of the most frequently used experimental methods in rodents is the subcutaneous implantation of pellets containing morphine (Collier, 1984), which then result in the development of tolerance/dependence. The withdrawal syndrome exhibited by drug-addicted rodents may be brought about either by administering a narcotic antagonist (e.g. naloxone in the case of opiates) or by the abrupt termination of chronic morphine administration. The non-antagonist precipitated withdrawal corresponds better to the condition encountered clinically. If, however, one must use drugs as tools to determine the role of various neurochemical mechanisms involved in the narcotic abstinence syndrome, as in the case of neuronal peptides, their analogues and fragments, the withdrawal syndrome precipitated by an antagonist might be more useful, mainly for technical reasons (withdrawal appears at the same time in each animal).

Particularly important for drug addiction processes are the reinforcing stimuli, which can be external or internal. Such stimuli maintain the occurrence of the responses at higher frequency than if the stimuli did not act. It is well known that the administration of some drugs (e.g. opiates) acts in a reinforcing way both in experimental animals and human patients. The experimental model of drug self-administration (Deneau *et al.*, 1969; Van Ree, 1979) is a technique based on the reinforcing effect of drugs, and is used as an attempt to demonstrate the development of psychological dependence in laboratory animals.

This chapter will describe the effects of posterior pituitary peptides (primarily that of oxytocin) on experimental drug addiction in rodents.

24.2. **The effect of oxytocin on opiate tolerance**

In recent years it has become increasingly evident that neurohypophyseal neuropeptides (oxytocin and vasopressin) might play a role in drug addiction processes (Krivoy *et al.*, 1974; Van Ree, 1983; Kovàcs, 1986; Kovàcs *et al.*, 1987). It has been demonstrated (Krivoy *et al.*, 1974) that desglycinamide^{9}-lysine^{8}-vasopressin (DG-LVP) facilitated the rate at which tolerance to morphine-induced analgesia has developed in mice. Van Ree and De Wied (1977a, b) found that the hormonally active nonapeptide (vasopressin) and a variety of hormonally inactive vasopres-

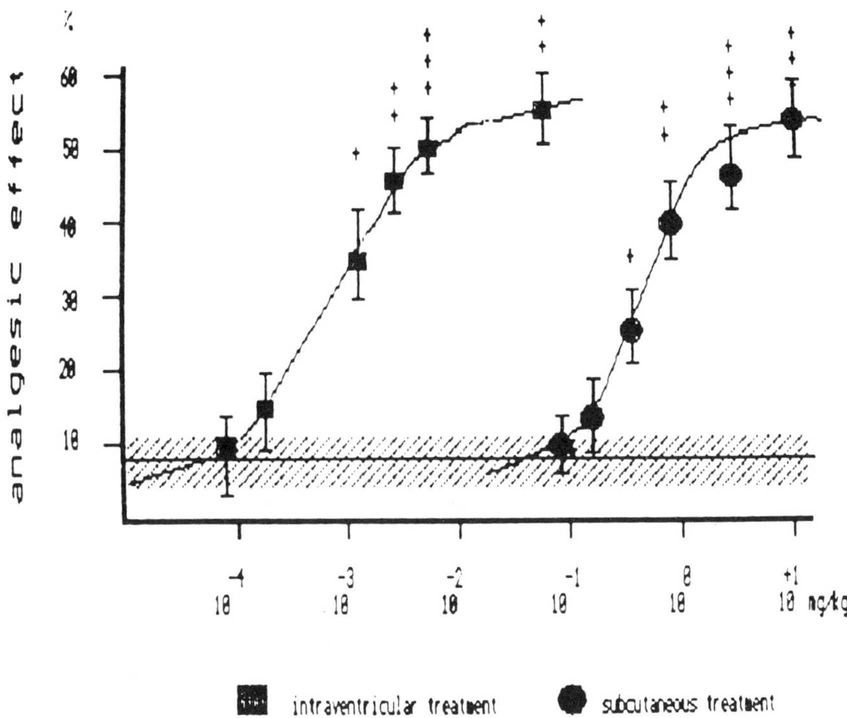

Fig. 24.1. The effect of oxytocin on the development of morphine tolerance. Mice were rendered tolerant to/dependent on morphine by the subcutaneous implantation of a morphine pellet (for details see Kovàcs *et al.*, 1981). A single injection of oxytocin in graded doses was given 2 h prior to the pellet implantation; 48 h later, a test dose of 5 mg/kg morphine was injected and the analgesic effect was measured 30 min later by the heat irradiant tail-flick method. Intraventricular injections were administered via chronically implanted plastic cannulae. Each point represents the mean (±SE) of eight experimental animals. The horizontal dotted area represents the analgesic effect of 5 mg/kg morphine in the tolerant control animals ($^{*}p < 0.05$; $^{**}p < 0.01$; $^{***}p < 0.001$ vs tolerant control animals).

sin-like peptide fragments increased the rate of tolerance (the analgesic action of opiates was investigated in these studies).

Oxytocin, on the other hand, dose-dependently attenuated the development of acute and chronic morphine tolerance in mice, i.e. in animals treated with this neuropeptide and then with a high, tolerance-inducing dose of morphine (Fig. 24.1) the analgesic effect of a second morphine injection was greater than in the control animals which did not receive oxytocin. Interestingly, oxytocin did not modify the magnitude or the duration of the analgesic effect of the first morphine challenge in drug-naive animals (Kovàcs *et al.*, 1985, 1987).

Of physiological interest might be the finding that tolerance to the antinociceptive effect of β-endorphin was also attenuated by graded doses of oxytocin (Fig. 24.2). The dose–response effect of oxytocin was U-shaped, because medium-high doses of the neurohypophyseal

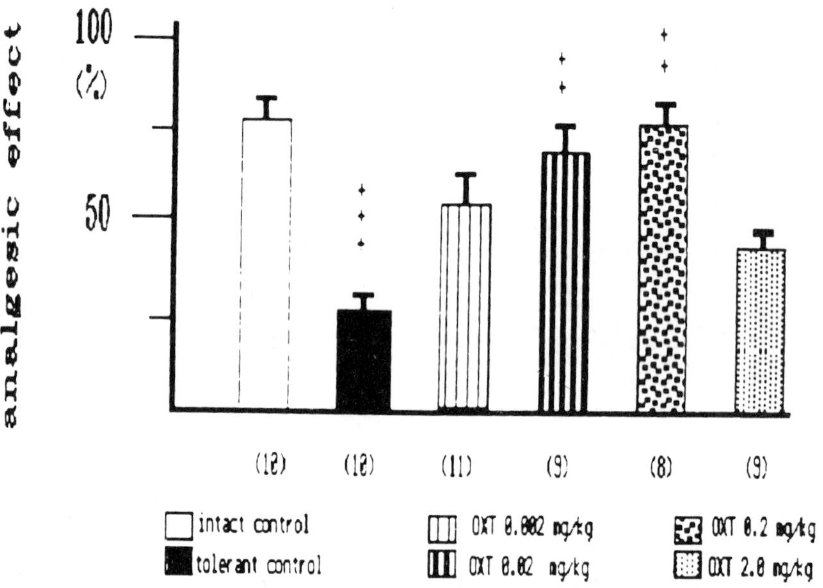

Fig. 24.2. Dose-related effect of oxytocin treatment on the development of tolerance to β-endorphin. Mice received daily injections of 0.005 mg β-endorphin i.c.v. for 3 consecutive days. On Day 4 the analgesic effect of i.c.v. β-endorphin treatment was measured 30 min following treatment, by the heat irradiant tail-flick method of D'Amour and Smith (1941) (tolerant control group). The intact control animals received artificial cerebrospinal fluid for 3 consecutive days, but they were given β-endorphin on day 4. Graded doses of oxytocin were injected s.c., 1 h prior to the daily injections of β-endorphin (for details see Kovàcs and Telegdy, 1987a). The columns indicate the mean ± SE of the analgesic effect of β-endorphin. (***$p < 0.001$ vs intact controls; **$p < 0.01$ vs tolerant controls). The numbers of experimental animals are shown in parentheses.

neuropeptide were more effective than high peptide doses. Since β-endorphin is an endogenous substance in the central nervous system, the results indicate the possibility of a physiological interaction between two different neuronal peptides (oxytocin and β-endorphin) in the organization of opiate tolerance (Kovàcs and Telegdy 1987a).

Intracerebroventricular (i.c.v.) injections of oxytocin were considerably more potent than were systemic peptide injections in attenuating analgesic morphine tolerance (Fig. 24.1). This finding suggests that central nervous, and not peripheral, mechanisms are involved in this effect of the neurohypophyseal neuropeptide. Local intracerebral microinjections of oxytocin were even more potent than i.c.v. injections (Kovàcs *et al.*, 1984; Sarnyai *et al.*, 1988; Ibragimov *et al.*, 1987). The most sensitive brain sites were the hippocampus and the basal forebrain (including the nucleus accumbens and the posterior olfactory nuclei). Of particular interest is the finding that oxytocinergic nerve terminals (Buijs, 1983; Sawchenko and Swanson, 1985) and binding sites (Ferrier *et al.*, 1983; De Kloet *et al.*, 1985) are present in these brain nuclei.

It is remarkable that the interaction of oxytocin with morphine tolerance could also be observed in non-analgesic effects of the opiate alkaloid. Accordingly, oxytocin treatment inhibited the development of tolerance that developed in mice in the locomotor hyperactivity following the administration of high doses of morphine (Kovàcs and Telegdy, 1987b). Since brain structures involved in the control of pain perception and those responsible for hyperlocomotion are not identical, these data suggest that the effect of oxytocin on morphine tolerance was not specifically related to the effector (output) mechanisms of these behavioural processes, but rather to more fundamental neuronal mechanisms responsible for the organization of tolerance.

24.3. **Oxytocin and morphine withdrawal**

The degree of physical dependence to a given narcotic drug is characterized by the severity of the withdrawal reactions (e.g. stereotyped jumpings due to extrapyramidal incoordination, salivation, loss of body weight as a consequence of intensive diarrhoea and urination, decrease in the colonic temperature, irritability, etc.). Injections of oxytocin in graded doses, given prior to the first morphine challenge, attenuated various signs of the naloxone-precipitated withdrawal reaction. Low doses of oxytocin were needed to attenuate the withdrawal-induced hypothermia, while higher doses of the neuropeptide were required to antagonize the withdrawal-induced loss in body weight and the latency of the onset of stereotyped jumpings (Fig. 24.3). It is therefore likely that this neurohypophyseal neuropeptide interfered with the development of physical

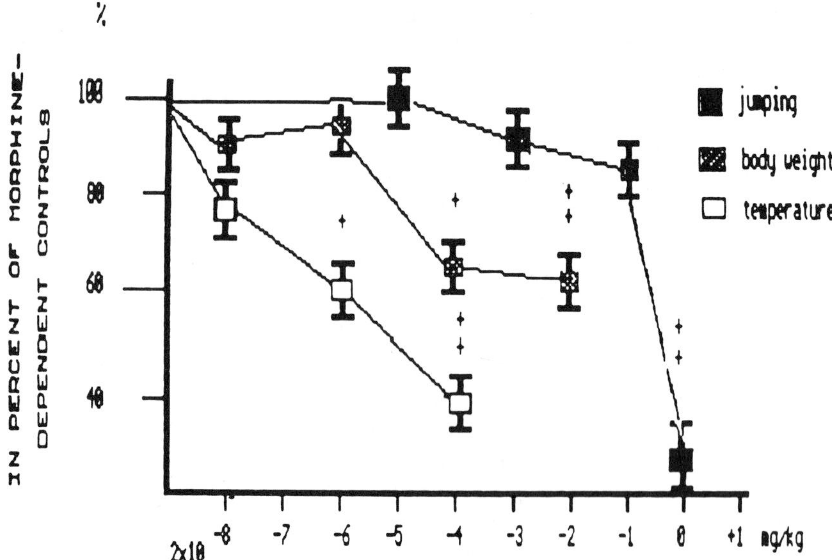

Fig. 24.3. Dose-related effect of oxytocin treatment on naloxone-precipitated withdrawal signs in mice. Mice were rendered physically dependent on morphine by the implantation of a morphine pellet; 72 h after pellet implantation, withdrawal was precipitated by the s.c. injection of naloxone (1 mg/kg). The latency of the onset of stereotyped jumpings, loss in body weight and changes in colonic temperature were used to quantify degree of physical dependence (for details see Kovàcs *et al.*, 1981, 1984). (100% = the full appearance of withdrawal signs in the dependent control animals.)

dependence and a lower degree of physical dependence resulted in a secondary manner in the appearance of less severe withdrawal signs.

24.4. **Heroin self-administration and neurohypophyseal peptides**

Since neurohypophyseal neuropeptides deeply affected the development of tolerance to, and dependence on, narcotic drugs (Kovàcs and Telegdy 1985, 1987b; Van Ree, 1983), their roles in drug-induced reinforcement processes have also been studied. Van Ree and De Wied (1977a, b) found that desglycinamide^{9}-arginine^{8}-vasopressin, a behaviourally active fragment of vasopressin, reduced intravenous self-administration of heroin in non-tolerant rats. Oxytocin exerted a different effect in this study: rats treated with this neuropeptide self-administered significantly more heroin (Van Ree and De Wied, 1977a) than did vasopressin-treated animals, but not more than the non-treated control animals. Basically different findings were published more recently (Kovàcs *et al.*, 1985, Kovàcs and Van Ree,

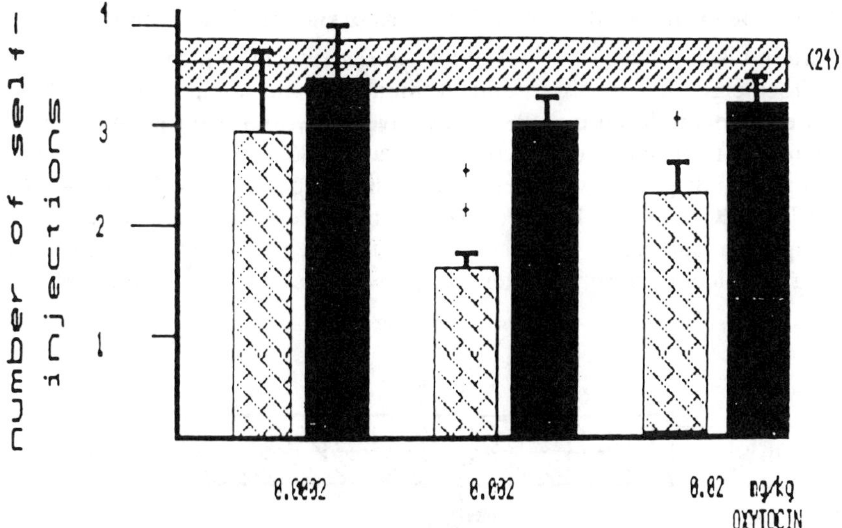

Fig. 24.4. The effect of graded doses of oxytocin on intravenous self-administration of heroin in rats. Rats were equipped with chronic jugular catheters and trained in an operant conditioning chamber to press a lever in order to activate a pump system and receive intravenous injections of heroin (for details see Kovàcs *et al.*, 1985). Once heroin self-administration was stable, rats were injected with graded doses of oxytocin. The columns indicate the mean ± SE of heroin self-injections (open columns 1 h, full columns 24 h after oxytocin). The horizontal dotted line indicates rate of heroin self-administration on the day preceding injection of oxytocin, when vehicle was administered instead of oxytocin (*$p < 0.05$; **$p < 0.01$ vs vehicle session).

1985; Ibragimov *et al.*, 1987), when the effect of oxytocin treatment on the development of heroin self-administration was investigated in heroin-tolerant rats. Oxytocin dose-dependently reduced the self-injection rate of heroin in heroin-tolerant animals (Fig. 24.4). The dose–response effect of oxytocin was U-shaped in this experimental situation, too. Interestingly, the analgesic effect of the self-injected lower heroin dose was not lower than the analgesic effect of higher heroin doses in the control animals. In non-tolerant rats, on the other hand, systemic injections of oxytocin did not modify the rate of heroin self-administration. Taken together, the neuropeptide reduced the higher self-administration rate of the heroin-tolerant/dependent rats to the lower self-injection level of the non-tolerant rats (Kovàcs and Van Ree, 1985). These data suggested the possibility that the primary action of oxytocin was not on the reinforcing efficacy of heroin, but rather on the degree of tolerance to, and dependence on, heroin.

24.5. **The involvement of central nervous oxytocinergic receptors**

Earlier results indicated that endogenous oxytocin, which is present in oxytocinergic nerve fibres and terminals in the brain (Buijs, 1983; Sawchenko and Swanson, 1985) may have a physiological role in adaptive components of narcotic addiction (Kovàcs, 1986). Since saturable oxytocinergic binding sites were described in limbic brain areas (Ferrier *et al.*, 1983; De Kloet *et al.*, 1985), the question has been studied as to whether receptor antagonists of oxytocin would effectively block the action of this neuropeptide on morphine tolerance (Kovàcs *et al.*, 1987). I.c.v. injections

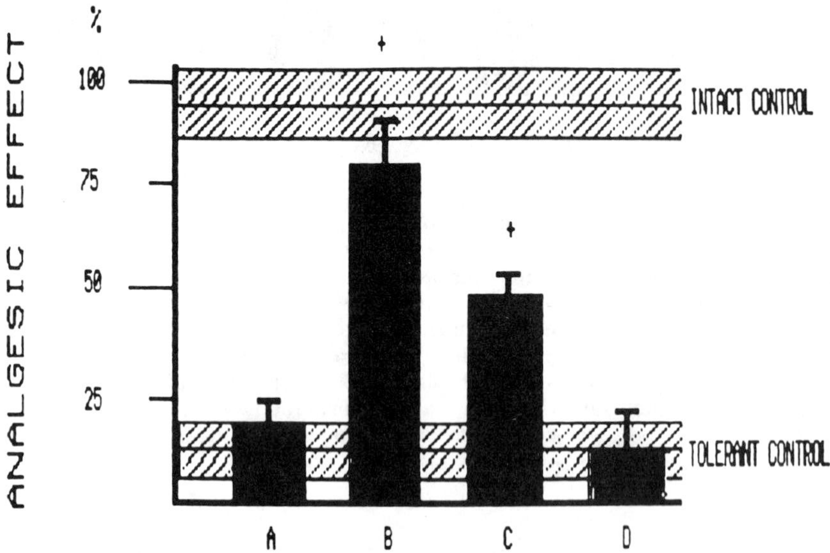

Fig. 24.5. Interaction of oxytocin with an oxytocin-antagonist on the development of acute morphine tolerance. Mice were equipped with an i.c.v. cannula in the lateral cerebral ventricle. Acute morphine tolerance was induced by the s.c. injection of 100 mg/kg morphine. Five hours later a test dose of 5 mg/kg morphine was injected, and the analgesic effect of this treatment was measured 30 min later (tolerant control group). A group of intact control animals was also investigated; this group did not receive the tolerance-inducing high dose of morphine, but later this group was also injected with the test dose of 5 mg/kg morphine. **A:** The effect of the oxytocin-antagonist: 90 min prior to the injection of the tolerance-inducing dose of morphine, mice received an i.c.v. injection of 50 pg *N*α-acetyl-[2-*O*-methyltyrosine]-oxytocin, an inhibitor of oxytocin receptors (Jost and Sorm, 1971; Krejci *et al.*, 1973). Thirty minutes later these animals were injected with 2 μl of artificial cerebrospinal fluid. **B:** The effect of oxytocin: Artificial cerebrospinal fluid was injected 90 min prior to the tolerance-inducing dose of morphine. Thirty minutes later, 50 pg oxytocin was administerd i.c.v. **C** and **D:** Combined effects of oxytocin and the oxytocin-antagonist: Mice received 5 (**C**) or 50 (**D**) pg of *N*α-acetyl-[2-*O*-methyltyrosine]-oxytocin. Thirty min later, both groups were injected with 50 pg oxytocin i.c.v. (*$p < 0.05$ vs tolerant control animals).

of N-α-acetyl-[2-O-methyltyrosine]-oxytocin, an antagonist of oxytocinergic receptors, dose-dependently antagonized the inhibitory effect of oxytocin on the development of acute morphine tolerance in mice (Fig. 24.5.). It has also been demonstrated that local microinjections of minute amounts of this receptor antagonist into limbic brain areas is highly effective in inhibiting the effects of oxytocin on morphine tolerance in mice (Sarnyai *et al.*, 1988), or heroin self-administration in rats (Ibragimov *et*

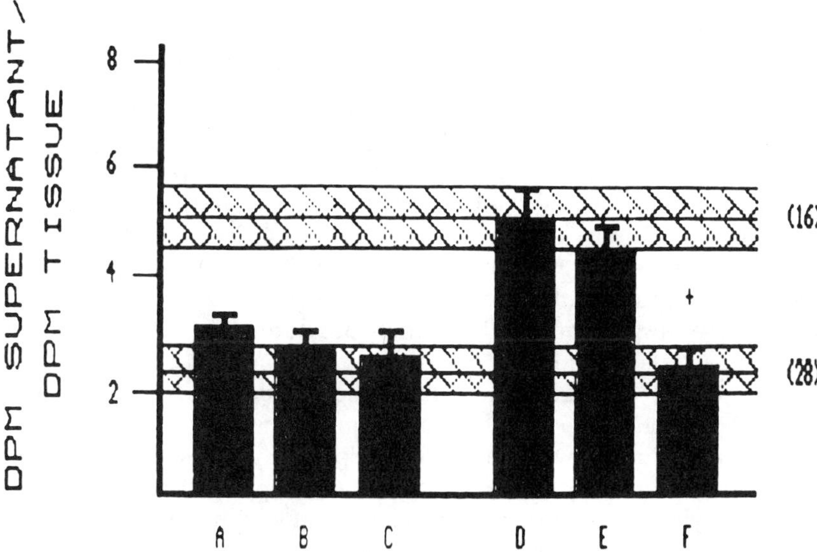

Fig. 24.6. The effect of oxytocin treatment on dopamine release in the mouse forebrain. The release of [^{3}H]dopamine was measured *in vitro* in brain slices of the basal forebrain. The experiments were performed with low potassium (4.2 mmol/l = control values are indicated by the lower horizontal dotted area) as well as high potassium (30 mmol/l = control values are indicated by the upper horizontal dotted area) concentrations (for details see Kovàcs *et al.*, 1986b). (The former gives information about the basal, the latter about stimulated release of dopamine.) Columns A, B, C: low potassium release groups, columns D, E, F: high potassium release groups. *Column A: in vitro* oxytocin treatment. The peptide was added to the incubation medium in a concentration of 1 nmol/l. *Column B:* Acute *in vivo* oxytocin treatment (0.2 mg/kg oxytocin s.c.). Mice were decapitated 1 h after peptide treatment. *Column C:* Chronic *in vivo* oxytocin treatment. The peptide was injected for 8 consecutive days in a dose of 0.2 mg/kg s.c. Mice were decapitated 1 h after the last peptide injection. *Column D:* Identical to group A, but the potassium concentration of the incubation medium was high. *Column E:* Identical to group B, but the concentration of potassium was high, *Column F:* Identical to group C, but the concentration of potassium was high. The *y*-axis indicates the ratio of the DPM of [^{3}H]dopamine in the supernatant, and that of the pellet fraction of the homogenate prepared from forebrain slices (*$p < 0.05$ vs high potassium control.) High ratio indicates increased *in vitro* release.

al., 1987). It is therefore likely that endogenous oxytocin in limbic brain areas inhibited adaptive central nervous mechanisms of experimental drug addiction through central nervous oxytocinergic receptors (Kovàcs *et al.*, 1986a).

24.6. **The role of forebrain dopamine in mediating the effect of oxytocin**

Several data in the literature (for summary see Redmond and Krystal, 1984) support the idea that forebrain dopamine might play an important, presumably causal, role in drug addiction processes. It was therefore of interest to investigate whether oxytocin was able to alter dopaminergic neurotransmission in these brain regions. In rats the effect of oxytocin was more pronounced on the nigrostriatal than on the mesolimbic dopamine system (Kovàcs and Telegdy, 1983; Versteeg, 1983; Van Heuven-Nolsen *et al.*, 1984). In mice, on the other hand, chronic oxytocin treatment decreased the utilization and the receptor binding (Kovàcs *et al.*, 1986b) of dopamine. In the basal forebrain structures, also involving the nucleus accumbens and the posterior olfactory nuclei, chronic treatment with oxytocin significantly inhibited the high potassium-induced stimulated *in vitro* release of dopamine in these brain structures (Fig. 24.6), while the basal (low potassium-induced) release of dopamine was not affected by the same treatment schedule. These and other findings (e.g. the fact that oxytocin interferes with the effect of apomorphine on locomotor activity and modulates the development of receptor supersensitivity following haloperidol treatment, Kovàcs *et al.*, 1986b) support the idea that the effect of oxytocin on narcotic addiction was – at least partly – mediated by dopamine receptors in the basal forebrain.

25.7 **Conclusions**

Recent results from our laboratories (for summary see Kovàcs, 1986), and other laboratories (Krivoy *et al.*, 1974; Van Ree and De Wied, 1977a, b; Van Ree, 1983) suggest that neurohypophyseal neuropeptides modulate the organism's response to drugs of abuse. In the case of oxytocin, adaptive components of drug addiction are affected primarily: the neuropeptide inhibits the development of tolerance to, and physical dependence on, morphine and reduces the self-administration of another frequently abused drug, heroin. These observations – made in rodents – might provide useful insights into the way in which an endogenous neuronal peptide modulates adaptive functions of the central nervous system by acting through its own receptors and by modifying the efficacy of a 'classical' neuronal transmitter system (dopamine) in the forebrain. Further investigations are needed to elucidate the significance of these findings in neuronal regulatory mechanisms of human opiate addiction.

Acknowledgements

At the time when the experimental part of these studies was done, the author was a co-worker of the Institute of Pathophysiology, Szent-Györgyi Medical University, Szeged, Hungary. The oxytocin antagonist was kindly donated by Dr Tomislav Barth (Prague, Czechoslovakia).

References

Buijs, R. M. (1983). Vasopressin and oxytocin – their role in neurotransmission. *Pharmacol. Ther.*, **22**, 127–41.

Collier, H. O. J. (1984). Cellular aspects of opioid tolerance and dependence. In: Hughes, J. *et al.* (eds), *Opioids: Past, Present and Future*. Taylor & Francis, London, pp. 109–25.

D'Amour, F. E. and Smith, D. L. (1941). A method for determining loss of pain sensation. *J. Pharmacol. Exp. Ther.*, **72**, 74–9.

De Kloet, E. R., Rotteveel, F., Voorhuis, T. A. M. and Terlou, M. (1985). Topography of binding sites for neurohypophyseal hormones in rat brain. *Eur. J. Pharmacol.*, **110**, 113–19.

Deneau, G., Yanagita, T. and Seevers, M. H. (1969). Self-administration of psychoactive substances by the monkey. *Psychopharmacology*, **16**, 30–48.

Ferrier, B. M., McClorry, S. A. and Cochrane, A. W. (1983). Specific binding of (^{3}H)oxytocin in female rat brain. *Can. J. Physiol. Pharmacol.*, **61**, 989–95.

Ibragimov, R., Kovàcs, G. L., Szabó, G. and Telegdy, G. (1987). Microinjection of oxytocin into limbic-mesolimbic brain structures disrupts heroin self-administration behavior: a receptor mediated event? *Life Sci.*, **41**, 1264–71.

Jost, K. and Sorm, F. (1971). The preparation of Nα-acetyl-[2-*O*-methyltyrosine]-oxytocin, a powerful inhibitor of the uterotonic activity of oxytocin. *Coll. Czechoslov. Chem. Commun.*, **36**, 297.

Kovàcs, G. L. (1986). Oxytocin and behaviour. In: Ganten, D. and Pfaff, D. (eds), *Current Topics in Neuroendocrinology*. vol. 6. Springer, Berlin, pp. 91–128.

Kovàcs, G. L. and Telegdy, G. (1983). Effects of oxytocin, desglycinamide-oxytocin and anti-oxytocin serum on the α-MPT-induced disappearance of catecholamines in the rat brain. *Brain Res.*, **268**, 307–14.

Kovàcs, G. L. and Telegdy, G. (1985). Role of oxytocin in memory, amnesia and reinforcement. In: Amico, J. A. and Robinson, A. G. (eds), *Oxytocin. Clinical and Laboratory Studies*. Elsevier, Amsterdam, pp. 359–71.

Kovàcs, G. L. and Telegdy, G. (1987a). β-Endorphin tolerance is inhibited by oxytocin. *Pharmacol. Biochem. Behav.*, **26**, 57–60.

Kovàcs, G. L. and Telegdy, G. (1987b). Neurohypophyseal peptides, motivated and drug-induced behaviour. *Front. Horm. Res.*, **15**, 138–74.

Kovàcs, G. L. and Van Ree, J. M. (1985). Behaviourally active oxytocin fragments simultaneously attenuate heroin self-administration and tolerance in rats. *Life Sci.*, **37**, 1895–900.

Kovàcs, G. L., Szontàgh, L., Balàspiri, L., Hódi, K., Bohus, P. and Telegdy, G. (1981). On the mode of action of an oxytocin derivative (*Z*-Pro-D-Leu) on morphine dependence in mice. *Neuropharmacology*, **20**, 647–51.

Kovàcs, G. L., Izbéki, F., Horvàth, Z. and Telegdy, G. (1984). Effects of oxytocin and a derivative (*Z*-prolyl-D-leucine) on morphine tolerance/dependence are mediated by the limbic system. *Behav. Brain Res.*, **14**, 1–8.

Kovàcs, G. L., Borthaiser, Z. and Telegdy, G. (1985). Oxytocin reduces intravenous heroin self-administration in heroin-tolerant rats. *Life Sci.*, **37**, 17–26.

Kovàcs, G. L., Sarnyai, Z., Szabó, G. and Telegdy, G. (1986a). Development of morphine tolerance is under tonic control of brain oxytocin. *Drug Alcohol Dep.*, **17**, 369–75.

Kovàcs, G. L., Faludi, M., Falkay, G. and Telegdy, G. (1986b). Peripheral oxytocin treatment modulates central dopamine transmission in the mouse limbic structures. *Neurochem. Int.*, **9**, 481–5.

Kovàcs, G. L., Sarnyai, Z., Izbéki, F., Szabó, G., Telegdy, G., Barth, T., Jost, K. and Brtnik, F. (1987). Effects of oxytocin-related peptides on acute morphine tolerance: opposite actions by oxytocin and its receptor antagonists. *J. Pharmacol. Exp. Ther.*, **241**, 569–74.

Krejci, I., Kupkovà, B. and Barth, T. (1973). Nα-acetyl-(2-*O*-methyltyrosine)-oxytocin: a specific antagonist of oxytocin. *Acta Physiol. Bohemoslov.*, **22**, 315–22.

Krivoy, W. A., Zimmermann, E. and Lande, S. (1974). Facilitation of development of resistance to morphine analgesia by desglycinamide^{9}-lysine vasopressin. *Proc. Natl. Acad. Sci. USA*, **71**, 1852–6.

Redmond, D. E. and Krystal, J. H. (1984). Multiple mechanisms of withdrawal from opioid drugs. *Ann. Rev. Neurosci.*, **7**, 443–78.

Sarnyai, Z., Viski, S., Krivàn, M., Szabó, G., Kovàcs, G. L. and Telegdy, G. (1988). Endogenous oxytocin inhibits morphine tolerance through limbic forebrain receptors. *Brain Res.*, **463**, 284–8.

Sawchenko, P. E. and Swanson, L. W. (1985). Relationship of oxytocin pathways to the control of neuroendocrine and autonomic functions. In: Amico, J. A. and Robinson, A. G. (eds), *Oxytocin. Clinical and Laboratory Studies*. Elsevier, Amsterdam, pp. 87–104.

Van Heuven-Nolsen, D., De Kloet, E. R. and Versteeg, D. H. G. (1984). Oxytocin affects noradrenaline utilization in distinct limbic-forebrain regions of the rat brain. *Neuropharmacology*, **23**, 1373–7.

Van Ree, J. M. (1979). Reinforcing stimulus properties of drugs. *Neuropharmacology*, **18**, 963–9.

Van Ree, J. M. (1983). Neuropeptides and addictive behavior. *Alcohol Alcoholism*, **18**, 325–30.

Van Ree, J. M. and De Wied, D. (1977a). Heroin self-administration is under control of vasopressin. *Life Sci.*, **21**, 315–20.

Van Ree, J. M. and De Wied, D. (1977b). Effect of neurohypophyseal hormones on morphine dependence. *Psychoneuroendocrinology*, 2, 35–41.

Versteeg, D. H. G. (1983). Neurohypophyseal hormones and brain neurochemistry. *Pharmacol. Ther.*, **19**, 297–325.

Part V

Neuro-immunology

25 *Eric M. Smith and Eve W. Johnson*

Neuropeptides and neuropeptide receptors in the immune system

25.1. Introduction

Traditionally, the immune system has been studied from the viewpoint that it is largely an autonomous unit, separate from the regulatory influences of other organ systems. While this view is advantageous for discerning *in vitro* the many mechanisms by which this complex network of cells and mediators operates to protect against foreign invasion or disease, it can also be misleading in understanding the larger *in vivo* picture. A relatively new and advancing area of research now explores one means by which immunomodulatory mechanisms may be derived from sources not classically considered to be a part of the immune system: neuroimmunoendocrinology defines an intercommunicative, interregulatory relationship that exists between cells of the nervous, endocrine and immune systems (see Ader, 1981; Smith, 1989). While such interactions have been suggested by anecdotal observations, and in studies of psychosocial factors in disease manifestations, it is only currently that the mechanisms by which these phenomena take place are becoming understood on a cellular and molecular level.

Early suggestions that the nervous system can influence immune function are found in correlations between 'state of mind' (i.e. anxiety, depression, stress) and the ability to adequately prevent disease. For example, correlations have been documented between certain personality traits and the incidence of autoimmune diseases, such as rheumatoid arthritis (reviewed in Solomon, 1981). Stress, both in humans and in carefully defined experiments with animals, has been reported to alter immune function profoundly. Marsh and Rasmussen (1960) have shown that immobilization or electric shock causes leukocytopenia, thymic involution, and loss of splenic tissue mass in mice. A greater incidence in mammary tumours has been documented in C3H mice subjected to chronic environmental stress, compared to control mice housed under conditions of low stress (Riley *et al.*, 1981). In humans, both bereavement (Schleifer *et al.*, 1983) and depression (Schleifer *et al.*, 1984) have been shown to correlate with suppressed immune function. Keller and colleagues (1983) have demonstrated in experimental animals that at least some of these stress-mediated changes in immune parameters cannot, as is established in the literature, be attributed to elevated glucocorticoid levels.

Neuroendocrine control of immune function can potentially occur via at least two non-mutually exclusive routes: directly by way of nerves and nerve endings, and indirectly through blood-borne mediators. Direct innervation of immune system tissues has been recognized and recently has been extensively characterized (Felten *et al.*, 1985; Bulloch and Pomerantz, 1984). Direct central nervous system control over immune responses has been demonstrated in conditioning studies and in lesioning studies, to be discussed further. Endocrinological maintenance of immunological integrity is witnessed in hypopituitary Snell–Bagg mice, which have a severe impairment of cell-mediated immunity that can be restored by growth hormone replacement therapy (Fabris *et al.*, 1971a, b).

More specifically our own group has shown that lymphocytes synthesize both neuroendocrine peptide hormones and their receptors (see Smith, 1989, and Blalock and Smith, 1985, for reviews). Various stimuli will induce lymphocytes to produce ACTH, endorphins (Smith and Blalock, 1981), enkephalins (Zurawski *et al.*, 1986), thyroid stimulating hormone (TSH) (Smith *et al.*, 1983), growth hormone (Weigent *et al.*, 1988), prolactin (Hiestand *et al.*, 1986), chorionic gonadotropin (Harbour-McMenamin *et al.*, 1986), and luteinizing hormone (Ebaugh and Smith, 1988). The stimuli include microorganisms, mitogens and hypothalamic releasing hormones (Smith and Blalock, 1981; Harbour-McMenamin *et al.*, 1985; Smith *et al.*, 1986). Conversely, we, and several other laboratories, have found evidence for specific neuroendocrine hormone receptors on lymphocytes, such as for: ACTH (Johnson *et al.*, 1982; Smith *et al.*, 1987), opiates (Carr, 1988), substance P (Payan and Goetzl, 1985), and vasoactive intestinal peptide (O'Dorisio *et al.*, 1985). Based on this evidence, plus many other examples, we feel that one way the immune and nervous systems may communicate is through soluble factors and receptors that are shared between the two systems. In this regard the purpose of this chapter is to demonstrate the similarity of these shared molecules by using an experiment with ACTH receptors on lymphoid cells as an example. This example will provide the basis for discussion of the implications and future directions of our work.

25.2. ACTH receptors on lymphocytes

Our previous studies studies have shown that ACTH will modulate immune responses (Johnson *et al.*, 1982, 1984), bind specifically with high affinity to leukocytes (Johnson *et al.*, 1982; Smith *et al.*, 1987), and induce cAMP production in lymphoid cells (Johnson *et al.*, 1988). This was certainly very strong evidence to show there is an ACTH receptor on lymphoid cells. But the question remained, is it the *same* receptor as on adrenal cells? To answer this question we chose to see if an antiserum that recognized the adrenal ACTH receptor binding site (Bost *et al.*, 1985) could activate the receptor on leukocytes and thereby show identity of the receptor structure.

The system for this study was the S49A T-cell lymphoma cell line (Johnson *et al.*, 1988). We found these cells to bind [^{125}I]ACTH and respond with an elevation in cAMP levels. In order to determine whether the S49A cAMP response induced by ACTH (Johnson *et al.*, 1988) is mediated directly through the ACTH receptor, an antibody known to recognize the ACTH receptor binding site was employed in an experiment illustrated by Table 25.1. The anti-HTCA antibody is a DEAE cellulose-purified fraction from an antiserum generated to the 'complementary peptide' of ACTH, and has been extensively characterized in terms of its ability to block binding of [^{125}I]ACTH to Y-1 adrenal cells and to stimulate steroidogenesis in Y-1 adrenal cells (Bost *et al.*, 1985). Anti-HTCA is thought to recognize the ACTH receptor binding site and is related to the anti-ACTH receptor antiserum in that immunoaffinity-purified material from anti-HTCA columns has been used to prepare another anti-ACTH receptor antiserum (Bost and Blalock, 1986). The experiment shown in Table 25.1 was conducted in a manner that would detect either antibody blockage of ACTH-mediated cAMP production, or antibody stimulation of this response. S49A cells preincubated with anti-HTCA, or simply treated with anti-HTCA, increased their intracellular production of cAMP (76%) to levels within the range of that seen for treatment with ACTH alone (97%). In addition, an even greater increase (155%) in intracellular cAMP was observed for cells preincubated with anti-HTCA, compared with those treated with ACTH, suggesting that ACTH treatment fills any unoccupied receptors to additively increase production of cAMP. Thus it

Table 25.1. Effect of ACTH and anti-HTCA antibody on cAMP induction in S49A, T-cell, lymphoma cells

Pretreatment	*Treatment*	*cAMP (pmol/7 × 10⁶ cells/25 min)*
Media	Media	3.85 ± 0.64
Media	Forskolin	69.32 ± 1.93
Anti-HTCA	Media	6.78 ± 0.91
Anti-HTCA	ACTH	9.85 ± 0.46
Media	ACTH	7.60 ± 0.14
Media	Anti-HTCA	6.63 ± 1.05

Cells (7×10^6/tube) were preincubated as shown, either with media alone or anti-HTCA (1:12 dilution) for 1 h. Cells were then washed and treated with forskolin (10^{-5} M), ACTH (5×10^{-5} M), anti-HTCA (1:12 dilution), or media alone. Both pretreatments and treatments were carried out in the presence of 0.5 mM IBMX. Reactions were stopped by addition of 7.5% TCA to cell pellets, and intracellular cAMP levels were measured as previously described (Peterson *et al.*, 1983).

appears that ACTH mediates activation of adenylate cyclase directly through S49A cell ACTH receptors.

25.3. Implications and future directions

On an antigenic basis these results suggest that the ACTH binding site on lymphoid cells is the same as the prototype ACTH receptor on adrenal cells. The similarities include specific binding of ACTH (Johnson *et al.*, 1982; Smith *et al.*, 1987), activation of adenylate cyclase (Johnson *et al.*, 1988), and now antigenicity. Superficially, one might say that this finding is predictable, based on the previously noted similarities between the lymphoid binding site and the adrenal ACTH receptor. This may be true, but it is a very essential finding to support our hypothesis that the neuropeptide receptors in the immune system are identical to their prototypes. This then has implications for using lymphoid receptors as models or windows into the neuroendocrine system (Smith *et al.*, 1987). The only way to validate such a model is by structural studies as reviewed and presented above, which has primarily been done for the lymphoid VIP receptor (O'Dorisio *et al.*, 1985) before this study.

These results are essential to our future directions. We have shown that a T lymphocyte mitogen, concanavalin A (Con A), will increase the expression of ACTH receptors on T cells (E. W. Johnson and E. M. Smith, manuscript in preparation) and therefore believe that ACTH receptors may be classified as an activation antigen. If true, this means that this neuroendocrine receptor can serve as a marker of immune functions. This may then be a specific link between the immune and neuroendocrine systems that could mediate regulatory effects such as those proposed for stress-associated suppression of immune responses. It is in this direction that we plan to proceed, to determine what immune function(s) might be modulated by ACTH during activation of the immune system, and to see if the ACTH receptor is a marker for physiological and psychological stresses.

Acknowledgements

E. W. J. was supported by a James W. McLaughlin Predoctoral Fellowship. This work was also funded in part by the Office of Naval Research (N000-14-89-J-1095) and the National Institutes of Health (DK 41034-01). The authors wish to thank Ms Karen Goodwin for typing the manuscript.

References

Ader, R. (1981). *Psychoneuroimmunology*. Academic Press, Orlando, FL.
Blalock, J. E. and Smith, E. M. (1985). Lymphocyte production of endorphin and

ACTH-like peptides: A complete regulatory loop between the immune and neuroendocrine systems. *Fed Proc.*, **44**, 108–11.

Bost, K. L. and Blalock, J. E. (1986). Molecular characterization of a corticotropin (ACTH) receptor. *Molec. Cell. Endocrinol.*, 44, 1–9.

Bost, K. L., Smith, E. M. and Blalock, J. E. (1985). Similarity between the corticotropin (ACTH) receptor and a peptide encoded by an RNA that is complementary to ACTH mRNA. *Proc. Natl. Acad. Sci.*, **82**, 1372–5.

Bulloch, K. and Pomerantz, W. (1984). Autonomic nervous system innervation of thymic-related lymphoid tissues in wild type and nude mice. *J. Comp. Neurol.*, **228**, 57–68.

Carr, D. J. J. (1988). Opioid receptors on cells of the immune system. *Prog. Neuroendocrinol. Immunol.*, **1**(2), 8–14.

Ebaugh, M. J. and Smith, E. M. (1988). Human lymphocyte production of immunoreactive luteinizing hormone. *FASEB J.*, **2**, A1642.

Fabris, N., Pierpaoli, W. and Sorkin, E. (1971a). Hormones and the immunologic capacity. III. The immunodeficiency disease of hypopituitary Snell-Bagg dwarf mice. *Clin. Exp. Immunol.*, **9**, 209.

Fabris, N., Pierpaoli, W. and Sorkin, E. (1971b). Hormones and the immunologic capacity. IV. Restorative effects of the developmental hormones or of lymphocytes on the immunodeficiency syndrome of the dwarf mouse. *Clin. Exp. Immunol.*, **101**, 1036.

Felten, D. L., Felten, S. Y., Carlson, S. L., Olschowka, J. A. and Livnat, S. (1985). Noradrenergic and peptidergic innervation of lymphoid tissue. *J. Immunol.*, **135**, 755s–765s.

Harbour-McMenamin, D., Smith, E. M. and Blalock, J. E. (1985). Bacterial lipopolysaccharide induction of leukocyte derived ACTH and endorphins. *Infect. Immun.*, **48**, 813–17.

Harbour-McMenamin, D., Smith, E. M. and Blalock, J. E. (1986). Production of immunoreactive chorionic gonadotropin during mixed lymphocyte reactions: A possible selective mechanism for genetic diversity. *Proc. Natl. Acad. Sci. USA.*, **83**, 6834–8.

Hiestand, P. C., Meker, P., Nordmann, R., Alfons, G. and Permmongkol, G. (1986). Prolactin as a modulator of lymphocyte responsiveness provides a possible mechanism of action for cyclosporine. *Proc. Natl. Acad. Sci. USA*, **83**, 2599–603.

Johnson, E. W., Blalock, J. E. and Smith, E. M. (1988). ACTH receptor-mediated induction of leukocyte cAMP. *Biochem. Biophys. Res. Commun.*, **157**, 1205–11.

Johnson, H. M., Smith, E. M. Torres, B. A. and Blalock, J. E. (1982). Regulation of *in vitro* antibody responses by neuroendocrine hormones. *Proc. Natl. Acad. Sci. USA*, **79**, 4717–4.

Johnson, H. M., Torres, B. A., Smith, E. M., Dion, L. D. and Blalock, J. E. (1984). Regulation of lymphokine (γ-interferon) production by corticotropin. *J. Immunol.*, **132**, 246–50.

Keller, S. E., Weiss, J. M., Schliefer, S. J., Miller, N. E. and Stein, M. (1983). Stress-induced suppression of immunity in adrenalectomized rats. *Science*, **221**, 1302–4.

Marsh, J. T. and Rasmussen, A. F., Jr (1960). Response of adrenals, thymus, spleen, and leukocytes to shuttle box and confinement stress. *Proc. Soc. Exp. Biol. Med.*, **104**, 180–3.

O'Dorisio, M. S., Wood, C. L. and O'Dorisio, T. M. (1985). Vasocative intestinal peptide and neuropeptide modulation of the immune response. *J. Immunol.*, **135**, 792s–796s.

Payan, D. G. and Goetzl, E. J. (1985). Modulation of lymphocyte function by sensory neuropeptides. *J. Immunol.*, **135**, 783s–786s.

Peterson, J. W., Molina, N. C., Houston, C. W. and Fader, R. C. (1983). Elevated cAMP in intestinal epithelial cells during experimental cholera and salmonellosis. *Toxicon*, **21**, 761–76.

Riley, V., Fitzmaurice, M. A. and Spackman, D. H. (1981). Psychoneuroimmunologic factors in neoplasia. In: Ader, R. (ed.), *Psychoneuroimmunology*. Academic Press, Orlando, FL, pp. 31–102.

Schleifer, S. J., Keller, S. E., Camarino, M., Thornton, J. C. and Stein, M. (1983). Suppression of lymphocyte stimulation following bereavement. *J. Am. Med. Assoc.*, **250**, 375–377.

Schleifer, S. J., Keller, S. E., Meyerson, A. T., Raskin, M. J., Davis, K. L. and Stein, M. (1984). Lymphocyte function in major depressive disorder. *Arch. Gen. Psychiatry*, **41**, 484–6.

Smith, E. M. (1989). Neuropeptide gene expression in lymphoid tissue. In: Inoue, S. and Krueger, J. M. (eds), *Endogenous Sleep Factors*. Bouma Text, Wassenaar (In press).

Smith, E. M. and Blalock, J. E. (1981). Human lymphocyte production of ACTH and endorphin-like substances. Association with leukocyte interferon. *Proc. Natl. Acad. Sci. USA*, **78**, 7530–4.

Smith, E. M., Phan, M., Coppenhaver, D., Kruger, T. E. and Blalock, J. E. (1983). Human lymphocyte production of immunoreactive thyrotropin. *Proc. Natl. Acad. Sci. USA*, **80**, 6010–13.

Smith, E. M., Morrill, A. C., Meyer, W. J. and Blalock, J. E. (1986). Corticotropin releasing factor production of leukocyte-derived immunoreactive ACTH and endorphins. *Nature*, **321**, 881–2.

Smith, E. M., Brosnan, P., Meyer, W. J. and Blalock, J. E. (1987). An ACTH receptor on human mononuclear leukocytes. *N. Engl. J. Med.*, **317**, 1266–9.

Solomon, G. F. (1981). Emotional and personality factors in the onset and course of autoimmune disease, particularly rheumatiod arthritis. In: Ader, R. (ed.), *Psychoneuroimmunology*. Academic Press, Orlando, FL, pp. 159–82.

Weigent, D. A., Baxter, J. B., Wear, W. E., Smith, L. R., Bost, K. L. and Blalock, J. E. (1988). Production of immunoreactive growth hormone by mononuclear leukocytes. *FASEB J.*, **2**, 2812–18.

Zurawski, G., Benedik, M., Kamb, B. J., Abrams, J. S., Zurawski, S. M. and Lee, F. D. (1986). Activation of mouse T-helper cells induces abundant proenkephalin in RNA synthesis. *Science*, **232**, 772–5.

26 *Michael K. Leung and Jonathan Lundy*

Opioid neuropeptides in invertebrate haemolymphs

26.1. Introduction

Shortly after the discovery of opioid neuropeptides in nerve tissues they were reported to be present in the plasma (Yamaguchi *et al.*, 1980; Clement-Jones *et al.*, 1980a). At the time little was known about their source and target tissues. It was generally assumed they were released by an endocrine gland and were on transit in the plasma to the target organ. It has been suggested that at least some of these opioid neuropeptides are released from the opioid-rich adrenal medulla in response to specific nerve signals (Schultzberg *et al.*, 1978). The actual source of plasma opioids is still not well understood. However, recent findings have shown that opioid neuropeptides have specific function in the plasma and they are not simply on transit to another part of the body. It is now well established that opioid neuropeptides such as Met- and Leu-enkephalin and β-endorphin play a significant role in the neuroimmune system (Sibinga and Goldstein, 1988). They exert powerful influence on plasma leukocytes and appear to function as modulators of the neuroimmune system.

Our recent studies have shown that a similar neuroimmune mechanism is also likely to exist in invertebrates. Previously, we have demonstrated the presence of Met- and Leu-enkephalin and Met-enkephalin-Arg-Phe in the pedal ganglia of the marine mussel, *Mytilus edulis* (Leung and Stefano, 1984; Stefano and Leung, 1984). Our subsequent studies also showed the synthetic Met-enkephalin analog, D[Ala^2]-Met-enkephalinamide (DAMA), is capable of inducing immune related activities such as cell adhesion, cell aggregation and chemotaxis in the immunocompetent granulocytes of *Mytilus* and *Leucophaea* hemolymphs (Stefano *et al.*, 1989). In this chapter we will review our current findings on Met-enkephalin-related neuropeptides in the haemolymphs of *Mytilus* and *Leucophaea*.

26.2. Chemical changes of Met-enkephalin induced by extraction process

Many different procedures have been used in the extraction of opioid neuropeptides from biological tissues. However, the question of whether any of these procedures are capable of causing chemical alteration to Met-enkephalin has not been studied in detail. It is well known that the thiol group of methionine is easily oxidized. Met-enkephalin is readily oxidized to Met-enkephalin sulphoxide (MEO) when extractions were carried out without the protection of reducing agent (Clement-Jones *et al.*, 1980b). Thus, the presence of MEO in tissue extracts has generally been assumed to be simply an artifact of the extraction procedure.

To better understand the chemical alteration of Met-enkephalin during extraction we studied the effects of several organic solvents which have been widely used in the extraction of Met-enkephalin. In our investigation lyophilized Met-enkephalin was exposed directly to each of the organic solvents individually in the absence of any protective agent. The solvents that were used included methanol, ethanol, acetone, acetonitrile and ether. All the solvents were obtained commercially in HPLC or ACS grade and they were used without additional purification. Thus, any observed alterations are likely to be caused by the trace contaminants in the solvents instead of by the solvents themselves. The Met-enkephalin was analysed for any chemical alteration by HPLC after it was exposed to each of the organic solvents. Our results showed that Met-enkephalin is oxidized to the sulphoxide form when exposed to ether, while it is altered to a yet-undefined structure with acetone. At this point nothing is known about the chemical structure of the altered Met-enkephalin after its exposure to acetone except that it has a HPLC R_t greater than that of Met-enkephalin, suggesting it is less polar than Met-enkephalin. Of the other solvents tested none appear to induce any chemical change to Met-enkephalin.

26.3. Extraction of invertebrate haemolymphs

Based on our findings, it is clear any chemicals that are to be used in an extraction solution must be tested carefully. This is particularly true of ether and acetone, as they have been widely used in many extraction procedures. Accordingly, a new extraction procedure is designed for the extraction of haemolymphs in our current study. The extraction solution is composed of 1 M acetic acid containing 1% 2-mercaptoethanol and 1 μg/ml each of pepstatin A, leupeptin and phenylmethylsulphonyl fluoride. Cell-free *Mytilus* haemolymph fluid was homogenized in an equal volume of the extraction solution. After centrifugation the supernatant is lyophilized, dissolved in a solution of 10% CH_3CN in 0.1% trifluoroacetic acid (TFA) and clarified by centrifugation before being fractionated by HPLC. An identical test run carried out with a solution of Met-enkephalin standard in artificial sea water showed no detectable level of MEO or any other chemically altered Met-enkephalin when analysed by HPLC. Thus, we believe the extraction procedure as described does not introduce any chemical change to Met-enkephalin.

A similar procedure was used for the extraction of *Mytilus* haemocytes. Haemocytes were collected from the haemolymph by centrifugation at 900

× *g* for 10 min. They were then homogenized with a small volume of extraction solution and processed as described for haemolymph. The haemocyte extract was not fractionated and was used only for estimation of total Met-enkephalin activity.

26.4. **Presence of Met-enkephalin and its related neuropeptides in haemolymphs**

The most effective biochemical assay currently available for the detection of neuropeptides in biological samples is that of the HPLC-RIA technique. In our laboratory this technique has been used in the studies of Met- and Leu-enkephalin-related materials in invertebrate neural tissues (Leung *et al.*, 1987; Stefano *et al.*, 1989). This procedure was modified for our study of invertebrate haemolymphs. The HPLC fractionations of haemolymph extracts were carried out with a reverse-phase column. It was eluted at a flow rate of 1 ml/min with a gradient using 0.1% TFA as solvent A and 80% CH_3CN in 0.1% TFA as solvent B. Figure 26.1 shows a typical separation of Met-enkephalin and MEO standards by HPLC. Under the described conditions the R_t values of Met-enkephalin and MEO are 10.6 and 15.8 min, respectively. One-millilitre fractions were collected for RIA.

An aliquot was removed from selected HPLC fractions for quantification

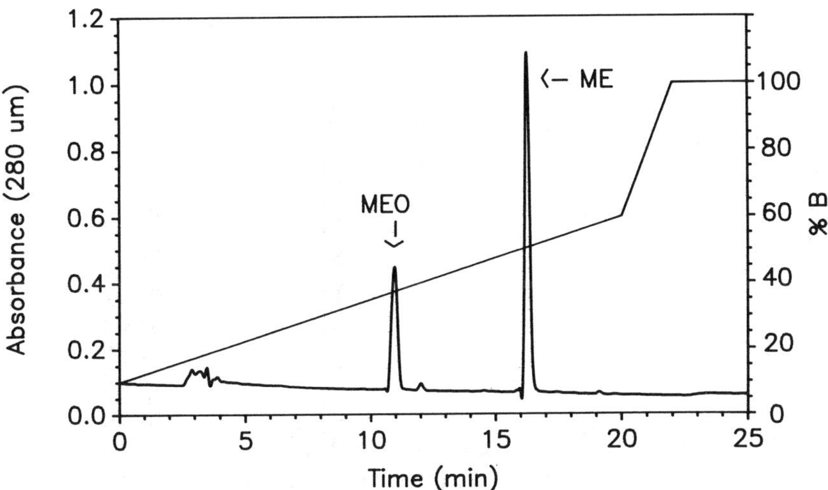

Fig. 26.1. HPLC chromatogram of Met-enkephalin and MEO standards. The standards were injected into a Brownlee Aquapore RP-300 C_8 reverse-phase column. The column was eluted at a flow rate of 1 ml/min with a binary solvent system consisting of 0.1% TFA and 80% CH_3CN in 0.1% TRA. The gradient used is as illustrated in the figure. ME = Met-enkephalin; MEO = Met-enkephalin sulphoxide.

of Met-enkephalin activity with RIA kits purchased from Incstar (Stillwater, MN, USA). Duplicated samples from each fraction were assayed and the results were averaged. Similar assays were performed with serially diluted unfractionated extracts for estimation of total activities. The antiserum used in these kits has a cross-reactivity of 72% for MEO and <3% for Met-enkephalin-Lys, Met-enkephalin-Lys-Arg, Met-enkephalin-Arg-Phe, Met-enkephalinamide, Leu-enkephalin, β-endorphin or dynorphin. Standard curves were established for both Met-enkephalin and MEO.

Figure 26.2 shows the RIA results for the HPLC fractions of *Mytilus* haemolymph. The major activities are found in fractions corresponding to the R_t values of Met-enkephalin and MEO. The estimated haemolymph Met-enkephalin level is 1.03 ng/ml while that of MEO is 1.07 ng/ml after making adjustment for the lower reactivity (72%) of the antiserum for MEO. Minor activities are also found in a few other fractions. The nature of the materials which are responsible for these activities is unknown at present. Similar results were obtained with studies carried out with whole *Leucophaea* haemolymph. RIA results obtained from unfractionated cell-free haemolymph fluid and haemocyte extracts from *Mytilus* showed nearly all of the Met-enkephalin reactivities are associated with the fluid

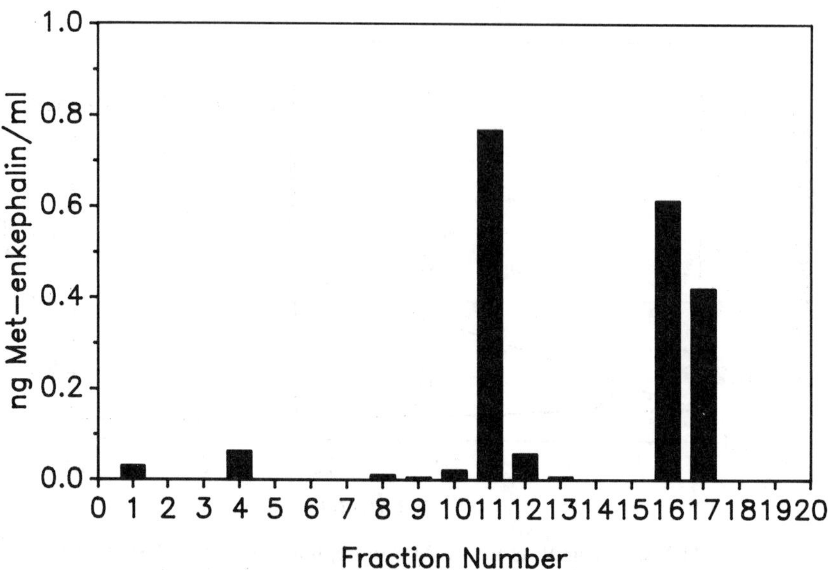

Fig. 26.2. Met-enkephalin RIA activities in HPLC fractions of *Mytilus* haemolymph extract. The extraction procedures are as described in the text. The HPLC fractionation procedures are as described in Fig. 26.1. The activities in the MEO fractions are expressed as Met-enkephalin equivalent.

portion of the haemolymph and only 4% of the reactivities are associated with the haemocytes. Since the haemocytes were not washed before extraction the actual amount contained in the cells should be less than 4%. Distribution of Met-enkephalin activities between the fluid and cell portions of *Leucophaea* haemolymph was not performed.

Our research findings showed Met-enkephalin and its related materials are present in the haemolymphs of invertebrates. The levels of Met-enkephalin in haemolymphs of *Mytilus* and *Leucophaea* are in the same range as in mammalian blood (Clement-Jones *et al.*, 1980a). In mammals the main source of plasma Met-enkephalin is generally considered to be that of the adrenal medulla which is rich in proenkephalin-related materials (Schultzberg *et al.*, 1978). In invertebrates Met-enkephalin has been shown to be present in several ganglionic tissues (Leung and Stefano, 1987). It is not yet clear if any ganglia are involved in the release of Met-enkephalin into the haemolymph. This will be one of the goals of our future research efforts. Results from our recent studies have demonstrated Met-enkephalin and its synthetic analogue, DAMA, exert strong influence on haemolymph immunocompetent cells. The types of Met-enkephalin-induced activities found in the haemocytes appear to parallel those found in plasma leukocytes (Falke and Fischer, 1985; Van Epps and Salant, 1984). The overall picture which emerged from our current findings strongly suggested a neuroimmune mechanism which involves Met-enkephalin as a modulator is also operating in the invertebrate. Thus, additional studies with invertebrate haemolymph could provide us with an uncomplicated path to the understanding of the complex neuroimmune mechanism.

26.5. **The origin of Met-enkephalin sulphoxide (MEO)**

The presence of MEO in biological extract has in the past been generally assumed to be an artifact of the extraction process. To determine if this is indeed the case we analysed in detail both the chemicals and the procedures used in our study. In the design of our extraction solution the use of both acetone and ether was avoided. Tests were carried out to verify if the integrity of Met-enkephalin is preserved by the procedures used in our studies (see above). When extraction of haemolymph was performed with the described extraction solution and procedure, MEO was still detected in the extract. This finding leads us to think that MEO may exist endogenously.

Even though there is no direct evidence for the hypothesis that MEO is a natural product, it is, however, supported by a number of published findings. First, it has been known for a long time that methionine residues in proteins can be oxidized to methionine sulphoxide endogenously. In the case of the extensively studied α_1-proteinase inhibitor, it has been shown that *in vivo* oxidation of methionine residues within this protein can occur readily under certain physiological conditions (Carp *et al.*, 1982). This oxidation renders the inhibitor ineffective in its protection of the alveolar lining. The destruction of the lining, in turn, leads to emphysema. Cigarette smoking appears to increase the oxidation of this inhibitor, presumably by increasing endogenous oxidant generated by superoxide (Carp *et al.*, 1982). Second, studies carried out with rat and human PMNL and macrophage have demonstrated that Met-enkephalin is capable of inducing superoxide production by these cells (Foris *et al.*, 1984; Sharp *et al.*, 1985). This increase has a bell-shaped dose–response curve and is reversible by naloxone. Finally, latex bead-induced phagocytic leukocytes have been reported to be capable of rapidly oxidizing free intracellular and extracellular methionine (Tsan and Chen, 1980). This oxidation of methionine has been linked to production of superoxide by phagocytic leukocytes. This oxidation phenomenon was not observed in similar experiments carried out with leukocytes isolated from patients suffering from chronic granulomatous disease. Chronic granulomatous disease is a genetic disorder in which the leukocytes from individuals with the disease are defective in superoxide production. Furthermore, superoxide production inhibitors such as azide and cyanide are capable of abolishing the oxidation of methionine by phagocytic leukocytes.

Endogenous chemical modifications of opioid neuropeptides are well documented phenomena (Akil *et al.*, 1984). Amidation and acetylation are two of the most commonly observed modifications. Interestingly, the biological activities of β-endorphin can be abolished by acetylation of the N-terminal amine (Smyth *et al.*, 1979). Since MEO is also reported to have no biological activity the oxidation of Met-enkephalin may provide an endogenous means of inactivating Met-enkephalin. The evidence presented by our present study, although not conclusive, does lend strong support for the endogenous existence of MEO. Assuming future studies would prove this to be true, we would like to propose here a hypothesis for a Met-enkephalin-mediated phagocytic mechanism. In this hypothesis we propose immune activities including superoxide production in immunocompetent cells are activated when Met-enkephalin is released, in response to neural stimulation, into the plasma of mammals or the haemolymphs of invertebrates. However, in order to prevent injury to the organism by an overproduction of superoxide a regulatory mechanism analogous to the feedback inhibition mechanism of the biosynthetic pathways is needed. The control of superoxide production can be achieved via the oxidation of Met-enkephalin with oxidant generated by superoxide. In brief, as Met-enkephalin induces the increase of superoxide production, the superoxide, along with destroying foreign particles, will also oxidize the Met-enkephalin to MEO; this in turn will slow down the production of superoxide. Experiments are in progress to provide confirmation for this proposed hypothesis.

26.6. **Degradation of enkephalin in haemolymph**

Met-enkephalin is rapidly degraded in mammalian plasma (Hambrook *et al.*, 1976). It appears that three classes of enzymes are involved in the degradation of enkephalin (Roscetti *et al.*, 1985). They are dipeptidylcarboxypeptidase, dipeptidylaminopeptidase and aminopeptidase. These same enzymes are also found in the nervous tissues. There is, at present, no concrete evidence indicating if any or all of these enzymes are truly responsible for the degradation of Met-enkephalin *in vivo*. The activities of these enzymes in some mammalian plasma have been separated into different chromatographic fractions. The dipeptidylcarboxypeptidase is inhibited by metalloenzyme inhibitors such as phosphoramidon and thiorphan, and the aminopeptidase is inhibited by bestatin (Lynch and Synder, 1986). Although some of these enzymes were studied more than others, none of them have been fully characterized.

As part of our study we investigated how Met-enkephalin is degraded in the invertebrate haemolymph as compared to mammalian plasma. Fresh haemolymph was removed from *Mytilus* as previously described (Stefano *et al.*, 1989). The haemolymph was pooled and an equal volume of whole haemolymph was aliquoted into different vials. Met-enkephalin was added to each of these vials to a final concentration of 500 μM from a stock solution of 5mM Met-enkephalin in artificial sea water. They were then incubated at 23°C with gentle agitation. At 5-min intervals a vial was removed and the incubation mixture was inactivated by heating in boiling water for 5 min. The control for this experiment was prepared by heating the sample immediately after the addition of the Met-enkephalin. A parallel experiment was carried out with MEO instead of Met-enkephalin. The inactivated mixtures were clarified by centrifugation and an aliquot was removed for analysis by HPLC.

HPLC analysis of the incubation mixtures at 280 μm showed Met-enkephalin and MEO are rapidly degraded by *Mytilus* haemolymph. Figure 26.3. is a plot showing the degradation of both substances by haemolymph with time. With an initial concentration of 500 μM, 50% of Met-enkephalin and MEO were degraded in about 9–10 min. This rate of degradation is comparable to that determined in human plasma. There is very little difference between the degradation curves of both Met-enkephalin and MEO. With increasing incubation time at least two peaks in the HPLC chromatograms appeared to increase in coordination with the decrease of Met-enkephalin. These peaks probably represent the degradative products of Met-enkephalin. Interestingly, a significant increase was observed in only one of these peaks with MEO, indicating that

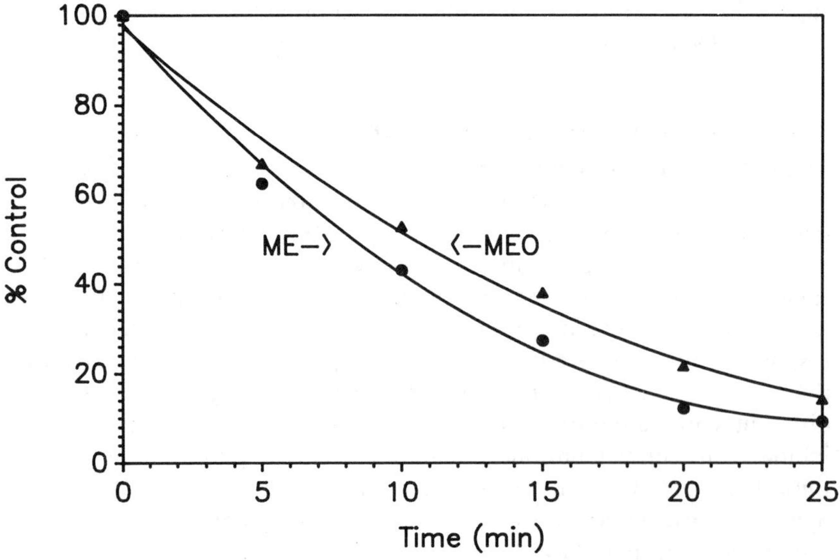

Fig. 26.3. Degradation of Met-enkephalin and MEO by *Mytilus* haemolymph. Controls were prepared by heat-inactivating the samples immediately after the addition of substrates. Each sample was analysed by HPLC as described in Fig. 26.1. The amount of undegraded substrate left at the end of each incubation period was calculated from the area under the peak and is expressed as percentage of control. The percentage of control is plotted against the time of incubation. ME = Met-enkephalin (●); MEO = Met-enkephalin sulphoxide (▲)

Met-enkephalin and MEO are degraded at different rates by different enzymes.

Overall, our data showed that, similar to mammalian plasma, Met-enkephalin and MEO are rapidly degraded in invertebrate haemolymph. Insufficient information is available at present to allow for a detailed comparison of the degradative enzymes found in haemolymph and plasma. Experiments are currently in progress to analyse the degradative products in greater detail. Their identifications should allow us to determine the types of enzymes involved in the degradation. Additional information can also be obtained from inhibition studies using inhibitors specific for the different degradative enzymes.

26.7. **Conclusion**

Our studies have shown the presence of Met-enkephalin and MEO in the haemolymphs of *Mytilus* and *Leucophaea*. Their presence in haemolymph – together with our earlier finding that the stable Met-enkephalin analogue, DAMA, exerts a strong influence on the immunocompetent cells in haemolymph – provide strong support for a Met-enkephalin-mediated neuroimmune mechanism in invertebrates. It appears MEO may have an endogenous origin. Hypothetically, the oxidation of Met-enkephalin may serve as a way of preventing Met-enkephalin from generating an excessive level of superoxide which could lead to injury to the organism. The presence of enzymes in invertebrate haemolymph, which could rapidly degrade both Met-enkephalin and MEO, suggests that the levels of Met-enkephalin and MEO in haemolymph are in a dynamic flux. Their concentration at any moment is controlled by the rate of release against that of degradation. In a situation analogous to neurotransmitter in the synapse, it appears Met-enkephalin is designed to produce an immediate response with short duration. Finally, the study of the neuroimmune mechanism in invertebrates serves more than just to provide us with an understanding of the evolutionary aspect of this mechanism. More importantly, the use of an invertebrate model would provide us with a simpler method of obtaining essential information which will assist our study of the more complex situation in higher organisms.

Acknowledgements

This work was supported by grants NIH-MBRS RR 18180, and ADAMHA-MARC MH 17138. Mr Jonathan Lundy is a student participant of the National Institutes of Health, Minority Biomedical Research Program at the State University of New York at Old Westbury.

References

Akil, H., Watson, S. J., Young, E., Lewis, M. E., Khachaturian, H. and Walker, J. M. (1984). Endogenous opioids: biology and function. *Ann. Rev. Neurosci.*, **7**, 223–55.

Carp, H., Miller, F., Hoidal, J. R. and Janoff, A. (1982). Potential mechanism of emphysema: α-proteinase inhibitor recovered from lungs of cigarette smokers contains oxidized methionine and has decreased elastase inhibitory capacity. *Proc. Natl. Acad. Sci. USA*, **9**, 2041–5.

Clement-Jones, V., Lowry, P. J., Rees, L. H. and Besser, G. M. (1980a). Met-enkephalin circulates in human plasma. *Nature* (*Lond.*), **283**, 295–7.

Clement-Jones, V., Lowry, P. J., Rees, L. H. and Besser, G. M. (1980b). Development of a specific extracted radioimmunoassay for methionine enkephalin in human plasma and cerebrospinal fluid. *J. Endocrinol.*, **86**, 231–43.

Falke, N. E. and Fischer, E. G. (1985). Cell shape of polymorphonuclear leukocytes is influenced by opioids. *Immunobiology.*, **169**, 532–9.

Foris, G., Medgyesi, G. A., Gyimesi, E. and Hauck, M. (1984). Met-enkephalin-induced alterations of macrophage functions. *Mol. Immunol.*, **21**, 747–50.

Hambrook, J. M., Morgan, B. A., Rance, M. J. and Smith, C. F. C. (1976). Mode of deactivation of the enkephalins by rat and human plasma and rat brain homogenates. *Nature* (*Lond.*), **262**, 782–783.

Leung, M. K. and Stefano, G. B. (1984). Isolation and identification of enkephalins in pedal ganglia of *Mytilus edulis* (Mollusca). *Proc. Natl. Acad. Sci. USA*, **81**, 955–8.

Leung, M. K. and Stefano, G. B. (1987). Comparative neurobiology of opioids in invertebrates with special attention to senescent alterations. *Prog. Neurobiol.*, **28**, 131–59.

Leung, M. K., Kessler, H., Whitfield, K., Murray, M., Martinez, E. A. and Stefano, G. B. (1987). The presence of enkephalin-like substances in the eyestalk and brain of the land crab, *Gecarcinus lateralis*. *Cell. Mol. Neurobiol.*, **7**, 91–6.

Lynch, D. R. and Snyder, S. H. (1986). Neuropeptides: multiple molecular forms, metabolic pathways, and receptors. *Ann. Rev. Biochem.*, **55**, 773–99.

Roscetti, G., Possenti, R., Bassano, E. and Roda, L. G. (1985). Mechanism of Leu-enkephalin hydrolysis in human plasma. *Neurochem. Res.*, **10**, 1393–404.

Schultzberg, M., Lundberg, J. M., Hökfelt, T., Terenius, L., Brandt, J., Elde, R. P. and Goldstein, M. (1978). Enkephalin-like immunoreactivity in gland cells and nerve terminals of the adrenal medulla. *Neuroscience*, **3**, 1169–86.

Sharp, B. M., Keane, E. F., Suh, H. J., Gekker, G., Tsukayama, D. and Peterson, P. K. (1985). Opioid peptides rapidly stimulate superoxide production by human polymorphonuclear leukocytes and macrophages. *Endocrinology*, **117**, 793–5.

Sibinga, N. E. S. and Goldstein, A. (1988). Opioid peptides and opioid receptors in cells of the immune system. *Ann. Rev. Immunol.*, **6**, 219–49.

Smyth, D. G., Massey, D. E., Zakarian, S. and Finnie, M. D. A. (1979). Endorphins are stored in biologically active and inactive forms: isolation of alpha-N-acetyl peptides. *Nature* (*Lond.*), **279**, 252–4.

Stefano, G. B. and Leung , M. K. (1984). Presence of Met-enkephalin-Arg^6-Phe^7 in molluscan neural tissues. *Brain Res.*, **298**, 362–5.

Stefano, G. B., Leung, M. K., Zhoa, X. and Scharrer, B. (1989). Evidence for the involvement of opioid neuropeptides in the adherence and migration of immunocompetent invertebrate hemocytes. *Proc. Natl. Acad. Sci. USA*, **86**, 626–30.

Tsan, M. and Chen, J. W. (1980). Oxidation of methionine by human polymorphonuclear leukocytes. *J. Clin. Invest.*, **65**, 1041–50.

Van Epps, D. E. and Salant, L. (1984). Endorphin and met-enkephalin stimulate human peripheral blood mononuclear cell chemotaxis. *J. Immunol.*, **132**, 3046–53.

Yamaguchi, H., Liotta, A. S. and Krieger, D. T. (1980). Simultaneous determination of human plasma immunoreactive β-lipotropin, gamma-lipotropin, and β-endorphin utilizing immune-affinity chromatography. *J. Clin. Endocrinol. Metab.*, **51**, 1002–8.

27 *George B. Stefano, Patrick Cadet, Juan Sinisterra and Berta Scharrer*

Comparative aspects of the response of human and invertebrate immunocytes to stimulation by opioid neuropeptides

27.1. Introduction

A major recent advance in the analysis of the diverse roles played by neuropeptides in intercellular communication concerns their involvement in immunoregulatory processes. They participate in the bidirectional exchange of information among the nervous, the endocrine and the immune systems (Smith and Blalock, 1981; Blalock, 1989) as well as autoregulatory processes within the immune system (Brown *et al.*, 1986; Stefano, 1989). Comparable regulatory phenomena also occur in the immune system of invertebrates (Stefano *et al.*, 1989a, b). In the mollusc *Mytilus edulis* and the insect *Leucophaea maderae*, exogenous and endogenous opioid signal molecules stimulate cellular adherence and migratory activity of immunocompetent haemocytes (immunocytes).

The operation of stereospecific receptors in these processes is demonstrated by the fact that they are markedly reduced in the presence of the opiate antagonist naloxone (Stefano *et al.*, 1989a).

27.2. Image analysis of immunocytes

This chapter is primarily concerned with a comparative analysis of the kinetic effects of opioids on invertebrate and vertebrate immunocytes (stimulation of locomotory behaviour and conformational changes). *In vitro* tests were carried out with haemolymph of *Mytilus* and human blood. The opioid of choice used in these experiments was the synthetic enkephalin analogue DAMA (D-Ala^2, Met^5-enkephalin). For the examination of *Mytilus* immunocytes 0.1 ml of haemolymph was withdrawn, placed on a slide coated with bovine albumin and incubated with DAMA, naloxone, or DAMA plus naloxone, as described elsewhere (Stefano *et al.*, 1989a, b). An inverted coverslip placed over the slide on a ring of Vaseline (to prevent drying out) permitted observation for periods up to 2 days. The mixture was allowed to incubate at room temperature (23°C) for at least 15 min, before the results were recorded. Human blood, obtained from the Long Island Blood Service in Melville, New York, was handled in the same way (Stefano *et al.*, 1989b).

The cell preparations were examined by use of phase contrast and Nomarski optics coupled with a Zeiss Axiophot microscope. Measurements were taken with the Zeiss Videoplan/Vidas image analysis system. Before they were recorded specific images were converted to binary images following frame grabbing. Simultaneously, specific cells were photographed by an internal Zeiss automated program photography system (35 mm film) with video-photography with a time-lapse video synchronization system (JVC).

Changes in cellular conformation, based on measurements of cellular area and perimeter, were mathematically expressed by use of the Form-Factor-pe (FF) calculation of the Zeiss Vidas analysis system, whereby the equation $(4 \times \pi \times \text{area})/\text{perimeter}^2$ provides mean numerical values. The lower this number, the higher is the cellular perimeter and the more amoeboid the cellular shape. Velocity measurements were taken by determining the distance travelled in regard to time. In both preparations the presence of the opioid markedly stimulated the locomotor activity of the immunoresponsive cells, as well as the conformational changes associated with this activity. In the presence of naloxone these effects of DAMA were markedly reduced. Changes in cellular shape effected by DAMA ranged from rounded (inactive) to amoeboid (activated). In both groups the FF mean values were below 0.40 after stimulation.

27.3. **Latency of opioid effect and velocity of cellular movements**

The effects of opioid stimulation showed a major difference between vertebrate and invertebrate immunocytes with respect to their time-course, namely the speed of their onset as well as the velocity of cellular movements. The data obtained from preliminary velocity measurements showed that locomotion of *Mytilus* immunocytes was initiated approximately 20–30 min following their exposure to DAMA. By contrast, human cells responded within less than 5 min (average about 2 min). Stimulated *Mytilus* cells had a velocity of approximately 1.3 μm/min, whereas responsive human blood cells moved at a rate of 4 μm/min. By comparison the distances travelled by unstimulated control cells were 0.7 and 1.7 μm/min, respectively.

In summary, under the anaerobic conditions of these experiments, human blood cells respond to opioid challenge more quickly, and they move with greater speed than *Mytilus* immunocytes.

27.4. **Response to stress**

Some additional information on the nature of immunocyte stimulation by opioids was gained by comparing their effects in slide tests carried out with normal blood with those observed in blood taken from animals that had been subjected to stressful stimuli. In *Mytilus* a treatment found to be effective in eliciting responses considered to be attributable to 'stress' was a combination of successive brief electrical shocks of 1 V for 10 ms on the in-current siphon, and the presence of a wedge preventing closure of the siphon. The haemolymph of these five animals contained a considerably larger number of 'activated' amoeboid cells than that of controls (Stefano *et al.*, 1989b). The FF values were between 0.35 and 0.27. The velocity with which these immunocytes moved was 1.9 μm/min. Both of these values fall within the range of those recorded in 'unstressed' specimens activated by DAMA or endogenous neuropeptides (Stefano *et al.*, 1989b).

The fact that receptor-mediated activity of endogenous neuropeptides is involved in the response on *Mytilus* immunocytes to stress has been demonstrated by the use of naloxone. The administration of this antagonist, at a concentration of 10^{-8} M previously found to be effective (Stefano *et al.*, 1989a), prior to that of stressful stimuli yielded haemolymph preparations with cellular components, very few of which appeared to have become activated. By contrast, once activated, as a consequence of induced stress, immunocytes no longer responded to the administration of naloxone. Thus in stress-activated immunocytes, naloxone seems to be able to block the onset of mobilization and locomotion, but once it has occurred it cannot reverse it. Moreover, the observation that stress-activated immunocytes did not appear to respond to the administration of exogenous opioid indicates that, under the conditions of the experiment, these cells had reached the (maximal) degree of activation of which they seem to be capable.

27.5. **Comparison of the effects of DAMA with those of related opioid substances**

None of six additional neuropeptides tested with respect to their stimulatory effect on conformational changes and locomotory activity of invertebrate and human immunocytes was as potent as DAMA (Stefano *et al.*, 1989b). Whereas the effective concentration of these substances, including delta, mu, and kappa type ligands, were in the range of 10^{-9} M, DAMA was the only opioid that turned out to be most effective at a concentration of 10^{-11} M. What proved to be of particular interest is that the distinctly lower effectiveness (10^{-9} M) observed in tests with the closely related opioid analog DADLE (D-Ala^2, Leu^5-enkephalin) applies to both human and invertebrate immune reactions. This discrepancy in the potency of DAMA and DADLE does not occur in the classical situation in the mammalian nervous system (Itzhak, 1988). This fact, and the presence of an endogenous Met-enkephalin-like material demonstrated in the immunocytes of *Mytilus*, add substance to the concept that Met-enkephalin plays a distinctive role in the animal's immunoregulatory activities. Its activity may be mediated by a special subtype of delta-receptor responsive to this ligand (Stefano *et al.*, 1989b).

27.6. Discussion and outlook

The effects of opioid peptides reported here are a contribution to the large body of information on the role of chemical signals in immunoregulatory processes. A major activity in an organism's fight against inflammation or antigenic challenge is the directional movement of immunoactive cells towards such sites of disturbance, by following a chemical gradient (chemotaxis). The selective movement and accumulation of these activated cells is directed by cell-specific signal molecules. In addition to lymphokines and serotonin, immune cells dispatch opioid peptides which enhance several activities characteristic of immune responses. The participation of opioids in chemotactic activities is documented by a variety of data obtained in vertebrates and invertebrates (Stefano, 1989). For example, injections of, or chronic infusions of, β-endorphin or Met-enkephalin into the cerebrospinal fluid resulted in the directed migration of neutrophils, macrophages and lymphocyte/type cells (Van Epps *et al.*, 1983; Brown *et al.*, 1986; Fischer and Falke, 1986). In the mollusc *Mytilus* the injection of DAMA into the haemolymph elicited directed migration of immunocytes to the site of injection (Stefano *et al.*, 1989).

In addition to their influence on chemotactic movements, opioids have been shown to stimulate chemokinetic activities. They appear to be implicated in the early phase of an immune response by initiating random movements accompanied by conformational changes of immunocytes, and by enhancing their velocity, as compared with untreated controls (present report). Evidence for chemokinetic opioid effects has been reported in mammalian (Fischer and Falke, 1986) as well as invertebrate studies (Stefano *et al.*, 1989b).

As to the influence of opioids on conformational changes of immunocytes, once again the information obtained in invertebrates parallels that in vertebrates. In mammalian preparations the administration of β-endorphin induced an increase in the diameter of polymorphonuclear leukocytes by approximately 20% in at least 33% of the total cellular population examined (Fischer and Falke, 1984). In the haemolymph of the invertebrates *Mytilus* and *Leucophaea*, immunocompetent granulocytes likewise responded to DAMA by comparable flattening and increase in cellular area and perimeter (present report and Stefano *et al.*, 1989b).

In *Mytilus* the appearance of activated haemocytes is amoeboid; in *Leucophaea* as well as human blood cells the shape changes from rounded to elongate (Stefano *et al.*, 1989b). These conformational differences may be related at least in part to those observed with regard to latency period and cellular velocity in the *Mytilus* versus the mammalian experiment.

More specifically, the slower response of *Mytilus* haemocytes could be due to the need for a higher degree of mobilization of the intracellular cytoskeleton suggested by the lower FF values in this invertebrate as compared with those in human blood.

The responses of the immune system attributable to stress-related disturbances resemble those elicited by other types of irritation in that they appear to be enhanced by endogenous opioids.

In summary, receptor-mediated stimulatory effects of endogenous opioid peptides on immune responses (e.g. those following stress or surgical trauma) exist in a wide variety of animals. The 'blueprint' of intercellular communication and the similarity of the signal molecules involved in these responses appear to have developed at an early evolutionary stage, and to have remained relatively intact (Stefano, 1986, 1988). Even though detailed information is still scanty, further exploration of these phenomena on a broad comparative basis promises to be rewarding in several respects, among them the elucidation of the problem of receptor specificity and the pharmacology of neuropeptide functions.

Acknowledgements

This work was supported by NIH-MBRS MH-08180, ADAMHA-MARC 17138, NSF INT-8803664, Hoffmann LaRoche Pharmaceuticals, The Upjohn Company, and the State University of New York Research Foundation.

References

Blalock, J. E. (1989). A molecular basis for bidirectional communication between the immune and neuroendocrine systems. *Physiol. Rev.*, **69**, 1–32.

Brown, S. L., Tokuda, S., Saland, L. C. and Van Epps, D. E. (1986). Opioid peptides effects on leukocyte migration, In: Plotnikoff, N. P. and Good, R. (eds), *Enkephalins and Endorphins: Stress and the Immune System*. Plenum Press, New York, pp. 367–86.

Fischer, E. G. and Falke, N. E. (1984). β-Endorphin modulates immune functions – a review. *Psychother. Psychosom.*, **42**, 195–204.

Fischer, E. G. and Falke, N. E. (1986). The influence of endogenous opioid peptides on venous granulocytes. In: Plotnikoff, N. P. and Good, R. (eds), *Enkephalins and Endorphins: Stress and the Immune System*. Plenum Press, New York, pp. 263–70.

Itzhak, Y. (1988). Multiple opioid binding sites. In: Pasternak, G. W. (ed.), *The Opiate Receptors*. Humana Press, Clifton, NJ, pp. 95–142.

Smith, E. M. and Blalock, J. E. (1981). Human lymphocyte production of adrenocorticotropin and endorphin-like substances: association with leukocyte interferon. *Proc. Natl. Acad. Sci., USA*, **78**, 7530–4.

Stefano, G. B. (1986). Conformational matching: a determining force in maintaining signal molecules. In: Stefano, G. B. (ed.), *Comparative Opioid and*

Related Neuropeptide Mechanisms. CRC Press, Boca Raton, FL, vol. 2, pp. 271–7.

Stefano, G. B. (1988). The evolvement of signal systems: conformational matching a determining force stabilizing families of signal molecules. *Comp. Biochem. Physiol.*, **90C**, 287–94.

Stefano, G. B. (1989). Role of opioid neuropeptides in immunoregulation. *Prog. Neurobiol.*, **33**, 149–59.

Stefano, G. B., Leung, M. K., Zhao, X. and Scharrer, B. (1989a). Evidence for the involvement of opioid neuropeptides in the adherence and migration of immunocompetent invertebrate hemocytes. *Proc. Natl. Acad. Sci. USA*, **86**, 626–30.

Stefano, G. B., Cadet, P. and Scharrer, B. (1989b). Stimulatory effects of opioid neuropeptides on locomotory activity and conformational changes in invertebrate and human immunocytes: evidence for a subtype of delta receptor. *Proc. Natl. Acad. Sci. USA.*, **86**, 6307–11.

Van Epps, D. E., Durant, D. A. and Potter, J. W. (1983). Migration of human helper/inducer T cells in response to supernatants from Con A-stimulated suppressor/cytotoxic T cells. *J. Immunol.*, **131**, 697–703.

28 *Dennis E. Van Epps and Margaret M. Mason*

Modulation of leukocyte migration by α-melanocyte stimulating hormone

28.1. **Introduction**

Melanotropins are peptides that can be found in a broad spectrum of species ranging from insects to humans. Alpha-melanocyte stimulating hormone (MSH) is a 13 amino acid peptide which was originally defined by its ability to darken amphibian skin. This activity is apparently due to the direct action of MSH on the melanosomes of melanocytes resulting in dispersion and increased pigmentation. Although in lower vertebrates this alteration in skin pigmentation may serve as a defence mechanism, the function of MSH in mammals is not well understood. MSH was first isolated from porcine pituitary glands in 1955 (Lerner and Lee, 1955) and its structure was determined in 1957 (Harris and Lerner, 1957). Although it was isolated from the pituitary and is considered to be an intermediate pituitary lobe peptide, it has also been identified in the hypothalamus, thalamus, midbrain, amygdala, septum and cortex (Shizame, 1985). In addition, MSH-like immunoreactivity has been identified in both mucosal and muscular layers throughout the gastrointestinal tract (Fox and Kraicer, 1981) and has been found in the peripheral circulation. Several studies have demonstrated that circulating levels of MSH are elevated by stress, including immobilization (Khorram *et al.*, 1985) and transient ether exposure (Usategui *et al.*, 1976). MSH is also increased in the circulation during the third trimester of pregnancy (Clark *et al.*, 1978). Taken together these data suggest a systemic role for MSH in mammals.

Sequence analysis of MSH demonstrates that it is homologous to the first 13 N-terminal amino acids of adrenocorticotropic hormone (ACTH) and therefore is a component of the parent molecule, pro-opiomelanocortin (POMC). It is this polypeptide that also serves as a precursor for Met-enkephalin and β-endorphin as well as other neuroactive peptides (Shizame, 1985). It is thought that differential cleavage of the precursor POMC molecule in the pituitary or at other sites leads to the generation of many of these bioactive neuropeptides, including MSH. Many of these hormones, such as ACTH and β-endorphin, are considered to play a systemic regulatory function in the stress response where they are released into the peripheral circulation and act at distal organs (reviewed by Smith and Johnson in this book, and by Blalock, 1989).

Over recent years, many studies have been published indicating that

MSH has extensive antipyretic activity in mammals (Glyn-Ballinger *et al.*, 1983; Glyn and Lipton, 1982; Medzihradsky, 1982; Murphy *et al.*, 1983; Lipton, 1988). In fact, Lipton has shown that MSH is 25,000 times more effective than acetaminophen at suppressing fever induced by endotoxin or interleukin 1 (IL1) (Lipton, 1988). Other studies have also shown that MSH given systemically or intracranially can suppress fever, leukocytosis, elevation of circulating hepatic serum amyloid-P protein (SAP), and circulating ACTH levels mediated by injection of IL1 (Lipton, 1988; Robertson *et al.*, 1986; Daynes *et al.*, 1987). It is well known that IL1 is a major contributor to the inflammatory response, and these studies suggest that MSH may serve as a potent regulator of physiological and immunological events associated with the response to inflammatory mediators such as IL1 and endotoxin.

Many studies have documented the role of IL1 in the inflammatory response, and have shown that it can trigger the mobilization of leukocytes, particularly neutrophils, into tissues where it has been injected (Granstein *et al.*, 1985; Cybulsky *et al.*, 1986; Mason and Van Epps, 1989a, b). Although IL1 can indeed stimulate the migration of PMN into tissues, other endogenous peptides also trigger a similar response. The two that will be discussed in this chapter are tumour necrosis factor α (TNF) and C5a. TNF is a cytokine produced by macrophages, which share many of the functional properties of IL1, including its ability to induce fever and to stimulate leukocytosis and tissue inflammation (Beutler and Cerami, 1987). C5a is derived from C5 of the complement system, and upon activation of this system by antibody-antigen complexes or alternate complement pathway activators such as endotoxin, this peptide is released. It also acts as a stimulus to attract PMN to an inflammatory site *in vivo* (Mason and Van Epps, 1989a, b).

In this chapter we will provide evidence that MSH can serve as a potent regulator of the mobilization of PMN to an inflammatory site *in vivo*; in particular, to those sites where IL1, TNF, and C5a may serve as mediators of the inflammatory reaction.

28.2. **Methods**

28.2.1. *Murine model of PMN immigration*

C3H/OUJ mice were used in the model system to evaluate leukocyte immigration in response to inflammatory mediators *in vivo*, as previously described (Mason and Van Epps, 1989a, b). Briefly, 1 cm^2 collagen sponges were implanted subcutaneously in the flank and shoulder regions of the mice. Four sponges were implanted per mouse and a 48 h recovery period was allowed. A minimum of three mice were used for each data point. The response of leukocytes to inflammatory mediators was evaluated by injecting 0.1 ml of the test stimulus into two sponges and control saline into the other two sponges. Sponges were then explanted at various time-points after injection and digested with collagenase. The resultant cell supension was counted and cell differentials were performed to ascertain the nature of the cells migrating into the sponge in response to the stimulus. When evaluating the effects of MSH on the leukocyte response, 0.1 ml of MSH was injected intraperitoneally (details in Mason and Van Epps, 1989b).

28.2.2. *Reagents*

Several cytokines and inflammatory mediators were used to stimulate the immigration of leukocytes into the implanted collagen sponges. These included recombinant IL1 (Genzyme, Boston, MA), recombinant TNF (Amgen, Thousand Oaks, CA), and C5a purified from activated human plasma (Van Epps and Chenoweth, 1984). POMC peptides tested for modulation of *in vivo* leukocyte immigration included MSH, and β-endorphin (Bachem, Torrance, CA).

28.3. **Results**

28.3.1. *Leukocyte immigration into collagen sponges in response to IL1 and TNF*

Our previous studies have shown that the injection of optimal concentrations of IL1 (5 U) or TNF (0.2 μg) into subcutaneous sponge implants stimulates a marked immigration of neutrophils into the sponges as compared to saline injected controls from the same animal (Mason and Van Epps, 1989a, b). Comparative responses showing the number of PMN per sponge present in TNF, IL1 and saline-injected sponges with respect to time are shown in Fig. 28.1. As can be seen, the maximal PMN response is observed 6 h after injection of the inflammatory stimulus. This response is only observed in the test sponges, with only a slight increase in the PMN response in the control sponges.

Since differential cell counts were done on each cell sample, it was possible to evaluate the nature of the cellular response. As shown in Fig. 28.2, the response to IL1 or TNF was selective for PMN in that no significant differences in lymphocytes or monocytes were observed between control and test sample-injected sponges. Similar differential responses were observed when 0.1 ml of 10^{-7} M C5a was used as a stimulus for leukocyte immigration.

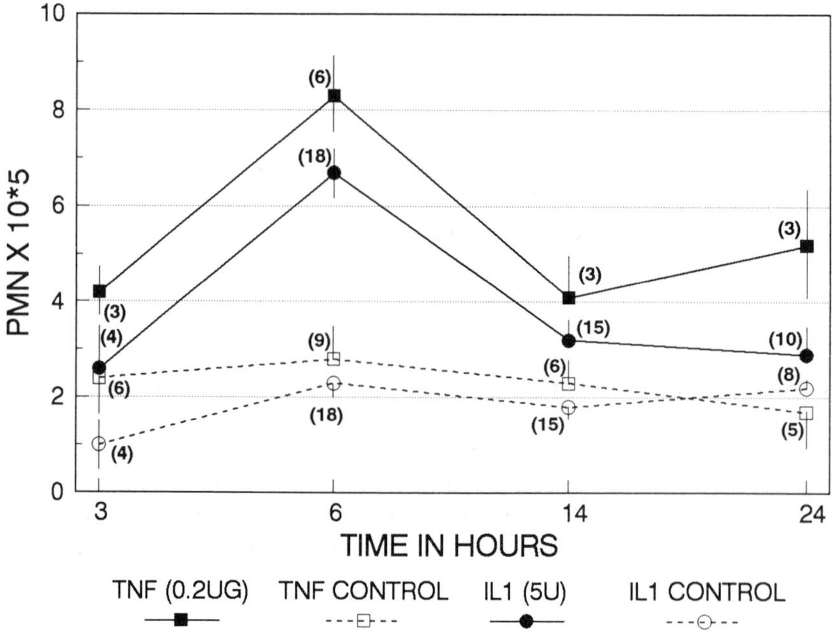

Fig. 28.1. Migration of PMN into collagen sponges in response to TNF or IL1. Each mouse was implanted with four sponges which were injected either with stimulus or control medium. 'Hours' indicates the time post-injection that the animals were sacrificed and the sponges removed and processed. Figures in parentheses represent the number of sponges that were retrieved and analysed at each time point.

28.3.2. *Alteration in IL1, TNF, and C5a-induced PMN immigration by intraperitoneal injection of MSH*

When 0.1 ml of 10^{-5} to 10^{-9} M MSH was injected intraperitoneally followed by an injection of 5 U of IL1 into the sponge, a dose-dependent inhibition of PMN immigration was observed with significant inhibition beginning with 10^{-7} M MSH. In these studies the difference between the PMN response to IL1 and control saline injection was 4×10^5 PMN per sponge without MSH injection. As shown in Fig. 28.3 this response was reduced by 13.5% with the i.p. injection of 10^{-9} M MSH and was inhibited by 65% when 10^{-5} M MSH was injected i.p.

In comparative studies where 10^{-5} M MSH was injected i.p. and 5 U of IL1, 0.2 μg of TNF or 0.1 ml of 10^{-5} M C5a was injected into the sponge, comparable suppression of PMN immigration in response to all inflammatory stimuli was observed. These results (Fig. 28.4.) indicate that the inhibition of PMN immigration by MSH is not specific to a single stimulus

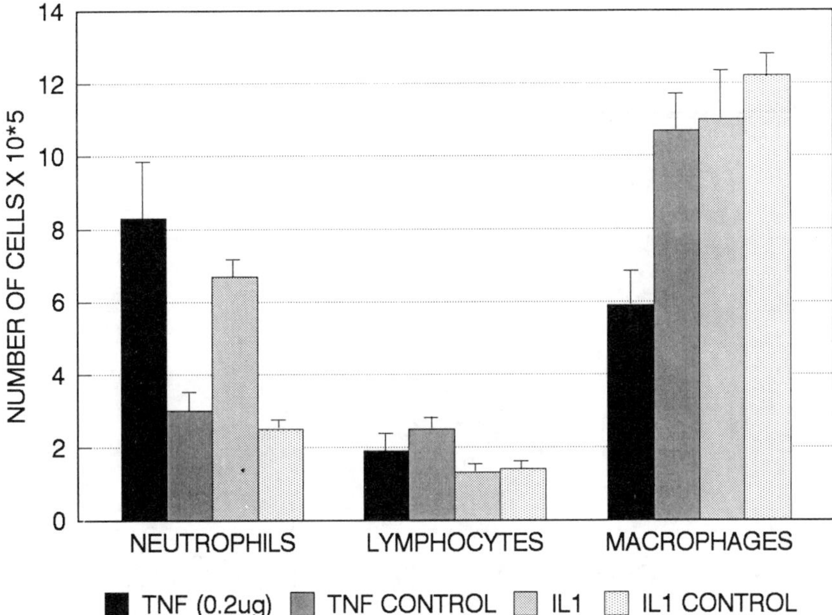

Fig. 28.2. Differential cell response to TNF and IL1 6 h after injection of sponge implants. In each case the specific cell type responding to IL1 or TNF is shown along with the cells present in sponges injected with control medium for each factor. Results represent the mean results of all sponges analysed for the data in Fig. 28.1

and represents a broad physiological effect on the inflammatory response.

In considering the specificity of action we have also compared the relative inhibitory activity of MSH versus the parent molecule, ACTH and another POMC peptide, β-endorphin. As shown in Fig. 28.5, the most effective inhibition of PMN migration was observed with MSH, with some activity also observed with ACTH. Beta-endorphin had little effect on the PMN response. These data indicate a specificity limited not only to MSH but extending to the ACTH molecule which contains the MSH sequence.

28.3.3. *Importance of timing of MSH exposure on the inhibition of PMN immigration*

If MSH truly serves as a regulator of the inflammatory response it would be expected to have short-term and reversible activity. In order to determine whether MSH fits these criteria, studies were performed to assess the relationship of the time of i.p. injection of MSH versus the timing of the injection of IL1. In these studies 0.1 ml of 10^{-5} M MSH was injected 30 min

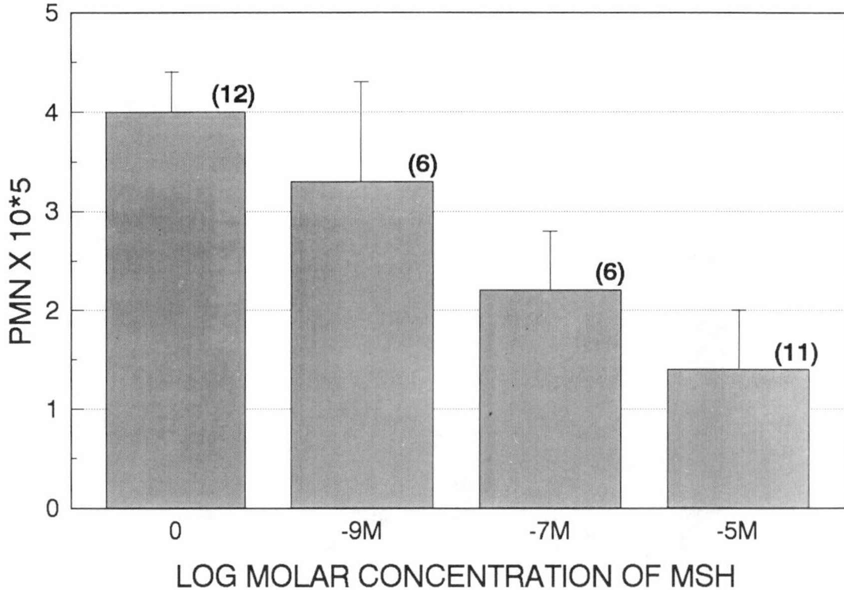

Fig. 28.3. Dose-dependence for MSH inhibition of PMN immigration into sponges in response to IL1. Data are results for sponges injected with 5 U of IL1 and analysed 6 h post-injection. Figures in parentheses indicate the number of sponges analysed. In each case the number of responding cells represents the difference between the number of PMN present in the IL1-injected sponges and the number of PMN present in the control medium-injected sponges. Significant ($p < 0.02$) inhibition of migration was observed with 10^{-7} M MSH and higher concentrations. MSH was injected i.p.

before, simultaneously with, and 1 h after the injection of IL1 into the collagen sponge. As shown in Fig. 28.6, the effects of MSH were observed only when it was injected simultaneously with IL1. These results indicate that the action of MSH is short-lived, and that it is necessary to have the MSH present at the time the animal is exposed to the stimulus.

28.4. **Discussion**

Although a clear understanding of the broad physiological role of MSH has been elusive, it is becoming quite clear that the function of MSH extends beyond its ability to stimulate melanocytes in amphibians. The evidence that MSH can prevent fever induced by endotoxin or IL1 (Glyn-Ballinger *et al.*, 1983; Glyn and Lipton, 1982; Medzihradszky, 1982; Murphy *et al.*, 1983; Lipton, 1988) implies that this peptide is an active participant in the regulation of the physiological and immunological response associated with infection. In the studies described here we have demonstrated that two

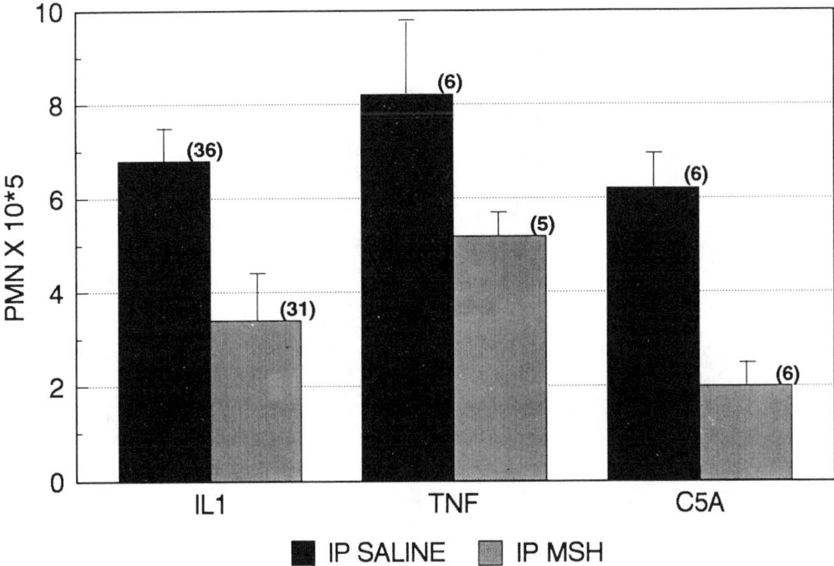

Fig. 28.4. Inhibition of PMN immigration in response to IL1, TNF, and C5a by intraperitoneal injection of MSH. In each case 0.1 ml of 10^{-5} M MSH or saline was injected i.p. simultaneously with the injection of sponges with 5 U of IL1, 0.2 μg of TNF or 0.1 ml of 10^{-7} M C5a. Sponges were analysed at 6 h. Media-injected control sponges resulted in an immigration of PMN ranging from 1.8 to 3 × 10^5 PMN per sponge.

cytokines which are released at inflammatory sites (IL1 and TNF) from monocytes, keratinocytes, endothelial cells, and several other cell types can stimulate the immigration of PMN into a region where these cytokines are focused (Granstein *et al.*, 1986; Cybulsky *et al.*, 1986; Mason and Van Epps, 1989a, b). This leukocyte immigration in response to a local production of cytokine represents a mechanism by which the host can mobilize its defence against infection.

The purpose of this study was to determine if MSH, which is known to be released under conditions of stress (Khorram *et al.*, 1985; Usategui *et al.*, 1976), could modulate the local cytokine-induced inflammatory response as it has been shown to inhibit IL1 and endotoxin-induced fever, leukocytosis and acute-phase reactant release (Lipton, 1988; Robertson *et al.*, 1986; Daynes *et al.*, 1987). It is apparent from these studies that low concentrations of MSH injected i.p. can indeed inhibit the mobilization of PMN to an inflammatory site in response to IL1, TNF and C5a. Although the exact mechanism of inhibition is still unknown, the fact that the injection site of MSH was the peritoneal cavity while the site of injection for the inflammatory mediator was subcutaneous implies that the action of

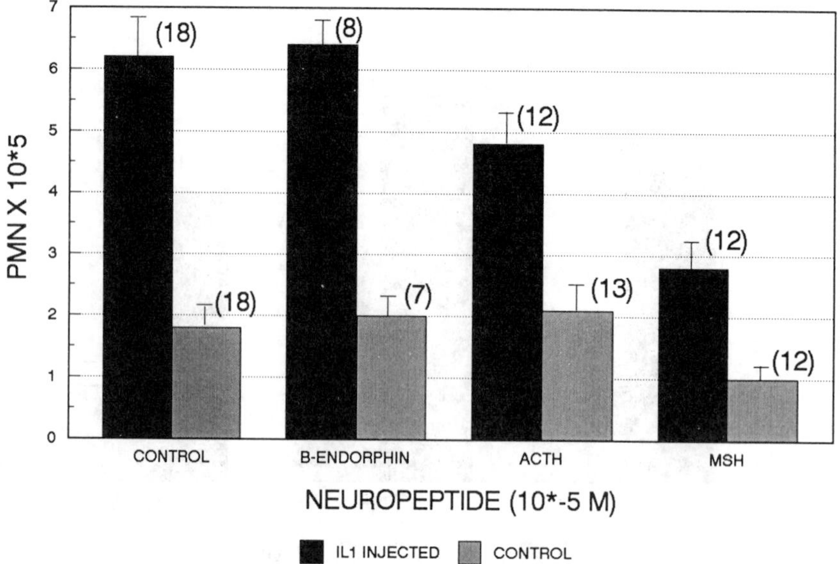

Fig. 28.5. Effect of other neuropeptides on the immigration of PMN in response to 5 U of IL1. Data were obtained 6 h after i.p. injection of saline or MSH with the simultaneous injection of sponges with the peptide or medium control indicated on the ordinate. Figures in parentheses indicate the number of sponges in each group. Significant inhibition ($p < 0.05$) was observed with ACTH and MSH.

MSH was systemic. One can only speculate as to how MSH works in this system. It may be that it affects the vascular response either directly or indirectly through an action on the central nervous system. Indeed, previous studies (Lipton, 1988) have shown that intracranial or i.v. injection of MSH both inhibit the febrile response to IL1 and endotoxin. Alternatively it is possible that MSH directly affects the ability of PMN to respond to an inflammatory stimulus, thus reducing the influx PMN to the site of injection.

Whatever the mechanism of MSH action in this system, it is clear that it is a short-lived event, since the injection of MSH must be done simultaneously with the injection of inflammatory stimulus as shown in Fig. 28.6. The reduced activity observed when MSH was injected 30 min prior to IL1 injection could be attributed to a very short half-life of MSH which is lost from the peripheral circulation in a matter of minutes (Kastin *et al.*, 1976). This, however, would not explain the lack of activity when the MSH was injected 1 h after the injection of the IL1. This latter phenomenon indicates that the initial events occurring within a few minutes of IL1 injection are the targets of the MSH activity, and not the more prolonged mobilization of PMN into the inflammatory site. Further studies will be

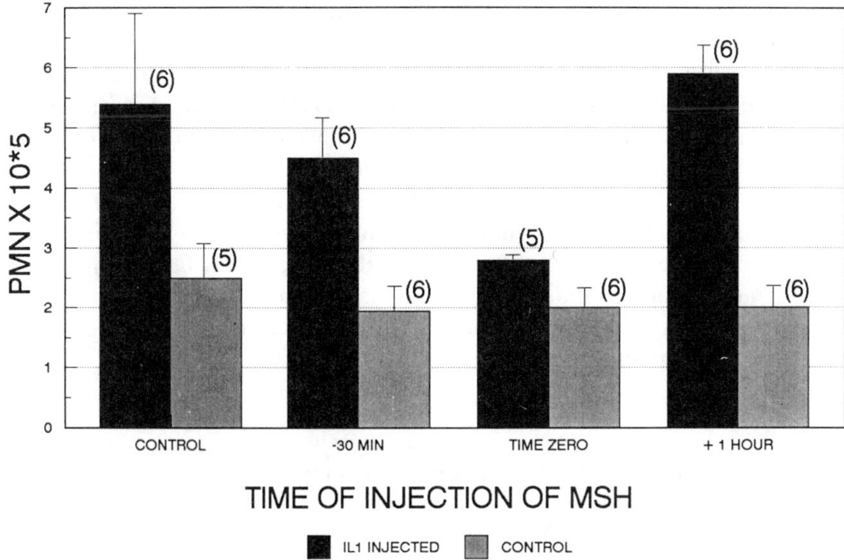

Fig. 28.6. Time-dependence for the effect of MSH on IL1-induced PMN immigration into collagen sponges. The time of injection refers to the time that 0.1 ml of 10^{-5} M MSH was injected i.p. with respect to the time that IL1 or control medium was injected into sponges. All sponges were harvested 6 h after IL1 injection. Figures in parentheses indicate the number of sponges analysed.

required to define the specific action of MSH on leukocyte mobilization *in vivo*.

An interesting observation with these studies is the selectivity of the action of MSH. It is apparent from Fig. 28.5 that MSH is the most potent of the peptides tested, but that the parent molecule ACTH which contains the MSH sequence also inhibits leukocyte immigration. In contrast, β-endorphin, also derived from POMC and shown to modulate the immune response and leukocyte migration *in vitro* (Blalock, 1989; Brown *et al.*, 1986), had little effect on PMN immigration *in vivo*. The comparative activity of MSH and ACTH in this system suggests that the N-terminal region of ACTH conveys the activity associated with this peptide, and that the addition of amino acids to the N-terminal region (at least those found in ACTH) reduces its biological activity. Because of the sequence homology between ACTH and MSH it is tempting to speculate that circulating ACTH, which is well known to increase following stress and stimulation with IL1, may be converted to MSH. This in turn would provide a mechanism to generate MSH locally or systemically, which then could serve to regulate the inflammatory response, fever and leukocytosis associated with infection and cytokine release.

Over recent years it has become increasingly apparent that there is a close interaction and bidirectional control of the immune system and the neuroendocrine system, as has been reviewed by Smith and Johnson in this book. This is particularly evident with the POMC peptide system where POMC fragments including β-endorphin, Met-enkephalin, ACTH, and α-endorphin have all been shown to modulate the function of cells of the immune system. The role of cytokines, including IL1 and IL2, in stimulating the release of these neurohormones further supports the interplay between the two systems. The data presented here with MSH provide additional evidence indicating a direct regulation of the inflammatory process by circulating neuropeptides which can be released in response to peripheral stimuli.

References

Beutler, B. and Cerami, A. (1987). Cachetin: more than a tumor necrosis factor. *N. Engl. J. Med.*, **316**, 379–85.

Blalock, J. E. (1989). A molecular basis for bidirectional communication between the immune system and neuroendocrine systems. *Physiol. Rev.*, **69**, 1–32.

Brown, S. L., Tokuda, S., Saland, L. C and Van Epps, D. E. (1986). Opioid peptide effects on leukocyte migration. In: Plotnikoff, N. and Good, R. (eds), *Enkephalins – Endorphins: Stress and the Immune System*. Plenum Press, New York, pp. 367–86.

Clark, D., Thody, A. J., Shuster, S., and Bowers, H. (1978). Immunoreactive alpha-MSH in human plasma in pregnancy. *Nature*, **273**, 163–4.

Cybulsky, M. I., Colditz, I. G. and Movat, H. Z. (1986). The role of interleukin-1 in neutrophile leukocyte emigration induced by endotoxin. *Am. J. Pathol.*, **124**, 367–72.

Daynes, R. A., Robertson, B. A., Cho, B., Burnham, D. K. and Newton, R. (1987). Alpha melanocyte stimulating hormone exhibits target cell selectivity in its capacity to affect interleukin 1 inducible responses *in vivo* and *in vitro*. *J, Immunol.*, **139**, 103.

Fox, J.-A., and Kraicer, J. (1981). Immunoreactive alpha-melanocyte stimulating hormone, its distribution in the gastrointestinal tract of intact and hypophysectomized rats. *Life Sci.*, **28**, 2127–32.

Glyn, J. R. and Lipton, J. M. (1982). Hypothermic and antipyretic effects of centrally administered ACTH (1–24) and alpha-melanotropin. *Peptides*, **2**, 177–87.

Glyn-Ballinger, J. R., Bernardini, G. L. and Lipton, J. M. (1983). Alpha-MSH injected into the septal region reduces fever in rabbits. *Peptides*, **4**, 199–203.

Granstein, R. D., Margolis, R. J., Mizel, S. B. and Sauder, D. N. (1985). *In vivo* chemotactic activity of epidermal cell-derived thymocyte activating factor (ETAF) and interleukin 1 in the mouse. *J. Leuk Biol.*, **37**, 709.

Harris, J .I and Lerner, A. B. (1957). Amino-acid sequence of alpha melanocyte stimulating hormone. *Nature*, **179**, 1346–7.

Kastin, A. J., Nissen, C., Nikolics, K., Medzihradszky, K., Coy, D. H., Teplan, I. and Schally, A. V. (1976). Distribution of [^{3}H]-MSH in rat brain. *Brain Res., Bull.*, **1**, 19–26.

Khorram, O., Bedran de Castro, J. C. and McCann, S. M. (1985). Stress-induced secretion of alpha-melanocyte-stimulating hormone and its physiological role in modulating the secretion of prolactin and luteinizing hormone in the female rat. *Endocrinology*, **117**, 2483–9.

Lerner, A. B. and Lee, T. H. (1955). Isolation of homogeneous melanocyte stimulating hormone from hog pituitary gand. *J. Am. Chem. Soc.*, **77**, 1066–7.

Lipton, J. M. (1988). MSH in CNS control of fever and its influence on inflammation/immune responses. In: Hadley, M. E. (ed.), *The Melanotropic Peptides*, vol. 2. CRC Press, Boca Raton, FL, pp. 97–113.

Mason, M. J. and Van Epps, D. E. (1989a). *In vivo* neutrophil emigration in response to interleukin-1 and tumor necrosis factor-alpha. *J. Leukocyte Biol.*, **45**, 62.

Mason, M. J. and Van Epps, D. E. (1989b). Modulation of IL-1, tumor necrosis factor, and C5a-mediated murine neutrophil migration by α-melanocyte-stimulating hormone. *J. Immunol.*, **142**,1646.

Medzihradszky, K. (1982). The bio-organic chemistry of alpha-melanotropin. *Med. Res. Rev.*, **2**, 247–70.

Murphy, M. T., Richards, D. B. and Lipton, J. M. (1983). Antipyretic potency of centrally administered alpha-melanocyte stimulating hormone. *Science*, **221**, 192–3.

Robertson, B. A., Gahring, L. C. and Daynes, R. A. (1986). Neuropeptide regulation of interleukin 1 activities: capacity of alpha-melanocyte stimulating hormone to inhibit interleukin 1 inducible responses *in vivo* and *in vitro* exhibits target cell selectivity. *Inflammation*, **10**, 371–85.

Shizame, K. (1985). Thirty-five years of progress in the study of MSH. *Yale J. Biol. Med.*, **58**, 561–70.

Usategui, R., Oliver, C., Vaudry, H., Lombardi, G., Rozenberg, I. and Mourre, A. M. (1976). Immunoreactive alpha-MSH and ACTH levels in rat plasma and pituitary. *Endocrinology*, **98**, 189–96.

Van Epps, D. E. and Chenoweth, D. E. (1984). Analysis of the binding of fluorescent C5a and C3a to human peripheral blood leukocytes. *J. Immunol.*, **132**, 2862–7.